Cloud Optics

by

Alexander A. Kokhanovsky

University of Bremen, Germany

Springer

A C.I.P. Catalogue record for this book is available from the Library of Congress.

ISBN-10 1-4020-3955-7 (HB)
ISBN-13 978-1-4020-3955-3 (HB)
ISBN-10 1-4020-4020-2 (e-book)
ISBN-13 978-1-4020-4020-7 (e-book)

Published by Springer,
P.O. Box 17, 3300 AA Dordrecht, The Netherlands.

www.springer.com

Printed on acid-free paper

Printed in the Netherlands.

To my parents

Since one must turn his eyes toward heaven to look at them, we think of them . . . as the throne of God . . . That makes me hope that if I can explain their nature . . . one will easily believe that it is possible in some manner to find the causes of everything wonderful about the Earth.

Rene Descartes

CONTENTS

FOREWORD

Clouds play an important role in the atmospheric radiative transfer and global water cycle. However, their properties are poorly understood. This is the main reason behind great efforts undertaken by the international research community to better understand cloud characteristics. Every year a lot of papers are published on various aspects of cloud research.

Optical remote sensing of clouds enables us to study cloud microphysical and geometrical characteristics using airborne, spaceborne, and ground-based optical instrumentation. Therefore, knowledge in the area of cloud optics is of great importance to any cloud physicist. Most important concepts of cloud optics and, in particular its theoretical basis, were formulated in 20th century. Time has come to summarize these concepts in a coherent way to establish a solid basis for cloud optics as a specific branch of physical optics in general.

Clouds are collections of droplets and crystals suspended in the air. Therefore, the main problem is to understand the laws of photon diffusion, scattering, absorption, and emission in a random collection of solid and liquid particles. The relationships of transmitted and reflected light fluxes with the geometrical and microphysical characteristics of clouds are of particular importance. Intuitively, one expects that the solar light transmittance by clouds decreases with cloud thickness. Also the cloud reflectance is larger for thicker clouds. Clearly, ice and liquid water absorption bands are present in spectral cloud reflectance as measured by the optical spectrometer orbiting the planet. However, what is the precise relationship of registered spectra and the cloud microstructure (e.g., the size of crystals and droplets)? The main aim of this book is to prepare the reader to deal with these and similar problems in a quantitative way. Therefore, we concentrate mostly on theoretical cloud optics. The description of optical instruments is outside the scope of this book.

Cloud optics is based on electromagnetic theory and statistical physics. In particular, Maxwell equations are used to establish laws of scattering and absorption of a light beam by a single droplet or an ice crystal. This allows us to calculate characteristics of single light scattering in a cloudy medium. Then the laws of statistical physics are applied to study diffusion, absorption, emission, and multiple scattering of photons in clouds. In particular, the Boltzmann transport equation, often used in statistical physics, forms a basis for studies of multiple scattering effects in clouds.

The book consists of four Chapters. Chapter 1 presents a review of cloud microphysical and geometrical properties. The interaction of an electromagnetic wave with a single liquid or solid particle is the main subject of Chapter 2. Here, starting from Maxwell equations, light scattering angular distributions and absorption characteristics are calculated depending on the size and shape of a single particle or shape/size statistical distributions of noninteracting scatterers. The problem of multiple light scattering is presented in Section 3. We introduce the radiative transfer equation in its most general form valid for three-dimensional radiative transfer with account for the change of light polarization due to scattering and transport processes. Chapter 4 is devoted to the selected applications.

The author is grateful to J. P. Burrows, A. Macke, B. Mayer, V. V. Rozanov, W. von Hoyningen-Huene, and E. P. Zege for many stimulating discussions and cooperation.

Chapter 1

MICROPHYSICS AND GEOMETRY OF CLOUDS

1.1 Microphysical Characteristics of Clouds

1.1.1 Droplet Size Distributions

Water clouds consist of small liquid droplets suspended in air. Cloud droplets have a spherical shape in most cases, although particles of other shapes can exist due to various external influences. For instance, the deformation of large particles due to the gravitational force is of importance for raining clouds with droplets having radii 1 mm and larger. The average radius of droplets in non-precipitating water clouds is usually around 0.01 mm and the approximation of spherical particles works quite well (see Fig. 1.1).

Natural clouds with droplets of uniform size throughout the cloud volume never occur due to the variability of physical properties of air both in space and in time domains. Thus, one must consider the radius of a droplet a as a random value, which is characterized by the probability distribution function $f(a)$. This function is normalized by the following condition:

$$\int_0^\infty f(a)da = 1. \tag{1.1}$$

The integral

$$F(a) = \int_{a_1}^{a_2} f(a)da \tag{1.2}$$

gives the fraction of particles with radii between a_1 and a_2 in a unit volume of a cloud. The probability distribution function $f(a)$ can be represented as a

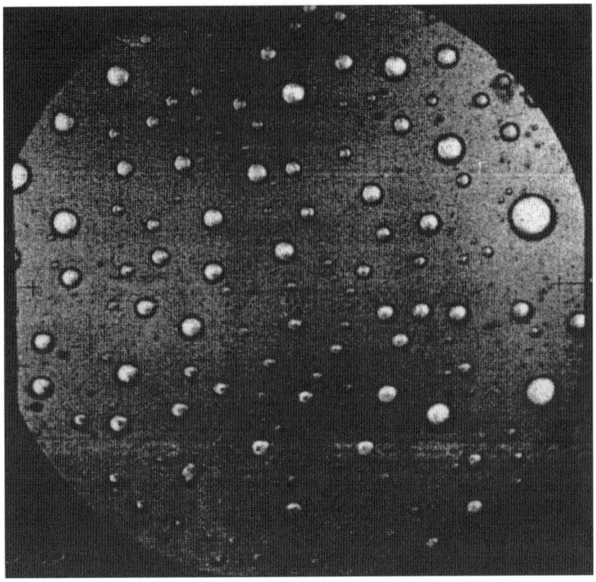

Fig. 1.1. A typical sample of cloud droplets caught on an oil slide and photographed under the microscope in an aircraft. The largest droplet has a radius of about 15 μm (Mason, 1975).

histogram, graphically or in a tabular form. However, it is most common to use an analytical form of this function, involving only one, two or three free parameters. This is, of course, a great simplification of real situations occurring in natural clouds, but most optical characteristics of a cloud only weakly depend on the fine structures of particle size distributions (PSDs) $f(a)$. McGraw et al. (1998) found that the local optical properties of polydispersions can be modelled with high accuracy by just the first six moments of the PSD. The usage of certain combinations of moments can reduce the number of parameters even further.

In most cases, the experimentally measured function $f(a)$ can be well represented by the gamma distribution (Deirmendjian, 1969):

$$f(a) = \mathbb{N} a^{\mu} e^{-\mu(a/a_0)}, \tag{1.3}$$

where

$$\mathbb{N} = \frac{\mu^{\mu+1}}{\Gamma(\mu+1) a_0^{\mu+1}} \tag{1.4}$$

is the normalization constant and $\Gamma(\mu+1)$ is the gamma function. It follows from Eq. (1.3) that the first derivative $f'(a_0) = 0$ and the second derivative $f''(a_0) < 0$. This means that the function $f(a)$ has the maximum at $a = a_0$. Equation (1.4) follows from Eqs. (1.1) and (1.3) and the definition of the gamma function:

$$\Gamma(\mu) = \int_0^\infty x^{\mu-1} e^{-x} dx. \tag{1.5}$$

In particular, one derives at integer $\mu \geq 1 : \Gamma(\mu) = (\mu - 1)!$. The parameter μ characterizes the width of the PSD $f(a)$, being smaller for wider distributions. Moments

$$\langle a^n \rangle = \int_0^\infty a^n f(a) da \tag{1.6}$$

of the PSD (1.3) can be found analytically:

$$\langle a^n \rangle = \left(\frac{a_0}{\mu}\right)^n \frac{\Gamma(\mu + n + 1)}{\Gamma(\mu + 1)}. \tag{1.7}$$

Equation (1.7) is used to find the average volume of spherical droplets

$$\langle V \rangle = \frac{4\pi}{3} \int_0^\infty a^3 f(a) da, \tag{1.8}$$

the average surface area

$$\langle \Sigma \rangle = 4\pi \int_0^\infty a^2 f(a) da, \tag{1.9}$$

and the average mass of droplets

$$\langle W \rangle = \rho \langle V \rangle, \tag{1.10}$$

where $\rho = 1\ \text{g/cm}^3$ is the density of water. It follows that

$$\langle V \rangle = \frac{\Gamma(\mu + 4)}{\mu^3 \Gamma(\mu + 1)} v_0, \tag{1.11}$$

$$\langle \Sigma \rangle = \frac{\Gamma(\mu + 3)}{\mu^2 \Gamma(\mu + 1)} s_0, \tag{1.12}$$

$$\langle W \rangle = \frac{\Gamma(\mu + 4)}{\mu^3 \Gamma(\mu + 1)} w_0, \tag{1.13}$$

where

$$v_0 = \frac{4\pi a_0^3}{3}, s_0 = 4\pi a_0^2, w_0 = \rho v_0 \tag{1.14}$$

are corresponding parameters for a droplet having the radius a_0. In the case of the most often employed cloud PSD given by Eq. (1.3) with $a_0 = 4\mu\text{m}$ and $\mu = 6$ [Cloud C1 model (Deirmendjian, 1969)] one can obtain:

$$\langle V \rangle = \frac{7}{3} v_0, \langle \Sigma \rangle = \frac{14}{9} s_0, \langle W \rangle = \frac{7}{3} w_0, \tag{1.15}$$

where $v_0 \approx 2.7 \times 10^{-16}\,\mathrm{m}^3$, $s_0 \approx 2 \times 10^{-12}\,\mathrm{m}^2$, $w_0 \approx 2.7 \times 10^{-10}\,\mathrm{g}$. Although parameters (1.14) are small, very large numbers of cloud droplets (typically, 100 particles in cm^3) create important factors for atmospheric processes.

Equation (1.3) allows to characterize the cloud droplet distribution by only two parameters: a_0 and μ. However, it should be remembered that neither a_0 nor μ is constant. They vary inside a cloud. Thus, a_0 and μ depend on the averaging scale, with large averaging scales producing more broad PSDs (with smaller values of μ). The value of $\mu = 2$ was found to be rather representative (Khrgian and Mazin, 1952) and this number is advised to be used in low resolution cloud satellite retrieval algorithms. It follows in this case: $f(a) = 8a_0^{-3}a^2\exp(-2a/a_0)$ or $f(a) = a^2\exp(-a)$ (in $\mu\mathrm{m}^{-1}$ if a is measured in μm) at $a_0 = 2\ \mu\mathrm{m}$. This function reaches a maximum at $a_0 = 2\ \mu\mathrm{m}$ and then decreases exponentially as $a \to \infty$.

The parameter $\mu = 6$ (Deirmendjian, 1969), used in the derivation of Eq. (1.15), is typical only for small averaging scales (Fomin and Mazin, 1998). General features of the droplet spectra in water clouds were studied experimentally in great detail by Warner (1973).

Parameters a_0 and μ are defined in terms of the specific unimodal cloud droplet distribution (1.3). It is more lucrative to characterize cloud PSDs by their moments. Moments can be retrieved from optical measurements without reference to specific distribution laws (McGraw et al., 1998).

The effective radius (Hansen and Travis, 1974)

$$a_{ef} = \frac{\langle a^3 \rangle}{\langle a^2 \rangle} \tag{1.16}$$

is one of the most important parameters of any PSD. It is proportional to the average volume/surface ratio of droplets. The parameter (1.16) can be defined for non-spherical particles as well. The coefficient of variance (CV) of the PSD

$$C = \frac{\Delta}{\langle a \rangle} \tag{1.17}$$

where

$$\Delta = \sqrt{\int_0^\infty (a - \langle a \rangle)^2 f(a)da}, \tag{1.18}$$

is also of importance, especially for narrow droplet distributions. The value of Δ is called the standard deviation. The CV, which is equal to the ratio of the standard deviation to the mean radius $\langle a \rangle$, is often expressed in percent.

It follows for the PSD (1.3):

$$a_{ef} = a_0\left(1 + \frac{3}{\mu}\right), \mathrm{C} = \frac{1}{\sqrt{1+\mu}} \tag{1.19}$$

and, therefore,

$$\mu = \frac{1}{C^2} - 1, a_0 = \frac{1 - C^2}{1 + 2C^2} a_{ef}. \quad (1.20)$$

The effective radius a_{ef} is always larger than the mode radius a_0. For instance, we obtain at $\mu = 3$: $a_{ef} = 2a_0$, $C = 0.5$, $\Delta = \langle a \rangle /2$. Therefore, the standard deviation is equal to half of average radius at $\mu = 3$.

Equation (1.20) gives the meaning of the parameter μ in the PSD (1.3). In particular, we have as $C \to 0 : \mu \to \infty$. In situ measurements show that the value of a_0 often varies from 4 to 20 μm (Mason, 1975) and that $\mu \in [2, 8]$ in most cases. It should be pointed out that clouds with smaller droplets are not stable due to coagulation and condensation processes. Larger particles cannot reside in the terrestrial atmosphere for a long time due to the gravity force. Thus, several physical processes lead to the existence of the most frequent mode radius range. One can obtain from Eq. (1.19) and inequality $2 \leq \mu \leq 8$ that the value of $C \in [0.3, 0.6]$. Thus, it follows that the standard deviation of the radius of particles in water droplets is usually 30–60% of an average radius. Smaller and larger values of C do occur but values of C smaller than 0.1 were never observed (Twomey, 1977). Larger values of C may indicate the presence of the second mode in the range of large particles (Ayvazyan, 1991).

Equation (1.19) and results for a_0 and μ just reported lead to the effective radius a_{ef} of water droplets being in the range from 5 to 50 μm, depending on the cloud type. Near-global survey of the value of a_{ef}, using satellite data, shows that typically 5 μm $\leq a_{ef} \leq$ 15 μm (Han et al., 1994). We see that water clouds with $a_{ef} >$ 15 μm are rare. This can be used to discriminate satellite pixels with ice crystals even at wavelengths where ice and water absorption coefficients are almost equal. Such a possibility of discrimination is due to much larger (e.g., in 5–10×) effective sizes of ice crystals as compared to droplets. The large size of ice crystals will reduce the reflection function in near infrared considerably as compared to droplets. This reduction can be easily detected. Note that clouds with $a_{ef} >$ 15 μm are often raining (Masunaga et al., 2002). Pinsky and Khain (2002) showed that the threshold of the occurrence of drizzle is around $a_{ef} >$ 15 μm. Then strong collisions of droplets start. The vertical extent and the thermodynamic state of clouds also influences the probability of precipitation.

Some authors prefer to use the representation of the PSD by the following analytical form (Ayvazyan, 1991):

$$f(a) = \frac{1}{\sqrt{2\pi}\sigma a} \exp\left(-\frac{\ln^2(a/a_m)}{2\sigma^2} \right), \quad (1.21)$$

which is called the log-normal distribution. The relations between values of a_{ef}, $\langle a \rangle$, Δ and parameters of the gamma and log-normal PSDs are presented

Table 1.1. Particle size distributions and their characteristics.

$f(a)$	B	$\langle a \rangle$	a_{ef}	C
Gamma distribution $Ba^{\mu}e^{-\mu(a/a_0)}$	$\frac{\mu^{\mu+1}}{a_0^{\mu+1}\Gamma(\mu+1)}$	$a_0\left(1+\frac{1}{\mu}\right)$	$a_0\left(1+\frac{3}{\mu}\right)$	$\sqrt{\frac{1}{\mu+1}}$
Log-normal distribution $\frac{B}{a}\exp\left(-\frac{\ln^2(a/a_m)}{2\sigma^2}\right)$	$\frac{1}{\sqrt{2\pi}\sigma}$	$a_m e^{0.5\sigma^2}$	$a_m e^{2.5\sigma^2}$	$\sqrt{e^{\sigma^2}-1}$

in Table 1.1. The value of Δ_{ef} in this table represents the effective variance, defined as:

$$\Delta_{ef} = \frac{\int_0^\infty (a - a_{ef})^2 a^2 f(a)da}{a_{ef}^2 \int_0^\infty a^2 f(a)da}. \tag{1.22}$$

This parameter is often used instead of the coefficient of variance C (Hansen and Travis, 1974) because of a special importance attached to the value of the effective radius of droplets a_{ef} as compared to the average radius $\langle a \rangle$ for the problems of cloud optics. For instance, light extinction in clouds is governed mostly by values of a_{ef} and liquid water content (LWC) independently of the type of PSD $f(a)$ (Kokhanovsky, 2004a). PSDs (1.3) and (1.21) at $a_{ef} = 6$ μm and $C = 0.38$ are shown in Fig. 1.2. Then it follows: $a_0 = 4$ μm, $\mu = 6$, $a_m = 5.6$ μm, $\sigma = 0.3673$.

The influence of μ on the PSD (1.3) is shown in Fig.1.3 at $\mu = 2, 6, 8$ and $a_{ef} = 6$ μm. Light extinction in media having droplet size distributions shown in Figs. 1.2 and 1.3 (and the same LWC) will be almost identical although PSDs are quite different.

Particle number concentration, N, in addition to particle size and shape, is of importance for the propagation, scattering and extinction of light in cloudy media. The value of N gives the number of particles in a unit volume and is expressed in cm^{-3}. It is usually in the range 50–1000 cm^{-3} (Fomin and Mazin, 1998). Clearly, the concentration of droplets depends on the concentration C_N of atmospheric condensation nuclei. The value of C_N is smaller over oceans than over continents, so the concentration of droplets in marine clouds is on average smaller than over continents. Generally, the smaller concentration of droplets over oceans means that they can grow larger, producing clouds with larger droplets over oceans, which is confirmed by the analysis of satellite optical imagery as well (Han et al., 1994). This influences the occurrence and the rate of precipitation.

Svensmark and Friis-Christensen (1997) and Marsh and Svensmark (2000) have speculated that cosmic ray ionization could influence the production of

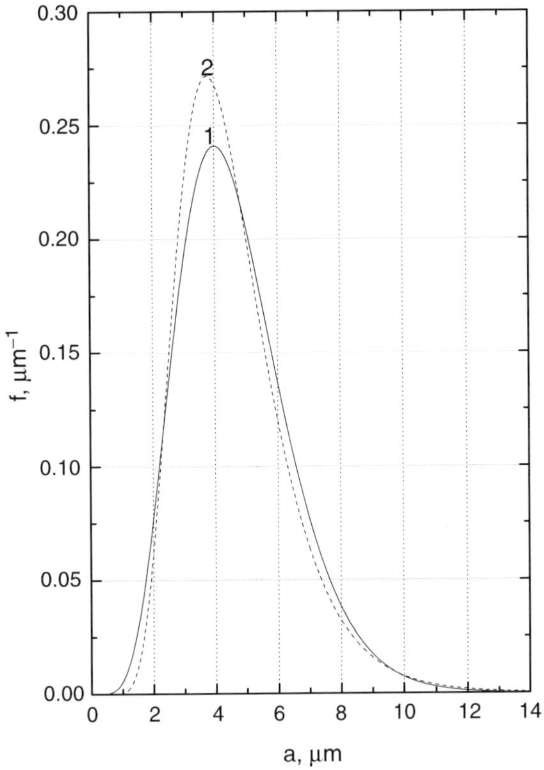

Fig. 1.2. Particle size distributions (1-gamma distribution, 2-log-normal distribution).

condensation nuclei and, therefore, cloud properties. This is of importance for climate change problems as discussed by Svensmark (1998).

The dimensionless volumetric concentration of droplets $C_v = N\langle V\rangle$ and LWC $C_w = \rho C_v$ or [see Eq. (1.10)] $C_w = N\langle W\rangle$ are often used in cloud studies as well. The value of C_w is usually in the range 0.01–1.0 gm^{-3} with typical values of 0.1 gm^{-3}. Therefore, the most frequent values of C_v are in the range of $[10^{-7}, 10^{-5}]$. This means that only a very small fraction of a cloud volume is occupied by droplets. This simplifies the solution of many cloud optics problems. In particular, local cloud optical characteristics can be calculated not accounting for the close-packed media effects (Kokhanovsky, 2004a). Numbers given above are representative, but it should be remembered that these values can change in a broader range in real situations. The LWC is not constant throughout a cloud but has larger values near the top of a cloud in most cases (Feigelson, 1981). This is illustrated in Fig. 1.4. Such a behaviour is characteristic for a_{ef} as well (see Fig. 1.5).

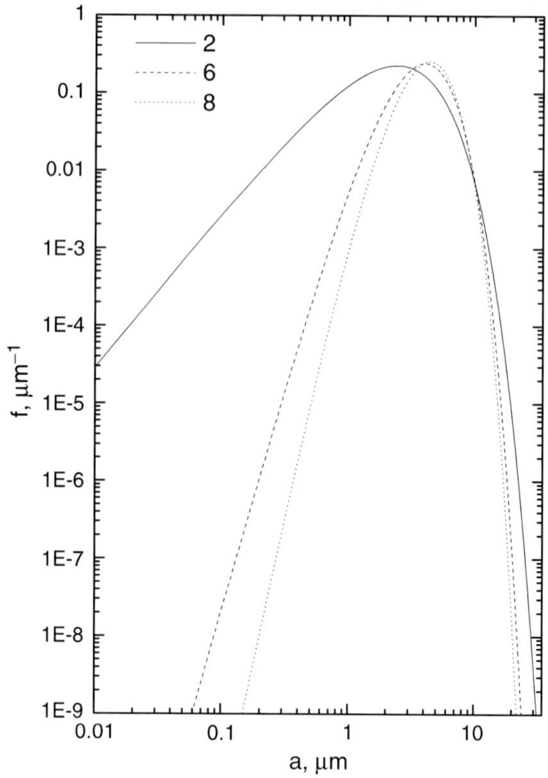

Fig. 1.3. Gamma particle size distributions for various μ at $a_{ef} = 6$ μm.

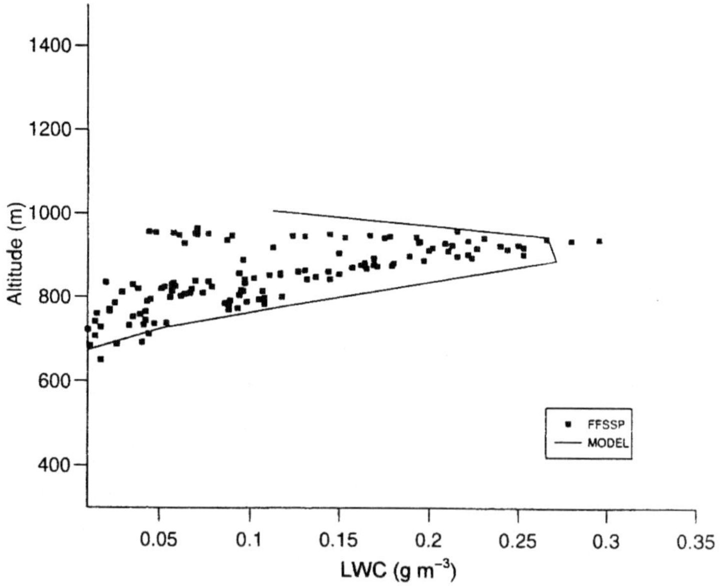

Fig. 1.4. Squares give measured values of the LWC as the function of altitude. Line gives the modelled LWC-profile. Further details are given by Ghosh et al. (2000).

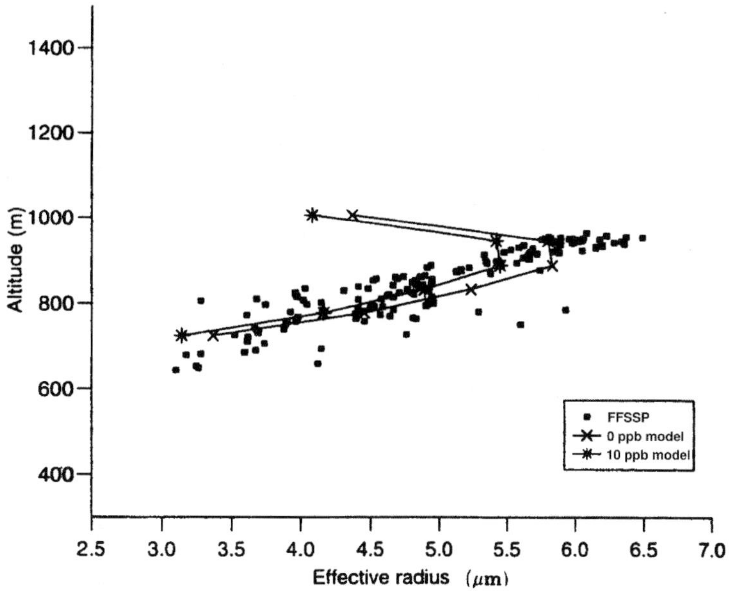

Fig. 1.5. Squares give measured values of the effective radius as the function of altitude. Line gives the modelled profile of the effective radius assuming different concentrations of HNO_3 acid in a cloud layer. Further details are given by Ghosh et al. (2000).

The liquid water path (LWP) w is defined as

$$w = \int_{z_1}^{z_2} C_w(z)dz, \tag{1.23}$$

where z_1 is the cloud bottom height and z_2 is the cloud top height, $l = z_2 - z_1$ is the geometrical thickness of a cloud. It follows at $C_w = \text{const}$:

$$w = C_w l. \tag{1.24}$$

The geometrical thickness of clouds varies, depending on the cloud type. It is in the range of 500–1000 m for stratocumulus clouds in most cases. Near-global data obtained by Han et al. (1994) from satellite measurements show that the LWP w is typically in the range of 50–150 g/m^2. The annual mean is equal to 86 g/m^2 (Han et al., 1994). Then we have at the cloud geometrical thickness $l = 860$ m: $C_w = 0.1$ g/m^3, which is a typical value of the LWC in water clouds (see Tables 1.2, 1.3 and Fig. 1.4). It should be underlined that remotely sensed values of the LWP and a_{ef} depend on the scale of horizontal averaging and also on the assumptions used in the retrieval procedures (see Figs. 1.6 and 1.7). Annual and monthly means of the LWP and a_{ef} over the globe in combination with many other cloud parameters derived from satellite data can be found at http://isccp.giss.nasa.gov/products/browsed2.html (Rossow and Schiffer, 1999).

Table 1.2. Geometrical and microphysical characteristics of marine water clouds obtained using experimental measurements (Miles et al., 2004).

B	T	H	N	LWC	a_{ef}	C	a_0	μ	a_m	σ
630	860	650	50	0.15	9.50	25.60	7.88	14.6	8.20	0.24
630	860	750	45	0.23	11.5	28.28	9.12	11.5	9.60	0.27
630	860	850	45	0.35	13.4	31.11	10.2	9.50	10.8	0.29
800	1000	900	75	0.28	11.2	49.69	6.53	4.20	7.55	0.40
800	1000	1000	100	0.49	13.0	44.00	6.16	2.70	7.70	0.46
240	760	240	23	0.01	5.75	42.50	2.97	3.20	3.60	0.43
240	760	440	56	0.13	8.95	32.00	6.82	9.60	7.25	0.29
240	760	620	111	0.37	9.80	27.59	8.29	16.5	8.55	0.23
—	—	—	91	0.18	7.95	15.13	7.44	43.6	7.50	0.15
—	—	1300	66	0.13	8.20	23.33	7.28	23.6	7.05	0.19
—	—	1300	22	0.03	7.60	30.53	6.10	12.2	6.40	0.26
610	960	700	160	0.09	6.00	43.68	3.50	4.20	4.05	0.40
610	960	800	182	0.31	8.00	29.20	6.27	10.9	6.60	0.28
610	960	930	158	0.47	9.60	28.31	7.63	11.6	8.00	0.27
1290	1460	1310	107	0.03	4.30	32.39	3.11	7.80	3.35	0.32
1290	1460	1340	142	0.06	5.10	30.59	3.87	9.40	4.15	0.29
1290	1460	1390	143	0.14	6.65	28.95	5.26	11.4	5.55	0.27
1290	1460	1430	140	0.22	7.65	98.54	6.47	16.5	6.70	0.23
—	—	310	12	0.02	7.75	44.25	4.69	4.60	5.30	0.39
—	—	380	5	0.004	7.20	47.92	3.48	2.80	4.35	0.45
420	730	460	28	0.08	9.90	42.48	6.74	6.40	7.40	0.34
420	730	530	39	0.13	11.2	55.70	5.95	3.40	7.20	0.42
420	730	610	59	0.2	11.3	60.14	5.46	2.80	6.80	0.45
420	730	690	73	0.17	11.0	69.23	3.48	1.40	5.30	0.54
380	830	480	79	0.16	8.90	40.00	5.68	5.30	6.35	0.37
380	830	730	94	0.59	12.5	33.65	9.58	10.0	10.1	0.24
408	684	575	296	0.22	6.45	41.67	4.00	4.90	4.65	0.35
408	684	682	228	0.29	7.75	41.74	4.77	4.80	5.35	0.39
410	740	410	116	0.04	4.75	22.35	3.95	14.8	4.10	0.24
410	740	580	123	0.29	9.00	29.61	6.86	9.60	7.30	0.29
410	740	740	149	0.35	9.80	33.75	6.86	7.00	7.60	0.32
1680	2350	1770	58	0.07	7.50	38.79	4.94	5.80	5.45	0.36
1680	2350	1890	40	0.08	9.00	40.74	5.63	5.00	6.35	0.38
1680	2350	240	30	0.10	12.0	60.00	4.50	1.80	6.30	0.51
1680	2350	2200	26	0.09	12.2	60.28	4.41	1.70	6.35	0.51
1310	1980	1250	40	0.12	10.4	42.48	6.24	4.50	7.15	0.38
1310	1980	1400	32	0.12	11.7	49.68	5.95	3.10	7.20	0.44
1310	1980	1620	52	0.26	12.2	40.98	7.63	5.00	8.60	0.37
1310	1980	1830	35	0.15	12.1	47.59	6.43	3.40	7.65	0.43

The meaning of columns as follows: B(m): cloud base height; T(m): cloud top height; H(m): height of measurements; N: droplet number concentration (cm^{-3}); LWC: liquid water content (gm^{-3}); a_{ef}: effective radius (µm); C: coefficient of variance; a_0: mode radius (µm) for the gamma PSD (1.3); μ: half-width parameter of the gamma PSD (1.3); and a_m (µm) and σ are parameters of the log-normal PSD (1.21).

Table 1.3. Geometrical and microphysical characteristics of continental water clouds obtained using experimental measurements (Miles et al., 2004).

B	T	H	N	LWC	a_{ef}	C	a_0	μ	a_m	σ
310	—	320	21	0.0038	4.15	46.55	2.29	3.7	2.70	0.41
350	—	360	59	0.0025	2.35	30.77	1.74	8.6	1.85	0.31
360	—	370	12	0.003	4.40	53.57	2.04	2.6	2.55	0.46
400	—	410	147	0.0093	2.75	32.65	1.77	5.4	2.00	0.36
400	—	410	228	0.0137	2.55	28.57	1.81	7.3	1.95	0.33
630	870	675	350	0.22	5.75	29.59	4.48	10.6	4.75	0.28
630	870	750	285	0.28	6.70	30.09	5.15	10.0	5.45	0.29
630	870	850	270	0.6	8.50	22.73	7.31	18.5	7.55	0.22
808	1040	908	480	0.12	4.20	27.03	3.49	14.8	3.60	0.24
808	1040	990	260	0.13	5.35	30.43	4.25	11.7	4.45	0.27
808	1040	908	370	0.13	4.65	31.25	3.68	11.5	3.90	0.27
808	1040	990	190	0.09	5.45	39.53	3.83	7.1	4.15	0.33
250	530	250	15	0.02	10.0	70.00	1.67	0.6	3.85	0.62
250	530	390	35	0.08	11.4	76.85	2.62	0.9	4.85	0.58
300	630	470	40	0.11	12.7	83.04	2.12	0.6	5.00	0.61
1350	2250	1800	148	0.4	9.15	26.38	7.73	16.3	8.00	0.23
—	—	800	215	0.01	2.60	58.06	1.65	5.2	1.85	0.37
—	—	800	418	0.12	2.25	46.88	1.32	4.3	1.50	0.40
800	1000	1000	250	0.2	6.50	37.25	4.38	6.2	4.80	0.35
380	790	440	60	0.003	3.05	38.46	1.18	1.9	1.65	0.50
380	790	550	366	0.06	4.20	48.21	1.20	2.7	2.45	0.46
380	790	650	401	0.11	4.85	48.48	2.50	3.2	3.05	0.43
380	790	750	396	0.15	5.40	50.68	2.83	3.3	3.40	0.43
1430	2010	1700	190	0.3	9.00	53.51	4.09	2.5	5.20	0.47
1430	2010	1700	450	0.7	7.70	27.61	6.20	12.5	6.50	0.26
1750	1870	1815	220	0.24	7.50	44.44	4.33	4.1	5.00	0.40
1750	1870	1815	200	0.035	4.00	41.67	2.48	4.9	2.80	0.38
450	673	571	693	0.2	5.25	57.14	2.1	2.0	2.80	0.50
450	673	673	575	0.26	6.25	61.97	2.17	1.6	3.25	0.51
750	1090	750	160	0.03	3.85	30.77	2.95	9.8	3.15	0.29
750	1090	870	225	0.1	5.05	26.67	4.21	15.1	4.35	0.24
750	1090	950	165	0.19	6.75	21.77	5.95	22.2	6.10	0.20
750	1090	1040	310	0.41	7.20	25.58	6.04	15.7	6.30	0.24
820	1010	840	680	0.08	3.30	26.32	2.68	13.0	2.80	0.26
820	1010	925	440	0.17	4.75	24.42	4.05	17.3	4.15	0.23
820	1010	980	297	0.16	5.40	28.42	4.41	13.3	4.60	0.25
400	1210	480	80	0.09	7.15	31.93	5.28	8.5	5.65	0.31
400	1210	890	320	0.67	8.25	21.71	7.21	20.8	7.45	0.21
400	1210	1120	360	1.00	9.05	20.96	8.00	22.9	8.20	0.20
2470	2990	2710	96	0.1	7.00	34.51	4.94	7.2	5.35	0.33
2470	2990	2790	103	0.18	8.40	37.12	5.72	6.4	6.30	0.34
2470	2990	2870	93	0.28	10.0	35.63	7.00	7.0	7.60	0.33

(continued)

Table 1.3. *Continued*

B	T	H	N	LWC	a_{ef}	C	a_0	μ	a_m	σ
2470	2990	2940	106	0.41	10.8	34.29	7.74	7.6	8.35	0.32
2470	2990	3020	96	0.45	12.0	41.57	7.38	4.8	8.35	0.38
200	500	430	139	0.14	7.75	57.73	3.52	2.5	4.45	0.47
1460	1940	1750	494	0.3	6.45	49.41	3.11	2.8	3.85	0.45
1460	1940	1840	432	0.33	6.80	48.91	3.51	3.2	4.25	0.43
1460	1940	1940	139	0.14	8.00	60.22	3.00	1.8	4.20	0.51

The meaning of columns as follows: B(m): cloud base height; T(m): cloud top height; H(m): height of measurements; N: droplet number concentration (cm^{-3}); LWC: liquid water content (gm^{-3}); a_{ef}: effective radius (μm); C: coefficient of variance; a_0: mode radius (μm) for the gamma PSD (1.3); μ: half-width parameter of the gamma PSD (1.3); and a_m (μm) and σ are parameters of the log-normal PSD (1.21).

Miles et al. (2000) summarized *in situ*—derived experimental data for various cloud parameters including N, a_{ef}, C_w, a_0, μ, a_m and σ. Some of them are shown in Tables 1.2–1.4 separately for marine and continental clouds. Correspondent frequency histograms are shown in Figs. 1.8–1.15. Also the correlation plot of LWC and N is given in Fig. 1.16 both for marine and continental clouds. We see that there is no important correlation between these parameters although continental

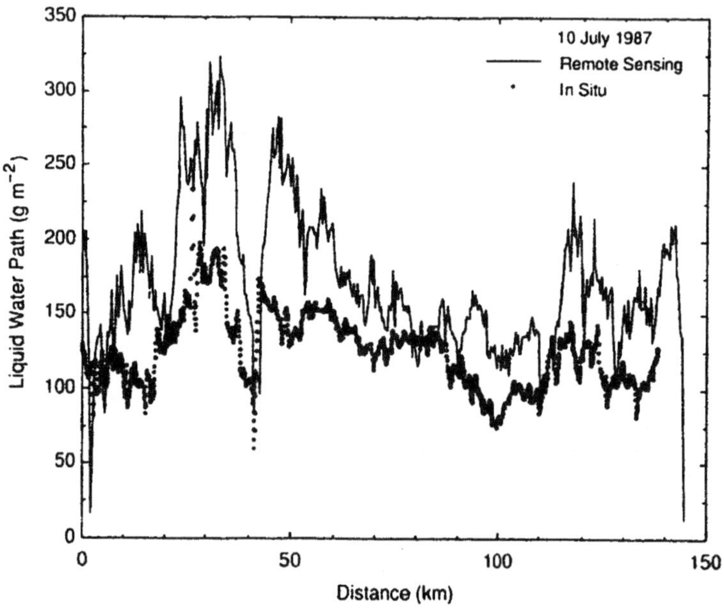

Fig. 1.6. Comparison of the liquid water path as a function of distance along the nadir track of the ER-2 as derived from remote sensing (solid line) and in situ measurements (solid circles) (Nakajima et al., 1991).

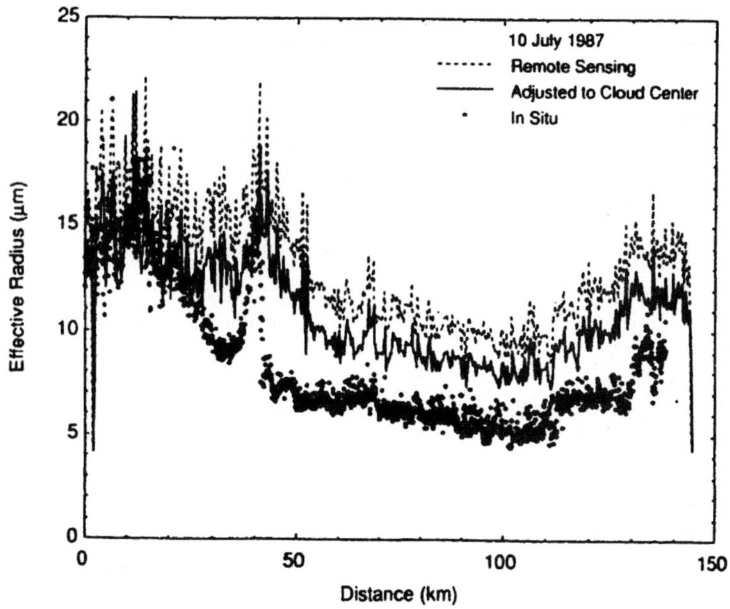

Fig. 1.7. Comparison of the effective radius as a function of distance along the nadir track of the ER-2 as derived from remote sensing (dashed line) and in situ measurements (solid circles) (Nakajima et al., 1991).

clouds have larger values of N. Similar results for values of a_{ef}, C are shown in Fig. 1.17. It follows that marine clouds have larger droplets, which is consistent with results shown in Fig. 1.16. Clearly, droplets have more chances to attract water molecules for smaller N. The frequently used cloud model C1 of Deirmendjian (1969) with $a_{ef} = 6$ μm and $\mu = 6(C = 1/\sqrt{7})$ is given by a circle in Fig. 1.17. We see that this model represents average cloud properties quite well. The analysis of Table 1.4 shows that the model is less accurate for marine clouds. So in addition

Table 1.4. Summary of results given in Tables 1.2 and 1.3 for average values of correspondent cloud parameters.

The cloud parameter	Continental clouds	Marine clouds	Average
a_{ef}, μm	6.0	9.0	7.5
C, %	44.0	43.0	43.5
N, cm^{-3}	254.0	91.0	172.5
LWC, gm^{-3}	0.20	0.17	0.185
a_0, μm	4.0	6.0	5.0
μ	7.0	8.0	7.5
a_m, μm	4.0	6.0	5.5
σ	0.4	0.4	0.4

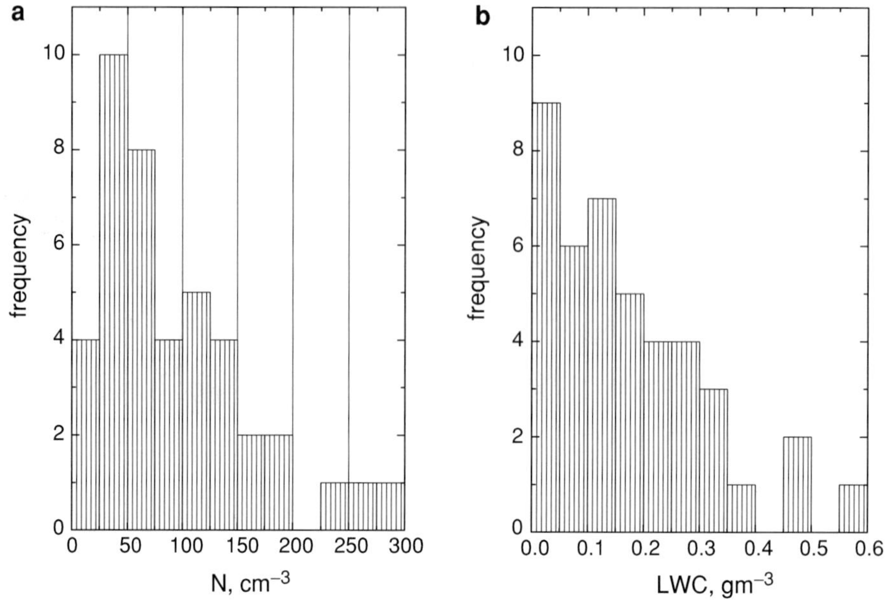

Fig. 1.8. Frequency distribution of N (a) and LWC (b) for marine water clouds obtained analysing multiple experimental measurements at different places and by different instruments and research groups (Miles et al., 2000).

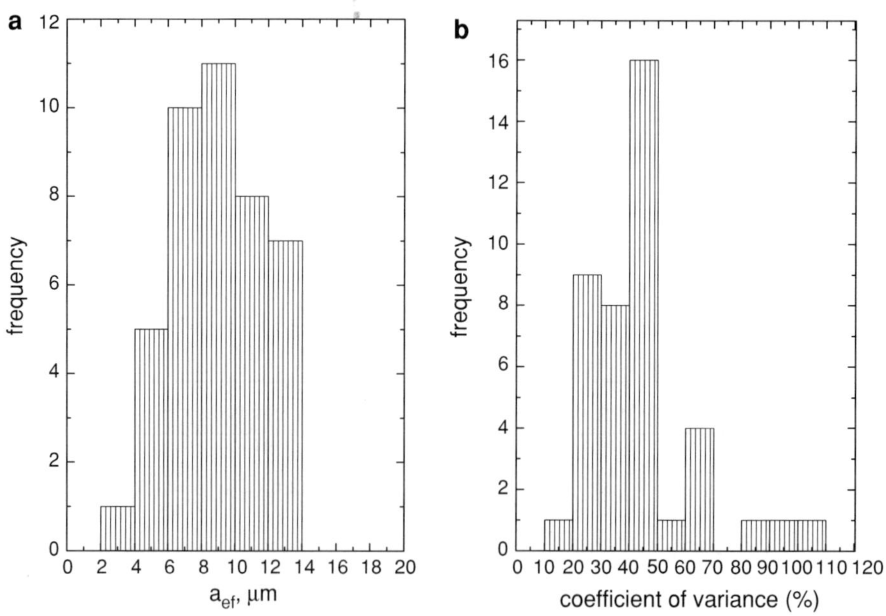

Fig. 1.9. The same as in Fig. 1.8 except for a_{ef} (a) and C (b).

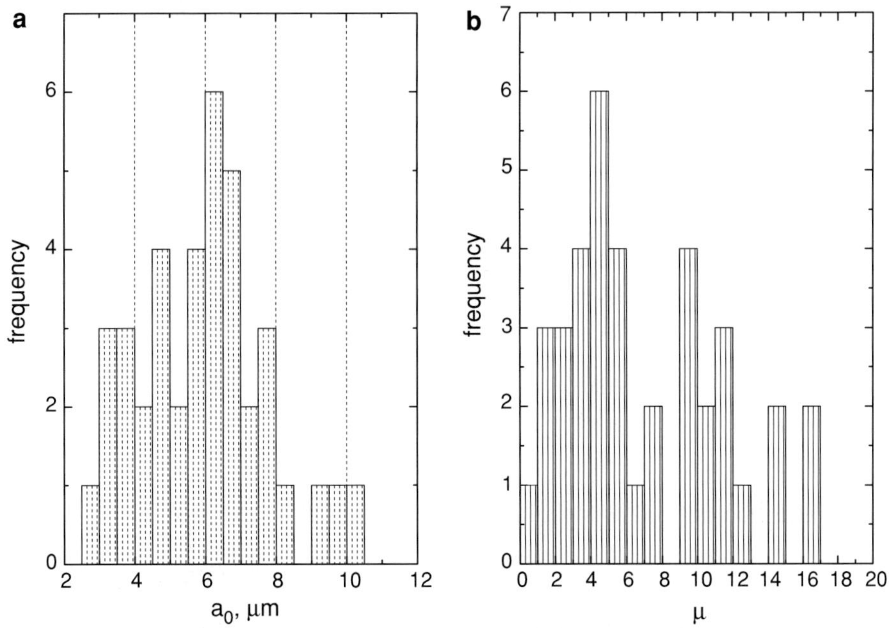

Fig. 1.10. The same as in Fig. 1.8 except for a_0 (a) and μ (b).

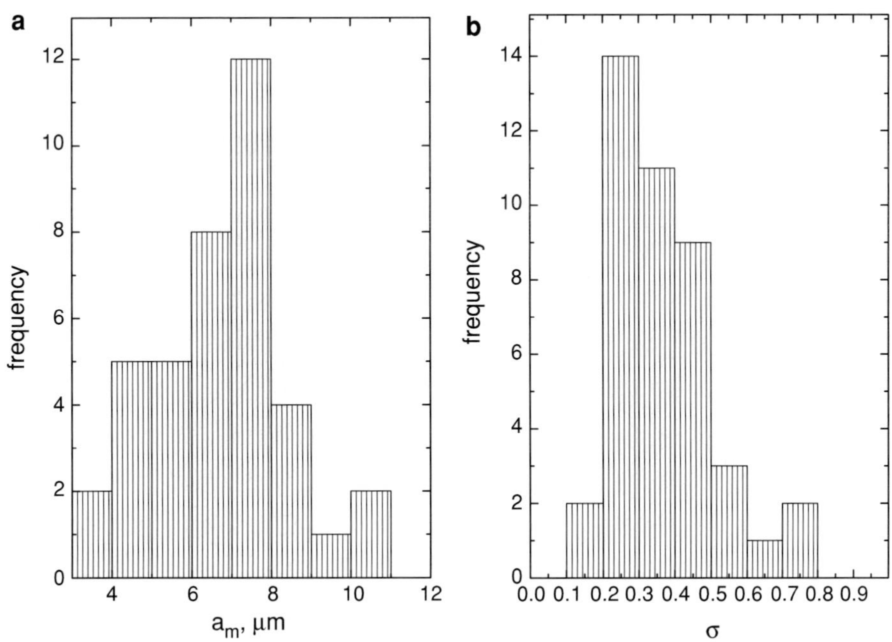

Fig. 1.11. The same as in Fig. 1.8 except for a_m (a) and σ (b).

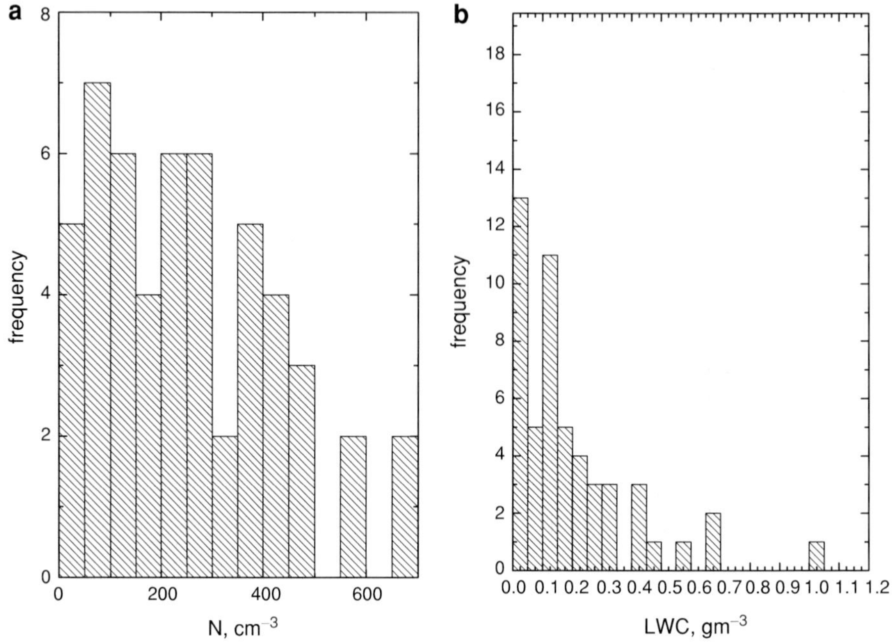

Fig. 1.12. Frequency distribution of N (a) and LWC (b) for continental water clouds obtained analysing multiple experimental measurements at different places and by different instruments and research groups (Miles et al., 2000).

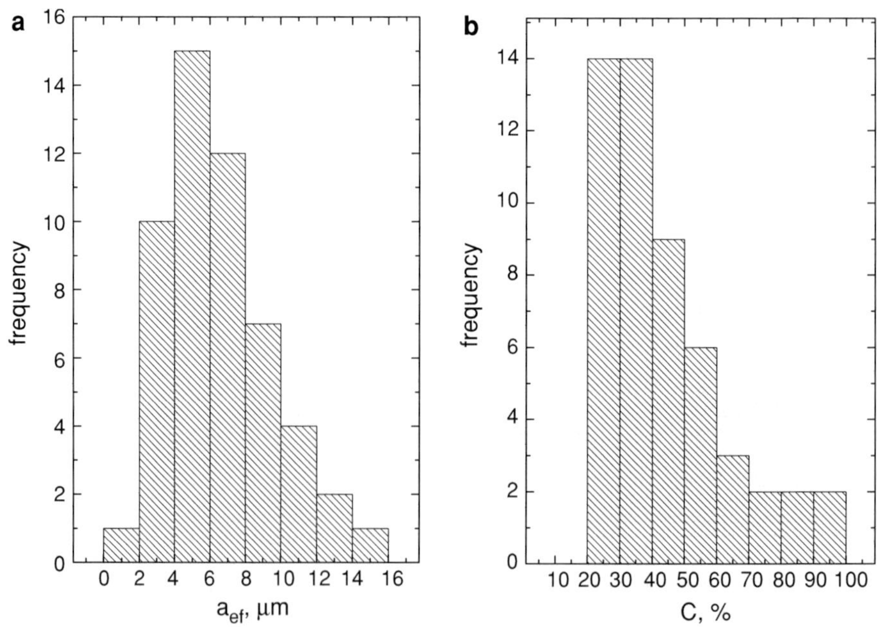

Fig. 1.13. The same as in Fig. 1.12 except for a_{ef} (a) and C (b).

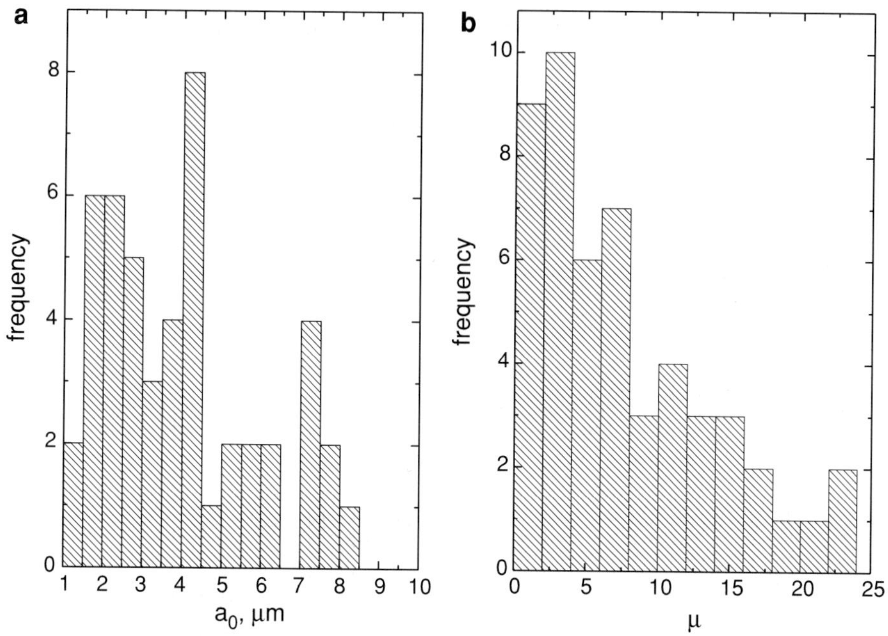

Fig. 1.14. The same as in Fig. 1.12 except for a_0 (a) and μ (b).

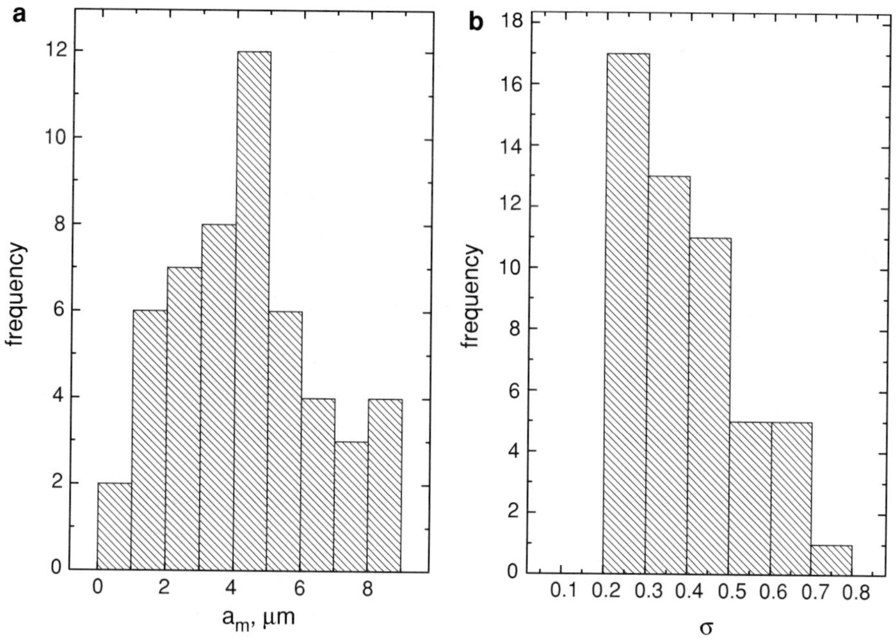

Fig. 1.15. The same as in Fig. 1.12 except for a_m (a) and σ (b).

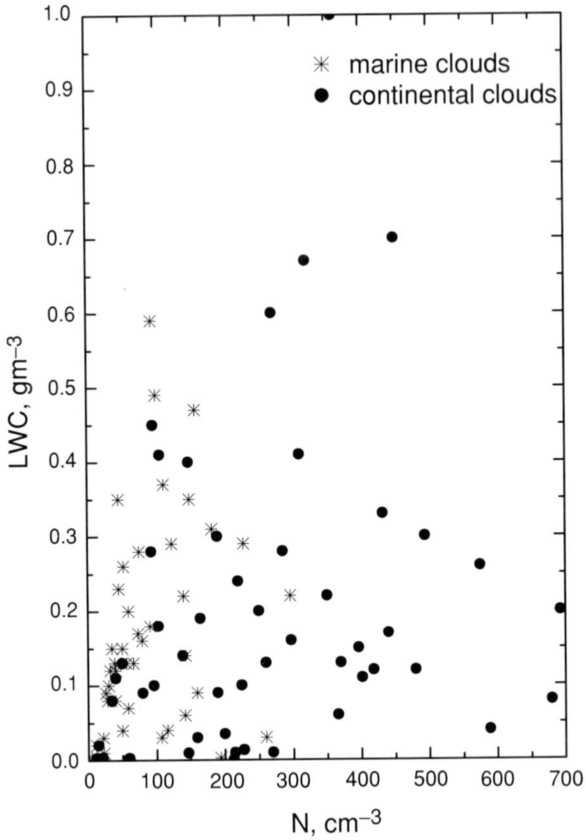

Fig. 1.16. The correlation between values of LWC and N for marine and continental clouds.

to the Deimendjian's continental cloud model (CCM), we propose also the marine cloud model (MCM) with the radius $a_0 = 6\ \mu\text{m}$ and $N = 100\ \text{cm}^{-3}$ (see Table 1.5). We choose the parameter $\mu = 6$ for the marine model because it provides physically plausible dependence of the PSD on V at small values of $V : f(V) \sim V^2$, where V is the volume of a droplet. Also, this value is close to values of μ equal to 7 and 8 given in Table 1.4.

Cloud systems can easily cover an area $S \approx 10^3\ \text{km}^2$ (Kondratyev and Binenko, 1984). So the total amount of water $W = wS$ (for idealized clouds having the same LWP for the whole cloudy area) stored in such a water cloud system is equal approximately to 10^8 kg, if we assume that $w = 100\ \text{g/m}^2$, which is a typical value for cloudy media (see Fig. 1.6). This underlines the importance of clouds both for climate problems and human activity (e.g., crops production, flooding, etc.).

Table 1.5. The continental cloud model (CCM) and the marine cloud model (MCM) proposed to be used in combination with Eq. (1.3).

The cloud parameter	CCM	MCM
a_0, μm	4	6
μ	6	6
N, cm^{-3}	250	100

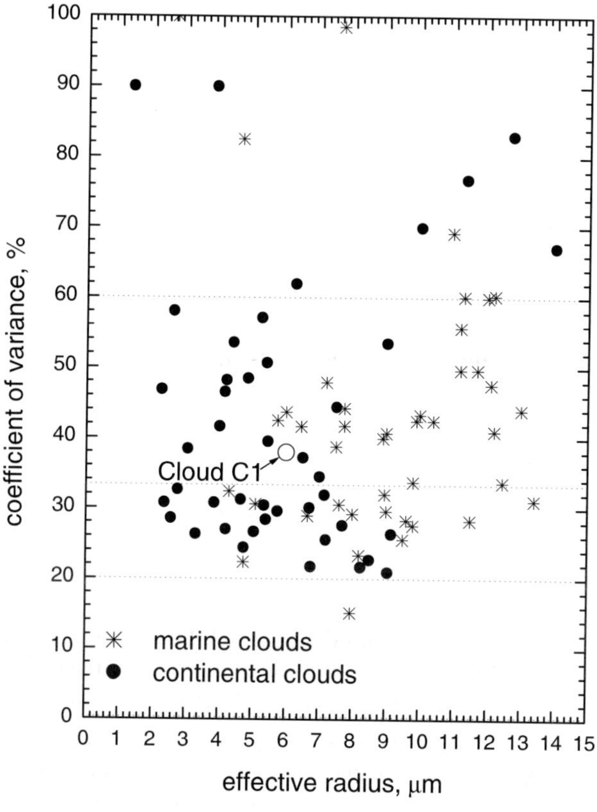

Fig. 1.17. Correlation between values of C and a_{ef} for marine and continental clouds.

1.1.2 Sizes and Shapes of Crystals

Microphysical properties of ice clouds cannot be characterized by a single PSD curve as in the case of liquid clouds even if one considers relatively small volumes of a cloudy medium. This is due to the extremely complex shapes of ice particles in crystalline clouds. Major shapes of ice crystals are plates, columns, needles, sheaths, dendrites, stars and bullets. Combinations of bullets and needles in one ice cluster are also common. The Magano–Lee classification (see Table 1.6) of natural crystals includes 80 shapes (Magano and Lee, 1966), ranging from the elementary needle (classification index N1a) to the irregular germ (classification index G6). The predominating shape of crystals depends on temperature and pressure. Therefore, it varies inside a cloud. In particular, smaller and more irregular shapes are usually found near cloud tops.

The concentration of crystals N varies with height, often in the range of 50–50,000 crystals per cubic meter. The ice water content

$$C_i = N\langle W\rangle, \tag{1.25}$$

where $\langle W\rangle$ is the average mass of crystals, is usually in the range of $10^{-4}-10^{-1}$ g/m^3. Therefore, $\langle W\rangle$ is in the range of $2\times10^{-9}-2\times10^{-3}$ g. Crystals have a bulk density, ρ, less than that of bulk ice ($\rho = 0.3-0.9\mathrm{g/cm^3}$)) due the presence of impurities and bubbles inside ice particles (Landolt-Börnstein, 1988). The size of crystals is usually characterized by their maximal dimension H, which is related to the effective size (Yang et al., 2000). H is usually in the range of 0.1–6 mm for single crystals and of 1–15 mm for snow crystal aggregates. Smaller crystals (e.g., with maximal sizes around 20 μm) also are present in ice clouds (Yang et al., 2001). The mode parameter H_0 of size distribution curves $f(H)$ depends on the shape of crystals, with characteristic values of H_0 being 0.5 mm for plates and columns, 1 mm for needles, sheaths and stars and 2 mm for dendrites. Distribution curves can be modelled by gamma distributions with half-widths of distributions $\Delta_{1/2}$ being larger for larger values of H. Frequently, it follows: $\Delta_{1/2}\approx\bar{H}$, where $\bar{H}$ is the average size of crystals.

Simple shapes of ice crystals (e.g., hexagonal prisms) can be characterized by two dimensions: the length of the prism L and the diameter $D = a\sqrt{3}/2$, where a is the side of a hexagonal cross-section. Even in this most simple case two-dimensional distribution functions $f(D, L)$ should be used. Note that $f(D, L)$ can be approximately reduced to one-dimensional functions $f(D)$ due to the existence of empirical relationships between the length of crystals and their diameter in natural clouds (Auer and Veal, 1970). For instance, it follows for hexagonal columns with semi-width ζ and length L(Mitchel and Arnott, 1994) $\zeta = 0.35\,L$ at $L < 100$ μm and $\zeta = 3.48\sqrt{L}$ otherwise. Pruppacher and Klett (1978) give the following empirical relationship for plates: $L = 2.4483\zeta^{0.474}$ at $\zeta\in[5\ \mu\mathrm{m}, 1500\ \mu\mathrm{m}]$. Similar relationships can be obtained for other shapes (Mitchell and Arnott, 1994).

It should be pointed out, however, that the whole concept of a single distribution function breaks down for ice clouds. Recall that the microphysical properties of most water clouds (as far as their optical properties are concerned) can

Table 1.6. Meteorological classification of snow crystals (Magano and Lee, 1966).

	N1a Elementary needle	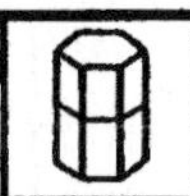	C1f Hollow column		P2b Stellar crystal with sectorlike ends
	N1b Bundle of elementary needles		C1g Solid thick plate		P2c Dendritic crystal with plates at ends
	N1c Elementary sheath	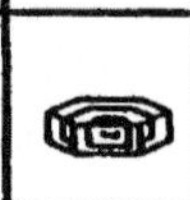	C1h Thick plate of skelton form		P2d Dendritic crystal with sectorlike ends
	N1d Bundle of elementary sheaths		C1i Scroll	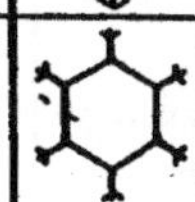	P2e Plate with simple extensions
	N1e Long solid column		C2a Combination of bullets	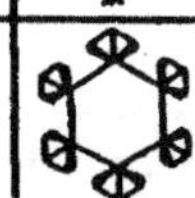	P2f Plate with sectorlike extensions
	N2a Combination of needles	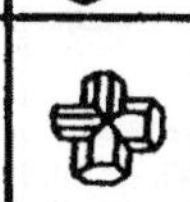	C2b Combination of columns		P2g Plate with dendritic extensions
	N2b Combination of sheaths	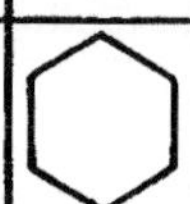	P1a Hexagonal plate	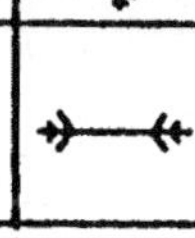	P3a Two-branched crystal
	N2c Combination of long solid columns		P1b Crystal with sectorlike branches	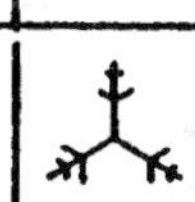	P3b Three-branched crystal
	C1a Pyramid		P1c Crystal with broad branches		P3c Four-branched crystal
	C1b Cup	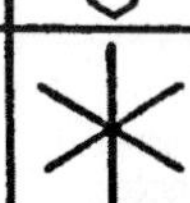	P1d Stellar crystal		P4a Broad branch crystal with 12 branches
	C1c Solid bullet		P1e Ordinary dendritic crystal		P4b Dendritic crystal with 12 branches
	C1d Hollow bullet		P1f Fernlike crystal		P5 Malformed crystal
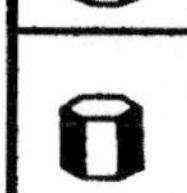	C1e Solid column	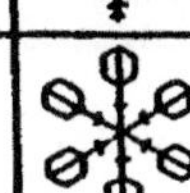	P2a Stellar crystal with plates at ends	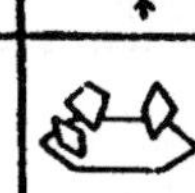	P6a Plate with spatial plates

(continued)

Table 1.6. *Continued*

	P6b Plate with spatial dendrites		CP3d Plate with scrolls at ends		R3c Graupellike snow with nonrimed extensions
	P6c Stellar crystal with spatial plates		S1 Side plates		R4a Hexagonal graupel
	P6d Stellar crystal with spatial dendrites		S2 Scalelike side planes		R4b Lump graupel
	P7a Radiating assemblage of plates		S3 Combination of side planes, bullets and columns		R4c Conelike graupel
	P7b Radiating assemblage of dendrites		R1a Rimed needle crystal		I1 Ice particle
	CP1a Column with plates		R1b Rimed columnar crystal		I2 Rimed particle
	CP1b Column with dendrites		R1c Rimed plate or sector		I3a Broken branch
	CP1c Multiple capped column		R1d Rimed stellar crystal		I3b Rimed broken branch
	CP2a Bullet with plates		R2a Densely rimed plate or sector		14 Miscellaneous
	CP2b Bullet with dendrites		R2b Densely rimed stellar crystal		G1 Minute column G2 Germ of skeleton form
	CP3a Stellar crystal with needles		R2c Stellar crystal with rimed spatial branches	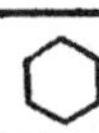	G3 Minute hexagonal plate
	CP3b Stellar crystal with columns		R3a Graupellike snow of hexagonal type	 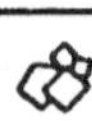	G4 Minute stellar crystal G5 Minute assemblage of plates
	CP3c Stellar crystal with scrolls at ends		R3b Graupellike snow of lump type		G6 Irregular germ

be characterized by just three numbers (e.g., the effective radius of droplets, the CV of the droplet distribution function $f(a)$ and the concentration of droplets) (Deirmendjian, 1969).

By contrast, at least 80 multi-dimensional PSDs are needed if one would like to use the classification of crystals developed by Magano and Lee (1966). This calls for the introduction of a new way of particle characterization in the case of complex particulate systems such as ice clouds. The same problem arises in the optics of mineral aerosol, blown from the Earth's surface to atmosphere (Mishchenko et al., 2002; Volten et al., 2001).

One of the possible solutions to the problem is the characterization of ice crystals in an elementary volume of a cloudy medium by the function

$$f(\vec{a},\vec{b}) = \sum_{r=1}^{N} c_r f_r(\vec{a}) + \sum_{i=1}^{M} c_i f_i(\vec{b}), \tag{1.26}$$

where $f_r(\vec{a})$ is the size distribution of particles of a regular shape (e.g., hexagonal plates or columns), $f_i(\vec{b})$ is the statistical distribution of particles with random surfaces or so-called irregularly shaped particles. Values of c_i and c_r give concentrations of different crystal habits. Note that instead of single variables (e.g., the radius of particles a) we need to introduce vector-parameters $\vec{a}$ and $\vec{b}$ in the case of crystalline clouds. In particular, we have for the components of the two-dimensional vector-parameter $\vec{a}$ with coordinates a_1, a_2 in the case of an idealized cloud with hexagonal columns: $a_1 = D, a_2 = L$. Functions $f_i(\vec{b})$ are distributions of some statistical characteristics of particles (e.g., average radii, correlation lengths, etc.). Clearly, the simplest case is the function $f(\vec{a},\vec{b})$ given by a sum of just two functions:

$$f(\vec{a},\vec{b}) = c_1 f_1(\vec{a}) + c_2 f_2(\vec{b}), \tag{1.27}$$

where the function $f_1(\vec{a})$ represents particles of a regular shape (say, hexagonal cylinders) and the function $f_2(\vec{b})$ represents statistical parameters of a single particle of an irregular shape. This irregularly shaped particle can be presented, e.g., as a fractal (Macke and Tzschihholz, 1992; Macke et al., 1996). It should be pointed out that the function $f_2(\vec{b})$ in this case represents fictive particles, which do not exist in clouds at all. However, ice cloud optical characteristics, calculated using $f_2(\vec{b})$, indeed represent quite well the optical characteristics of particles with extremely diverse shapes (Kokhanovsky, 2004a; Macke et al., 1996).

An important approach dealing with optical properties of large irregularly shaped particles was developed by Peltoniemi et al. (1989) and Muinonen et al. (1996). It is based on Monte-Carlo calculations of light scattering by a large particle with a rough surface. The model of spheres with rough surfaces was successfully applied to the optical characterization of irregularly shaped aerosol particles (Volten

Table 1.7. The frequency of occurrence of ice particle habits for the size ranges >125, >250, and >500 μm (Korolev et al., 2000).

	Fraction of spheres, %			Fraction irregulars, %			Fraction of needles and columns, %			Fraction dendrites, %			Particle concentration, *l*		
T, degrees C	125	250	500	125	250	500	125	250	500	125	250	500	125	250	500
[−5,0]	13	11	10	69	63	52	13	15	18	5	11	20	3.2	1.5	0.9
[−10,−5]	14	10	9	69	65	58	10	9	8	8	16	24	2.8	1.5	0.9
[−15,−10]	6	5	4	75	71	64	7	4	3	12	20	28	2.8	1.8	1.2
[−20,−15]	8	9	10	84	83	80	6	3	3	3	4	6	2.6	1.8	1.1
[−25,−20]	7	7	9	87	88	84	5	3	3	1	2	4	2.5	1.7	0.9
[−30,−25]	6	3	6	90	94	88	4	2	4	0	1	2	3.3	1.7	0.6
[−35,−30]	3	2	8	93	96	86	4	2	4	0	0	2	3.0	1.5	0.2
[−40,−35]	1	0	3	95	97	85	4	2	10	0	0	2	3.7	1.1	0.04
[−45,−40]	0	0	0	91	89	80	9	11	19	0	0	1	0.9	0.3	0.04
[−45,0]	7	5	6	84	83	76	6	6	8	3	6	10	2.9	1.6	0.9

et al., 2001). The model as given by Muinonen et al. (1996) can be applied to the case of ice clouds as well. For this the parameters of the irregularity of so-called fictive particles (Kokhanovsky, 2004a) should be changed accordingly. This is due to the different morphology of ice crystals (Magano and Lee, 1966) as compared to mineral aerosol (Okada et al., 2001). Also note that the refractive index of aerosol in the visible is larger than that of ice.

The importance of irregular shaped particles for optics of ice clouds is emphasized by experimental data shown in Table 1.7. These data confirm that the most of particles in ice clouds are of irregular shape. Actually, the shape parameters are functions of temperature and height inside of a cloud (see Fig. 1.18). Also it follows from Fig. 1.18 that shapes are not evenly distributed in different size intervals. Generally, smaller particles are closer to the spherical shape assumption as compared to larger ones (see Fig. 1.18). Larger ice crystals are found at the bottom of a cloud.

1.1.3 Refractive Indices of Liquid Water and Ice

The complex refractive index of particles suspended in the atmosphere is another important parameter in atmospheric optics studies (Liou, 2002). This is due to the fact that not only the size, shape and concentration of particles influence the light propagation in the atmosphere. The chemical composition, thermodynamic phase, temperature, presence of impurities and inhomogeneity of particles are of

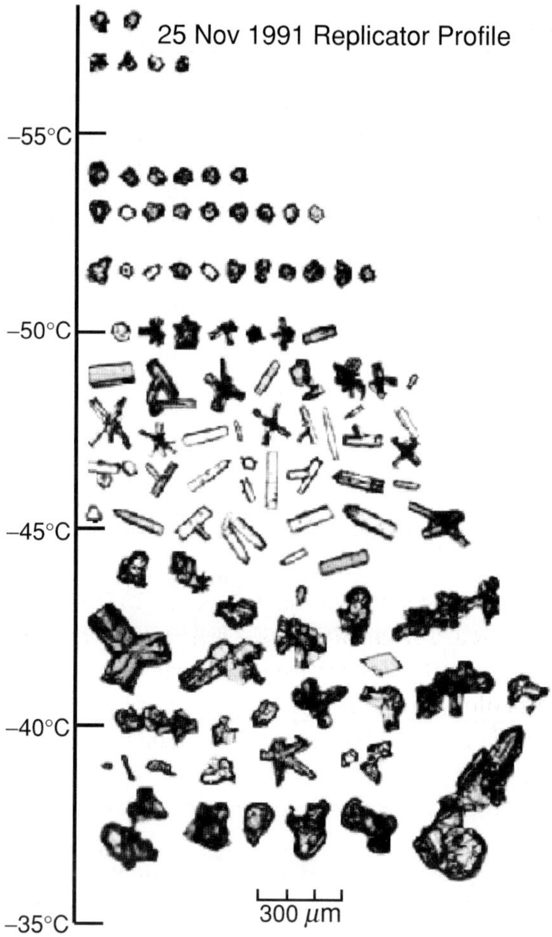

Fig. 1.18. Temperature profile of characteristic ice particles sampled by the NCAR balloon-borne replicator in a cirrus cloud on November 25th, 1991, near Coffeyville, Kansas, USA, during the NASA FIRE-II experiment (courtesy A. Heymsfield).

importance as well. The refractive index of water droplets and ice crystals varies with the temperature and has been tabulated by many authors [see e.g., Hale and Querry (1973) for liquid water and Warren (1984) for ice].

Spectral dependencies of real and imaginary parts of the complex refractive index of water and ice are presented in Appendix A and also in Figs. 1.19 and 1.20. The differences in light absorption by liquid and solid water are considerable, a fact that can be used for the retrieval of the cloud thermodynamic phase (liquid, ice or mixed-phase clouds) from ground (Dvoryashin, 2002) and satellite (Knap

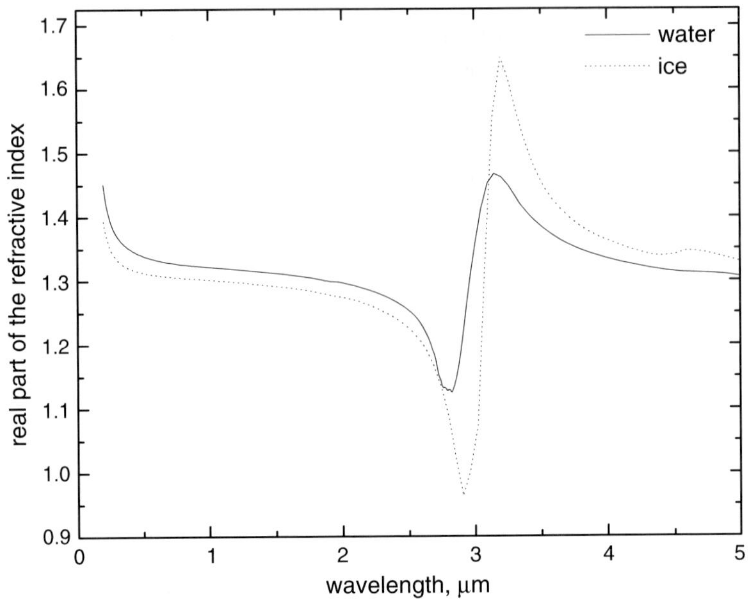

Fig. 1.19. Real part of the refractive index of water and ice as the function of the wavelength.

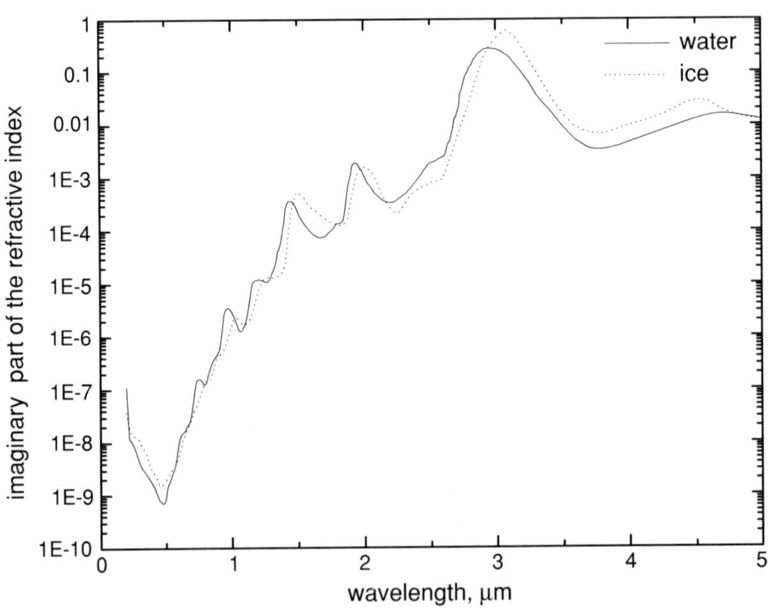

Fig. 1.20. The same as in Fig. 1.19 except for the imaginary part of the refractive index.

et al., 2002) measurements. In particular, we see that the absorption band around the wavelength $\lambda = 1.55$ μm for ice is shifted to larger wavelengths as compared to liquid water. This shift in the absorption band position can be easily detected with the use of modern spectrometers (see, e.g., Bovensmann et al., 1999).

The real part of the refractive indices of water n_w and ice n_i do not vary considerably in the visible and near-infrared regions of the electromagnetic spectrum. Generally $n_i(\lambda) < n_w(\lambda)$, with n_w in the range 1.33–1.34 for $\lambda = 0.4$–1 μm and n_i in the range 1.30–1.32 within the same spectral band. Somewhat larger values of refractive indices occur at shorter wavelengths.

The spectral variability of the imaginary part of the refractive index of water and ice, which is responsible for the level of absorption of solar radiation by clouds, is much higher (see Fig. 1.20). It changes by six orders of magnitude in the spectral range of 0.4–2 μm for both liquid water and ice.

Different impurities in water droplets, mainly soot (Markel, 2002) and various aerosol particles (Twomey, 1977), can change the imaginary part of the refractive index of droplets (especially in the visible, where water is almost transparent). This may influence the accuracy of modern cloud remote sensing techniques (Asano et al., 2001; Nakajima et al., 1991). Asano et al. (2001) found that the temperature dependence (Kuo et al., 1993) of the liquid water absorption influences the droplet sizing techniques.

1.2 Geometrical Characteristics of Clouds

Clouds are usually formed by the lifting of damp air. The lifting air parcel cools. This leads to an increase of the relative humidity. The excess vapour condenses on tiny particles (clouds nuclei), producing water droplets, which can grow due to coagulation processes. Drops at high altitudes freeze forming ice clouds. The ice phase is of great importance to precipitation processes. Some crystals originate on water-insoluble aerosol particles on which water vapour is deposited as ice. Formed clouds can travel large distances due to atmospheric circulation. Particles can evaporate. Also they can be deposited on the surface of our planet due to precipitation (e.g., snow and rain).

More than 50% of the Earth's surface is covered by clouds at any given time (see Fig. 1.21). This underlines an important role of clouds both in the hydrological cycle and planetary radiative budget. The collections of droplets and crystals can occupy large volumes of the troposphere at any place up to 20 km from the Earth's surface, depending on the latitude.

Clouds are classified not only in respect to their thermodynamic state (e.g., water, ice or mixed clouds) but also in respect to the cloud base height H_{bot}. In

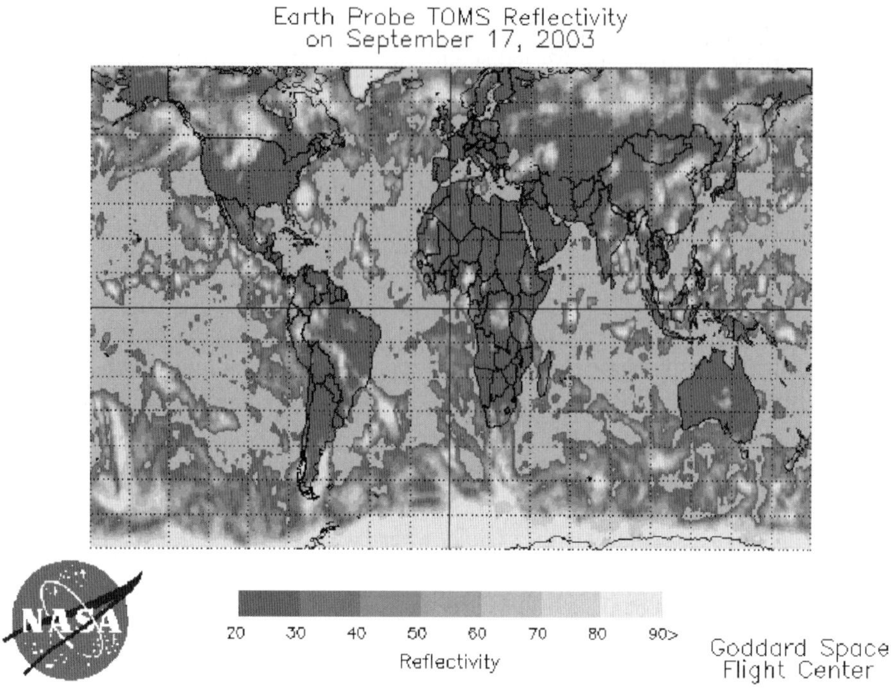

Fig. 1.21. Reflectivity of TOMS on September 17th, 2003. Bright areas correspond to clouds or ice/snow. Hurricane Isabel is clearly visible to the north-east of Florida (courtesy of NASA).

particular, Stratocumulus (Sc), Stratus (St) and Nimbostratus (Ns) are classified as low-level clouds ($H_{bot} < 2$ km). Typical values of H_{bot} for St are close to 0.4 km. They are a little bit higher for Sc and Ns (≈ 1 km). These estimations are at least 1 km higher in the areas close to the equator. The geometrical thickness ΔH of Sc is usually close to 0.6 km. St clouds have values of ΔH close to 1 km. Ns clouds have the largest thickness of all low-level clouds ($\Delta H \approx 1$–3 km and even larger than that). The position of the cloud top of low-level clouds depends on the season with higher cloud tops in summer. The latitude is also an important factor. Usually cloud top heights of St and Sc are close to 1 km and they are close to 2 km for Ns at mid-latitudes.

Cumulus (Cu) are usually classified as low-level clouds because their cloud bases are often below 2 km. Note that cumulus clouds belong to a category of clouds having considerable vertical extent. In summer in the middle latitudes these convective clouds have the following typical depths: Cu humilis, up to 1.0 km; Cumulus mediocris, 1–2 km; Cumulus congestus, up to 2–3.5 km; Cumulus fractus, 0.4–0.7 km. Cumulonimbus (Cb) has an extent more than 3–4 km. The principles of cloud classification, etymology, meaning of the Latin names and colour photos of different cloud types are given at http://users.pandora.be/skystef/clasclouds.htm.

Note that Cb clouds belong to low-level clouds as far as the classification with respect to H_{bot} is concerned. However, they have a typical geometrical thickness of more than 3–4 km and can reach the tropopause and even the stratosphere (up to 20 km in the tropics).

Middle-level clouds have cloud base heights in the region of 2–6 km (3–7 km in the tropics). Altostratus (As) and Altocumulus (Ac) clouds belong to this category. Usually Ac clouds are thinner (the geometrical thickness $\Delta H \approx 0.2$–0.7 km) as compared to As ($\Delta H \approx 1$ km). Median values of middle-level cloud top heights are in the range of 3–4 km in most of cases. High-level clouds are situated just below the tropopause. Due to low temperatures at such heights, they are composed of ice crystals. The values of H_{bot} are usually larger than 7 km. They are in the range of 11–13 km for tropics and of 6–8 km at middle latitudes. However, the height of high-level clouds is lower in the Arctic and the Antarctic (typically 5–6 km). This is due to the lower position of the tropopause near the poles as compared to the tropics. Cirrus (Ci), Cirrostratus (Cs) and Cirrocumulus (Cc) all belong to this category of clouds.

Ci are white, transparent and thin clouds. The values of H_{bot} for these clouds are in the range of 17–18 km in tropics. They are in the range of 7–10 km at middle latitudes. These clouds, although very thin in the majority of cases, may extend till the ground (e.g., in the Arctic) forming so-called "diamond dust". Cs clouds give an impression of a white or slightly bluish semi-transparent layers. They can have a great horizontal extent and cover all sky above the observer. The thickness of Cs is usually in the range of 0.1–3.0 km. Cc clouds are very thin non-precipitating clouds ($\Delta H \approx 0.2$–0.4 km). The frequency of appearance of such clouds is lower than that of Ci and Cs.

Cloud top heights for high-level clouds are typically above 12–14 km in the tropics, 9 km in middle altitudes and 8 km near poles. They can penetrate into the stratosphere (10–15% cases). In most cases, the cloud top height is situated in the 2–3 km layer below the tropopause.

We summarize the cloud classification described above in Table 1.8. Clouds can have various forms. Cumulus clouds may form clusters of about 3 km in diameter. The distances between clouds in a cluster range from about 200 to 300 m. Such groups can be 3–10 km apart. Fields of cumulus clouds typically occupy areas ranging from 10^4 to 10^5 km^2. Contrary to cumulus, stratocumulus clouds have horizontal dimensions of orders of magnitude greater than their vertical dimensions. In particular, it was found that such cloud systems can cover from 5 to 15 million km^2 on average. It should be remembered that all these different clouds are present in the atmosphere simultaneously and at different levels. The probability of having

Table 1.8. The cloud classification. Numbers in brackets give the most frequent altitudes, where correspondent clouds appear. The optical thickness τ of clouds in the column 1 is usually below 4. The sub-visual Cirrus clouds have optical thickness below 0.03. The optical thickness for clouds in the second column is usually in the range 4–25. Clouds in the last column are very thick ($\tau \approx 25 - 400$).

Thin clouds	Clouds having the moderate thickness	Thick clouds
Cirrus/6–6.5 km	Cirrostratus/7–7.5 km	Deep convection/0.5–6.5 km
Altocumulus/4–5 km	Altostratus/3–4 km	Nimbostratus/0.5–5 km
Cumulus/0.3–1 km	Stratocumulus/0.3–1.0 km	Stratus/0.5–1.0 km

just one single cloud layer is usually smaller (depending on the region and latitude) than of having a muti-layered (two or more layers) cloud system. In particular, it was found that the frequency of occurrence of single cloud layers over the territory of the former Soviet Union is only 30–50% (without account for high-level clouds above 6 km).

The typical cloudiness structure at a spring day over Cabauw (Netherlands) is shown in Fig. 1.22. These data were obtained using radar observations. We see Ci clouds around 8 km height, Cu—below 2 km and also middle-level clouds at

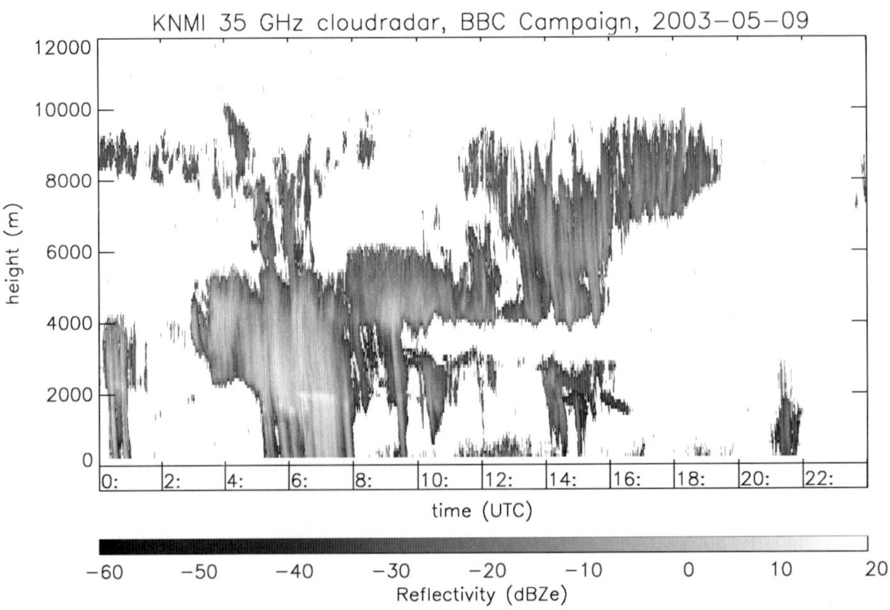

Fig. 1.22. The KNMI 35 GHz radar reflectivity as the function of time for May 9th, 2003 over Cabaw (The Netherlands) (Kokhanovsky et al., 2004).

4–6 km height. Most probably mixed-phase clouds also exist in the picture shown. This figure makes it easy to understand the complexity associated with calculations of light fields transmitted and reflected by cloud systems (Marshak and Davies, 2005). Because of this, most approaches (e.g., in operational cloud remote sensing) assume homogeneous cloud layers. This is very remote from reality and must be improved in future studies.

Chapter 2

OPTICS OF A SINGLE PARTICLE

2.1 Vector Wave Equation

The problem of light interaction with a single droplet or crystal lays in the very heart of cloud optics. Only if this interaction is fully understood further advances in studies of light multiple scattering in clouds are possible (e.g., using the radiative transfer theory).

A light beam consists of electromagnetic waves having large frequency ν (typically, 10^{15} oscillations per second). Therefore, we need to consider a general problem of an electromagnetic wave interaction with a macroscopic liquid or solid particle. The solution can be used at any ν (e.g., in the microwave region as well).

The electromagnetic field is fully described by Maxwell equations:

$$\vec{\nabla} \times \vec{\mathrm{E}} = -\frac{1}{c}\frac{\partial \vec{B}}{\partial t}, \tag{2.1}$$

$$\vec{\nabla} \times \vec{\mathrm{H}} = \frac{1}{c}\frac{\partial \vec{D}}{\partial t} + \frac{4\pi}{c}\vec{j}, \tag{2.2}$$

$$\vec{\nabla} \cdot \vec{D} = 4\pi\rho, \tag{2.3}$$

$$\vec{\nabla} \cdot \vec{B} = 0, \tag{2.4}$$

where $\vec{\mathrm{E}}$ is the electric vector, $\vec{B}$ is the magnetic induction, $\vec{D}$ is the electric displacement, $\vec{\mathrm{H}}$ is the magnetic vector, $\vec{j}$ is the electric current density, ρ is the density of charge, t is time, and c is the speed of light. Vectors in Maxwell equations satisfy the following relationships for the case of electromagnetic wave interaction with various isotropic symmetric substances (e. g., water)

$$\vec{j} = \sigma\vec{\mathrm{E}}, \tag{2.5}$$

$$\vec{D} = \varepsilon\vec{\mathrm{E}}, \tag{2.6}$$

$$\vec{B} = \mu\vec{\mathrm{H}}, \tag{2.7}$$

where σ is the specific conductivity, ε is the permittivity, μ is the magnetic permeability.

Equations (2.1)–(2.7) together with boundary conditions, which state that tangential components of $\vec{\mathrm{E}}$ and $\vec{\mathrm{H}}$ are continuous across the interface, allow us to consider a great variety of problems related to the interaction of electromagnetic waves with various substances. This is the main subject of the electrodynamics of continuous media.

Values of σ, ε, and μ are assumed to be known (e.g., from measurements). They depend on the substance under study. It follows for a vacuum: $\rho = \sigma = 0$, $\varepsilon = \mu = 1$, and therefore, $\vec{j} = 0$, $\vec{D} = \vec{\mathrm{E}}$, $\vec{B} = \vec{\mathrm{H}}$. It is known that $\rho = \sigma = 0$ and $\mu = 1$ for water at optical frequencies. So it follows from Eqs. (2.1)–(2.7) in this case:

$$\vec{\nabla} \times \vec{\mathrm{E}} = -\frac{1}{c}\frac{\partial \vec{\mathrm{H}}}{\partial t}, \tag{2.8}$$

$$\vec{\nabla} \times \vec{\mathrm{H}} = \frac{\varepsilon}{c}\frac{\partial \vec{\mathrm{E}}}{\partial t}, \tag{2.9}$$

$$\vec{\nabla} \cdot \vec{\mathrm{E}} = \vec{\nabla} \cdot \vec{\mathrm{H}} = 0. \tag{2.10}$$

We should find the solution of these equations, which are valid both inside and outside a spherical particle. Also boundary conditions should be satisfied. One possibility of the solution is the finite-difference time domain method. For this Eqs. (2.8) and (2.9) are discretized in the time domain. This is often used in studies of light scattering by nonspherical particles. We, however, would like to consider now the much more simpler problem of light scattering by a sphere. Then Eqs. (2.8) and (2.9) can be transformed into a single wave equation, which can be solved analytically in terms of special functions. Note that Eq. (2.10) follows from Eqs. (2.8) and (2.9), assuming that ε does not depend on spatial coordinates. The permittivity ε $\left(\varepsilon = m^2,\ m = n - i\chi \text{ is the refractive index}\right)$ depends on the frequency ω. This frequency dependence is determined by the electronic structure of a given substance and is the subject of quantum mechanics. We will assume that $\varepsilon(\omega)$ is known (e.g., from measurements). Also, we consider only plane monochromatic electromagnetic waves with the following time dependence: $\vec{\mathrm{E}} = \vec{E}e^{i\omega t}$, $\vec{\mathrm{H}} = \vec{H}e^{i\omega t}$. Note that only real parts of these fields have the physical sense. So, e.g., the product $\vec{\mathrm{E}} \times \vec{\mathrm{H}}$ means $Re(\vec{\mathrm{E}}) \times \mathrm{Re}(\vec{\mathrm{H}})$, where $\vec{\mathrm{E}}$ and $\vec{\mathrm{H}}$ can be substituted by their exponential forms given above.

Equations (2.8) and (2.9) simplify to the following relationships with account for exponential forms given above:

$$\vec{\nabla} \times \vec{E} = -\frac{i\omega}{c}\vec{H}, \vec{\nabla} \times \vec{H} = \frac{i\varepsilon\omega}{c}\vec{E}. \tag{2.11}$$

It follows from these expressions:

$$\vec{\nabla} \times \vec{\nabla} \times \vec{E} = \frac{\varepsilon\omega^2}{c^2}\vec{E} \tag{2.12}$$

or

$$\Delta\vec{E} + k^2 m^2 \vec{E} = 0, \tag{2.13}$$

where $k = \omega/c$ and we accounted for Eq. (2.10), the vector identity $\vec{\nabla} \times \vec{\nabla} \times \vec{E} = \vec{\nabla} \cdot (\vec{\nabla} \cdot \vec{E}) - \Delta\vec{E}$ and introduced the refractive index $m = \sqrt{\varepsilon}$ (see Appendix A). Also we have:

$$\Delta\vec{H} + k^2 m^2 \vec{H} = 0. \tag{2.14}$$

Equations (2.13) and (2.14) have the form similar to that of the scalar wave equation:

$$\Delta\Psi + k^2 m^2 \Psi = 0, \tag{2.15}$$

where Ψ is the scalar wave amplitude.

Vector and scalar wave equations have important applications in various branches of physics. Therefore, they have been studied in great detail. In particular, it was found that the solution of the vector wave equation for the arbitary vector $\vec{M}$, namely:

$$\Delta\vec{M} + k^2 m^2 \vec{M} = 0, \tag{2.16}$$

can be related to the solution of the scalar wave equation. In particular, it follows in spherical coordinates (r, θ, ϕ):

$$\vec{M} = \vec{\nabla} \times [\vec{a}_r(r\Psi)] = \vec{a}_\theta \frac{1}{r\sin\theta}\frac{\partial(r\Psi)}{\partial\phi} - \vec{a}_\phi \frac{1}{r}\frac{\partial(r\Psi)}{\partial\theta}, \tag{2.17}$$

where Ψ satisfies Eq. (2.15). Vectors $\vec{a}_r$, $\vec{a}_\theta$ and $\vec{a}_\phi$ are unit vectors in spherical coordinates. Note that the vector

$$\vec{N} = \frac{1}{mk}\vec{\nabla} \times \vec{M} \tag{2.18}$$

also satifies Eq. (2.16). This vector can be written in spherical coordinates as follows:

$$\vec{N} = \frac{\vec{a}_r}{mk}\left\{\frac{\partial^2(r\Psi)}{\partial r^2} + m^2k^2 r\Psi\right\} + \vec{a}_\theta \frac{1}{mkr}\frac{\partial^2(r\Psi)}{\partial r\partial\theta} + \vec{a}_\varphi \frac{1}{mkr\sin\theta}\frac{\partial^2(r\Psi)}{\partial r\partial\varphi}. \tag{2.19}$$

The soluton of Eq. (2.13) for the electric vector $\vec{E}$ can be presented as

$$\vec{E} = \vec{M}_v + i\vec{N}_u. \tag{2.20}$$

Also we have:

$$\vec{H} = m(-\vec{M}_u + i\vec{N}_v). \tag{2.21}$$

Equations (2.20) and (2.21) also satisfy Eq. (2.11). Subscripts u and v signify that $\vec{M}$ and $\vec{N}$ are related to two independent solutions (u and v) of the scalar wave Eq. (2.15).

Let us find the solution of Eq. (2.15) now. For this we follow the standard procedure. Namely, because the scatterer has a spherical shape, we write this equation in spherical coordinates:

$$\frac{1}{r^2}\frac{\partial}{\partial r}\left(r^2\frac{\partial\Psi}{\partial r}\right) + \frac{1}{r^2\sin\theta}\frac{\partial}{\partial\theta}\left(\sin\theta\frac{\partial\Psi}{\partial\theta}\right) + \frac{1}{r^2\sin\theta}\frac{\partial^2\Psi}{\partial\varphi} + k^2m^2\Psi = 0. \tag{2.22}$$

This insures an easy application of the boundary conditions at a later stage of derivations.

We represent the solution of Eq. (2.22) as

$$\Psi(r,\theta,\varphi) = A(r)B(\theta)C(\varphi). \tag{2.23}$$

The substitution of Eq. (2.23) into Eq. (2.22) gives:

$$\left[\sin^2\theta\frac{1}{A}\frac{\partial}{\partial r}\left(r^2\frac{\partial A}{\partial r}\right) + \sin\theta\frac{1}{B}\frac{\partial}{\partial\theta}\left(\sin\theta\frac{\partial B}{\partial\theta}\right) + k^2m^2r^2\sin^2\theta\right] = -\frac{1}{C}\frac{\partial^2 C}{\partial\varphi^2}. \tag{2.24}$$

Right and left parts of Eq. (2.24) are functions of different arguments. Therefore, the equality holds only if both parts are equal to the same constant j. Then it follows from Eq. (2.24):

$$\frac{1}{A}\frac{\partial}{\partial r}\left(r^2\frac{\partial A}{\partial r}\right) + k^2m^2r^2 = -\frac{1}{\sin\theta}\frac{1}{B}\frac{\partial}{\partial\theta}\left(\sin\theta\frac{\partial B}{\partial\theta}\right) + \frac{j}{\sin^2\theta} \tag{2.25}$$

and

$$\frac{1}{C}\frac{\partial^2 C}{\partial^2\varphi} + j = 0. \tag{2.26}$$

Eq. (2.26) can be easily solved. The answer is:

$$C(\varphi) = \varsigma_j\cos\sqrt{j}\varphi + \upsilon_j\sin\sqrt{j}\varphi, \tag{2.27}$$

where ς_j and υ_j are integration constants defined by boundary conditions.

Right and left parts of Eq. (2.25) depend on different arguments. So the equality can be satisfied if both parts are equal to the same constant, which we denote

$n(n+1)$ for further mathematical convenience:

$$\frac{\partial^2 (rA)}{\partial r^2} + \left\{k^2 m^2 - \frac{n(n+1)}{r^2}\right\} rA = 0, \tag{2.28}$$

$$\frac{1}{\sin\theta}\frac{d}{d\theta}\left(\sin\theta\frac{dB}{\partial\theta}\right) + \left\{n(n+1) - \frac{j}{\sin^2\theta}\right\} B = 0. \tag{2.29}$$

These two equations are well known in mathematical physics. In particular, Eq. (2.29) is the standard equation defining the associated Legendre polynomial $P_n^l(\cos\theta)$, where $l = \sqrt{j}$. Therefore, we have: $B \equiv P_n^l(\cos\theta)$.

Equation (2.28) can be solved in terms of the spherical Bessel functions. Namely, we have:

$$A(r) = c_n \frac{\psi_n(kmr)}{r} + d_n \frac{\chi_n(kmr)}{r}, \tag{2.30}$$

where

$$\psi_n(kmr) = \sqrt{\frac{\pi kmr}{2}} J_{n+\frac{1}{2}}(kmr), \tag{2.31}$$

$$\chi_n(kmr) = -\sqrt{\frac{\pi kmr}{2}} N_{n+\frac{1}{2}}(kmr) \tag{2.32}$$

and $J_{n+(1/2)}(kmr)$ and $N_{n+(1/2)}(kmr)$ are Bessel functions. Constants c_n and d_n are defined by boundary conditions. Using Eq. (2.23), we have for the general solution of the scalar wave

$$\Psi(r,\theta,\phi) = r^{-1}\sum_{n=0}^{\infty}\sum_{l=-n}^{n} P_n^l(\cos\theta)\left[c_n\psi_n(kmr) + d_n\chi_n(kmr)\right](\varsigma_l\cos l\phi + \upsilon_l\sin l\phi), \tag{2.33}$$

where any term in the sum satisfies Eq. (2.15). Equation (2.33) can be used to find the solution of the vector wave equations for $\vec{E}$ and $\vec{H}$ using Eqs. (2.17), (2.19)–(2.21). This completes the solution.

Let us apply the mathematical theory presented above to the solution of a particular problem, namely electromagnetic light scattering by an isotropic homogeneous sphere. The main issue is the derivation of arbitrary coefficients c_n, d_n, ς_n, υ_n in Eq. (2.33) as functions of the refractive index and the size of a spherical particle.

2.2 Mie Theory

We consider the problem of a plane monochromatic linearly polarized electromagnetic wave scattering by an isotropic homogeneous sphere with the refractive index m and the radius a. It is assumed that the sphere is surrounded by vacuum, which is a valid approximation for cloud droplets and crystals in air. Let us select the origin of a rectangular coordinate system in the center of the sphere, with the positive Z axis along the direction of propagation of the incident wave ($\vec{k} \parallel \vec{e_z}$). Here $\vec{e_z}$ is the unity vector along the positive Z axis and the wavevector $\vec{k}(|\vec{k}| = \omega/c)$ points in the direction of the wave propagation. The electric vector $\vec{E}^i$ of the incident wave is assumed to be parallel to the unit vector $\vec{e_x}$ along the coordinate axis OX:

$$\vec{E}^i = E_0^i e^{-ikz}\,\vec{e_x}. \tag{2.34}$$

Then we have for the magnetic vector $\vec{H}^i$ of the incident wave:

$$\vec{H}^i = H_0^i e^{-ikz}\vec{e_y}. \tag{2.35}$$

Here E_0^i and H_0^i are correspondent amplitudes and $\vec{e_y}$ is the unity vector in the direction of the positive axis OY. The unit vectors $\vec{a_r}$, $\vec{a_\theta}$ and $\vec{a_\varphi}$ in spherical coordinates are related to unit vectors $\vec{e_r}$, $\vec{e_\theta}$ and $\vec{e_\varphi}$ in Cartesian coordinates as

$$\vec{a_r} = \vec{e_x}\sin\theta\cos\phi + \vec{e_y}\sin\theta\sin\phi + e_z\cos\theta, \tag{2.36}$$

$$\vec{a_\theta} = \vec{e_x}\cos\theta\cos\phi + \vec{e_y}\cos\theta\sin\phi - e_z\sin\theta, \tag{2.37}$$

$$\vec{a_\phi} = -\vec{e_x}\sin\phi + \vec{e_y}\cos\phi. \tag{2.38}$$

Taking into account Eqs. (2.36)–(2.38), we obtain:

$$E_r^i = \vec{E}^i\vec{a_r} = E_0^i\sin\theta\cos\phi e^{-ikr\cos\theta}, \tag{2.39}$$

$$E_\theta^i = \vec{E}^i\vec{a_\theta} = E_0^i\cos\theta\cos\phi e^{-ikr\cos\theta}, \tag{2.40}$$

$$E_\phi^i = \vec{E}^i\vec{a_\phi} = -E_0^i\sin\phi e^{-ikr\cos\theta}, \tag{2.41}$$

where we accounted for the fact that $z = r\cos\theta$. Note that it also follows: $x = r\sin\theta\cos\phi$, $y = r\sin\theta\sin\phi$. We obtain in a similar way:

$$H_r^i = H_0^i\sin\theta\sin\phi e^{-ikr\cos\theta}, \tag{2.42}$$

$$H_\theta^i = H_0^i\cos\theta\sin\phi e^{-ikr\cos\theta}, \tag{2.43}$$

$$H_\phi^i = H_0^i\cos\phi e^{-ikr\cos\theta}. \tag{2.44}$$

Let us find potentials u and v for the incident field. It follows from Eqs. (2.17), (2.19) and (2.20):

$$\vec{E} = E_r\vec{a_r} + E_\theta\vec{a_\theta} + E_\phi\vec{a_\phi}, \tag{2.45}$$

where

$$E_r = \frac{i}{mk}\left\{\frac{\partial^2(ru)}{\partial r^2} + m^2k^2ru\right\}, \tag{2.46}$$

$$E_\theta = \frac{1}{r\sin\theta}\frac{\partial(rv)}{\partial\phi} + \frac{i}{mkr}\frac{\partial^2(ru)}{\partial r\partial\theta}, \tag{2.47}$$

$$E_\phi = -\frac{1}{r}\frac{\partial(rv)}{\partial\theta} + \frac{i}{mkr\sin\theta}\frac{\partial^2(ru)}{\partial r\partial\phi} \tag{2.48}$$

Table 2.1. Scattered and internal field components expressed via scalar potentials of the scattered field (u_s, v_s) and internal field (u_t, v_t). Here $\bar{m}$ is the refractive index of outer space assumed to be equal one in this work. The value of m gives the complex refractive index of a scatterer.

Field component	Scattered field	Internal field
E_r	$\frac{i}{\bar{m}k}\left\{\frac{\partial^2(ru_s)}{\partial r^2}+\bar{m}^2k^2ru_s\right\}$	$\frac{i}{mk}\left\{\frac{\partial^2(ru_t)}{\partial r^2}+m^2k^2ru_t\right\}$
E_θ	$\frac{1}{r\sin\theta}\frac{\partial(rv_s)}{\partial\phi}+\frac{i\partial^2(ru_s)}{\bar{m}kr\partial r\partial\theta}$	$\frac{1}{r\sin\theta}\frac{\partial(rv_t)}{\partial\phi}+\frac{i\partial^2(ru_t)}{mkr\partial r\partial\theta}$
E_ϕ	$-\frac{\partial(rv_s)}{r\partial\theta}+\frac{i\partial^2(ru_s)}{\bar{m}kr\sin\theta\partial r\partial\phi}$	$-\frac{\partial(rv_t)}{r\partial\theta}+\frac{i\partial^2(ru_t)}{mkr\sin\theta\partial r\partial\phi}$
H_r	$\frac{i}{k}\left\{\frac{\partial^2(rv_s)}{\partial r^2}+\bar{m}^2k^2rv_s\right\}$	$-\frac{i}{k}\left\{\frac{\partial^2(rv_t)}{\partial r^2}+m^2k^2rv_t\right\}$
H_θ	$-\frac{\bar{m}}{r\sin\theta}\frac{\partial(ru_s)}{\partial\phi}+\frac{i}{kr}\frac{\partial^2(rv_s)}{\partial r\partial\theta}$	$-\frac{m}{r\sin\theta}\frac{\partial(ru_t)}{\partial\phi}+\frac{i}{kr}\frac{\partial^2(rv_t)}{\partial\theta r\partial\theta}$
H_ϕ	$\frac{\bar{m}}{r}\frac{\partial(ru_s)}{\partial\theta}+\frac{i}{kr\sin\theta}\frac{\partial^2(rv_s)}{\partial r\partial\phi}$	$\frac{m}{r}\frac{\partial(ru_t)}{\partial\theta}+\frac{i}{kr\sin\theta}\frac{\partial^2(rv_t)}{\partial r\partial\phi}$

and similarly for $\vec{H}$. For convenience, we present electric and magnetic scattered and internal field components in Table 2.1.

It follows from Eqs. (2.39) and (2.46):

$$E_r^i=\frac{i}{k}\left(\frac{\partial^2(ru^i)}{\partial r^2}+k^2ru^i\right)=e^{-ikr\cos\theta}\sin\theta\cos\phi, \tag{2.49}$$

where we assumed that $E_0^i=1$ and accounted for the fact that $m=1$ outside the particle.

The next step is to determine u^i from Eq. (2.49). We use the Bauer's formula:

$$e^{-ikr\cos\theta}=\sum_{n=0}^{\infty}(-i)^n(2n+1)\frac{\psi_n(kr)}{kr}P_n(\cos\theta) \tag{2.50}$$

and also account for identities:

$$e^{-ikr\cos\theta}\sin\theta=\frac{1}{ikr}\frac{\partial}{\partial\theta}(e^{-ikr\cos\theta}),\ \frac{\partial}{\partial\theta}P_n(\cos\theta)=-P_n^1(\cos\theta),\ P_0^1(\cos\theta)=0. \tag{2.51}$$

Then it follows:

$$e^{-ikr\cos\theta}\sin\theta\cos\phi=\frac{1}{(kr)^2}\sum_{n=1}^{\infty}(-i)^{n-1}(2n+1)\psi_n(kr)P_n^1(\cos\theta)\cos\phi. \tag{2.52}$$

We apply the following trial solution of Eq. (2.49) for u^i:

$$u^i = \frac{1}{kr} \sum_{n=1}^{\infty} \alpha_n \psi_n (kr) P_n^1 (\cos\theta) \cos\phi, \tag{2.53}$$

which follows from Eq. (2.33) taking into account the form given by Eq. (2.52). Substitution of Eqs. (2.52) and (2.53) into Eq. (2.49) gives the following result:

$$\alpha_n \left[k^2 \psi_n (kr) + \frac{\partial^2 \psi_n (kr)}{\partial r^2} \right] = (-i)^n (2n+1) \frac{\psi_n (kr)}{r^2}. \tag{2.54}$$

This can be written as

$$\frac{\partial^2 \psi_n (kr)}{\partial r^2} + \left[k^2 - \frac{s}{r^2} \right] \psi_n (kr) = 0, \tag{2.55}$$

where

$$s = (-i)^n (2n+1) \alpha_n^{-1}. \tag{2.56}$$

It follows from Eq. (2.28) that Eq. (2.55) is satisfied if $s = n(n+1)$ or [see Eq. (2.56)]

$$\alpha_n = (-i)^n \frac{2n+1}{n(n+1)}. \tag{2.57}$$

Therefore, we obtain [see Eqs. (2.53) and (2.57)]:

$$u^i = \frac{1}{kr} \sum_{n=1}^{\infty} (-i)^n \frac{2n+1}{n(n+1)} \psi_n (kr) P_n^1 (\cos\theta) \cos\phi. \tag{2.58}$$

Using the same procedure and Eqs. (2.42), (2.17), (2.19) and (2.21), it follows:

$$v^i = \frac{1}{kr} \sum_{n=1}^{\infty} (-i)^n \frac{2n+1}{n(n+1)} \psi_n (kr) P_n^1 (\cos\theta) \sin\phi \tag{2.59}$$

and, therefore, $v^i = u^i \tan\phi$. Functions u^i, v^i can be used to calculate the components of the incident electric vector E_θ^i, E_ϕ^i (and also H_θ^i, H_ϕ^i). The general solution (2.33) should match u^i and v^i for the incident field at the boundary of the particle. It means that only the case $l = 1$ should be left in Eq. (2.33). Then we have for the internal wave potentials u^t and v^t:

$$u^t = \frac{1}{kr} \sum_{n=1}^{\infty} \bar{c}_n \psi_n \left(\tilde{k}r\right) P_n^1 (\cos\theta) \cos\phi, \tag{2.60}$$

$$v^t = \frac{1}{kr} \sum_{n=1}^{\infty} \bar{d}_n \psi_n \left(\tilde{k}r\right) P_n^1 (\cos\theta) \sin\phi, \tag{2.61}$$

where $\bar{c}_n = kc_n\varsigma_1, \bar{d}_n = kd_n\upsilon_1$, and we neglected the terms with $\chi_n(kmr)$ because they lead to nonphysical results as $r \to 0$ (e.g., an infinite value of an electric field in the center of the sphere). Similar equations are valid for potentials of the scattered wave outside the particle:

$$u^s = \frac{1}{kr}\sum_{n=1}^{\infty} \tilde{c}_n \xi_n (kr)\, P_n^1 (\cos\theta) \cos\phi, \tag{2.62}$$

$$v^s = \frac{1}{kr}\sum_{n=1}^{\infty} \tilde{d}_n \xi_n (kr)\, P_n^1 (\cos\theta) \sin\phi \tag{2.63}$$

where we used functions

$$\xi_n (kr) = \psi_n (kr) + i\chi_n (kr) \tag{2.64}$$

instead of $\psi_n(kmr)$ as in Eqs. (2.60) and (2.61) because functions $\xi_n (kr)$ (and u^s, v^s) vanish at infinity, which is the case for scattered fields.

Coefficients $\bar{c}_n, \bar{d}_n, \tilde{c}_n, \tilde{d}_n$ can be derived using boundary conditions. Namely, it follows:

$$E_\theta^i + E_\theta^s = E_\theta^t,\ H_\theta^i + H_\theta^s = H_\theta^t,\ E_\phi^i + E_\phi^s = E_\phi^t,\ H_\phi^i + H_\phi^s = H_\phi^t \tag{2.65}$$

at the surface of the sphere ($r = a$). Components E_θ, E_ϕ (and also H_θ, H_ϕ) depend on the following combinations: $\partial(rv)/\partial r$, $\partial(ru)/m\partial r, mu$, and also v. These combinations should satisfy boundary conditions at $r = a$. This ensures that boundary conditions for field components are satisfied as well. Therefore, we have

$$\frac{\partial}{\partial r}[r(u^i + u^s)] = \frac{1}{m}\frac{\partial}{\partial r}(ru^t), \tag{2.66}$$

$$\frac{\partial}{\partial r}[r(v^i + v^s)] = \frac{\partial}{\partial r}(rv^t) \tag{2.67}$$

$$u^i + u^s = mu^t,\ v^i + v^s = v^t. \tag{2.68}$$

Let us introduce amplitude coefficients a_n, b_n, p_n, q_n in such a way that

$$\tilde{c}_n = -\beta_n \cdot a_n,\quad \tilde{d}_n = -\beta_n b_n,\quad \bar{c}_n = \frac{\beta_n p_n}{m},\quad \bar{d}_n = \frac{\beta_n q_n}{m} \tag{2.69}$$

where

$$\beta_n = (-i)^n \frac{2n+1}{kn\,(n+1)}. \tag{2.70}$$

Then it follows from Eq. (2.66) at $r = a$:

$$\psi_n' (ka) - a_n \xi_n' (ka) = \frac{1}{m} p_n \psi_n' (kma). \tag{2.71}$$

The value of m in the dominator appears because the prime denotes differentiation with respect to the argument ($\partial\psi_n(mkr)/\partial r = mk\psi_n'$). Also we have from Eqs. (2.68) at $r = a$:

$$\psi_n(ka) - a_n\xi_n(ka) = p_n\psi_n(mka). \tag{2.72}$$

The combination of Eqs. (2.71) and (2.72) gives:

$$a_n = \frac{\psi_n'(mka)\psi_n(ka) - m\psi_n'(ka)\psi_n(mka)}{\psi_n'(mka)\xi_n(ka) - m\xi_n'(ka)\psi_n(mka)} \tag{2.73}$$

and

$$p_n = \frac{m\left(\xi_n(ka)\psi_n'(ka) - \psi_n(ka)\xi_n'(ka)\right)}{\psi_n'(mka)\xi_n(ka) - m\xi_n'(ka)\psi_n(mka)}. \tag{2.74}$$

It follows from remaining boundary conditions:

$$\psi_n'(ka) - b_n\xi_n'(ka) = q_n\psi_n'(kma), \tag{2.75}$$

$$\psi_n(ka) - b_n\xi_n(ka) = \frac{q_n}{m}\psi_n(kma), \tag{2.76}$$

where we used the properties

$$\frac{\partial\psi_n(kmr)}{\partial r} = km\psi_n', \quad \frac{\partial\psi_n(kr)}{\partial r} = k\psi_n'(kr), \quad \frac{\partial\xi_n(kr)}{\partial r} = k\xi_n'(kr). \tag{2.77}$$

This means that

$$b_n = \frac{m\psi_n'(mka)\psi_n(ka) - \psi_n(mka)\psi_n'(ka)}{m\psi_n'(mka)\xi_n(ka) - \psi_n(mka)\xi_n'(ka)} \tag{2.78}$$

and

$$q_n = \frac{m\left(\psi_n'(ka)\xi_n(ka) - \psi_n(ka)\xi_n'(ka)\right)}{m\psi_n'(mka)\xi_n(ka) - \psi_n(kma)\xi_n'(ka)}. \tag{2.79}$$

Now electric $\vec{E}$ and magnetic $\vec{H}$ fields inside and outside the spherical particle can be easily found (see Table 2.1). The only obstacle is the calculation of Legendre and Bessel functions. However, for this standard subroutines are available. These equations can be used to find the electromagnetic field in any point outside and inside a spherical particle. Internal field solutions are of a great importance to studies of laser beams propagation in clouds. Then, depending on the intensity of a beam, droplets can evaporate. Also other interesting effects are happening (see, e.g., Prishivalko, 1983, Prishivalko et al., 1984).

Droplets in clouds are separated by large distances ($kr \gg 1$). This means that only far-field solutions for a scattered field are of interest for most cloud optics

Table 2.2. Far field solutions and incident fields ($E_0 = H_0 = 1$). The angular functions are defined as follows: $\pi_n = \dfrac{P_n^1(\cos\theta)}{\sin\theta}$, $\tau_n = \dfrac{dP_n^1(\cos\theta)}{d\theta}$.

Field component	Far—field solutions and incident fields
$E_\theta^s = H_\phi^s$	$\dfrac{1}{ikr}e^{-ikr}\cos\phi\sum_{n=1}^{\infty}\dfrac{2n+1}{n(n+1)}[a_n\tau_n + b_n\pi_n]$
$E_\phi^s = -H_\theta^s$	$-\dfrac{1}{ikr}e^{-ikr}\sin\phi\sum_{n=1}^{\infty}\dfrac{2n+1}{n(n+1)}[a_n\pi_n + b_n\tau_n]$
$E_\theta^i = H_\theta^i\cot\phi$	$\dfrac{1}{kr}\cos\phi\sum_{n=1}^{\infty}\dfrac{2n+1}{n(n+1)}(-i)^n\left[\psi_n(kr)\pi_n + i\psi_n'(kr)\tau_n\right]$
$E_\phi^i = -H_\phi^i\tan\phi$	$-\dfrac{1}{kr}\sin\phi\sum_{n=1}^{\infty}\dfrac{2n+1}{n(n+1)}(-i)^n\left[\psi_n(kr)\tau_n + i\psi_n'(kr)\pi_n\right]$

studies. Then it follows: $\xi_n(kr) \to i^{n+1}\exp(-ikr)$ and expressions for $U_s \equiv ru_s$ and $V_s = rv_s$ are greatly simplified. Namely, we have:

$$U_s = -\frac{ie^{-ikr}\cos\phi}{k}\sum_{n=1}^{\infty}\frac{(2n+1)}{n(n+1)}a_n P_n^1(\cos\theta), \tag{2.80}$$

$$V_s = -\frac{ie^{-ikr}\sin\phi}{k}\sum_{n=1}^{\infty}\frac{(2n+1)}{n(n+1)}b_n P_n^1(\cos\theta). \tag{2.81}$$

The substitution of these expressions in formulae given in Table 2.1 allows us to obtain analytical equations for E_θ, E_ϕ, H_θ, H_ϕ (see Table 2.2) in terms of spherical coordinates (r, θ, φ) and parameters of spherical cloud droplets $(m,\ a)$. It follows that E_θ, E_φ, H_θ, $H_\phi \sim (kr)^{-1}$. It is easy to show that $E_r \sim (kr)^{-2}$. Therefore, the radial component can be neglected, if one concentrates on far field effects.

Note that scattered fields depend on the product ka or $2\pi a/\lambda$. This allows for experimental studies of light scattering by small particles using microwave techniques. Indeed scattered patterns for radiation of any wavelength coincide if the parameter $x = ka$ is kept constant. For this, however, the value of m should also be constant, which is difficult to achieve taking into account strong spectral dependencies of the refractive index of correspondent substances. Therefore, to model light scattering, e.g., by an ice crystal of a given shape, using microwave techniques one needs to prepare a sample having the same size/wavelength characteristics as a crystal from a substance, which has the same value of m at a given microwave frequency as, e.g., ice in visible light. Such experiments are of great importance in studies of light scattering by nonspherical particles, where theory still did not achieve the simplicity expressed in expressions given in Table 2.2 valid for spherical particles.

2.3 Differential and Integral Light Scattering Characteristics

2.3.1 Amplitude Scattering Matrix

Expressions for $\vec{E}$ and $\vec{H}$ obtained in the previous section give a complete solution of the problem of an electromagnetic wave interaction with a single spherical particle. Now we will introduce a number of important characteristics, which are used in studies of light scattering and propagation in cloudy media. We start from the definition of the amplitude scattering matrix.

The distances between particles in clouds are much larger than the wavelength of incident solar light. Therefore, we should use far-field ($kr \to \infty$) solutions of equations presented above. Then we have: $\xi_n(kr) \approx i^{n+1} \exp(-ikr)$ and solutions given in Table 2.2 can be presented in the following form:

$$E_\phi^s = \frac{i}{kr} e^{-ikr} \sin\phi S_1(\theta), \; E_\theta^s = -\frac{i}{kr} e^{-ikr} \cos\phi S_2(\theta), \tag{2.82}$$

where we introduced the amplitude functions:

$$S_1(\theta) = \sum_{n=1}^{\infty} \frac{(2n+1)}{n(n+1)} (a_n \pi_n + b_n \tau_n), \; S_2(\theta) = \sum_{n=1}^{\infty} \frac{(2n+1)}{n(n+1)} (b_n \pi_n + a_n \tau_n) \tag{2.83}$$

and

$$\pi_n = \frac{P_n^1(\cos\theta)}{\sin\theta}, \quad \tau_n = \frac{dP_n^1(\cos\theta)}{d\theta}. \tag{2.84}$$

Note that it follows at $\theta = 0$:

$$\pi_n(1) = \tau_n(1) = \frac{n(n+1)}{2} \tag{2.85}$$

and, therefore,

$$S(0) = S_1(0) = S_2(0) = \sum_{n=1}^{\infty} \left(n + \frac{1}{2}\right)(a_n + b_n). \tag{2.86}$$

Let us define components perpendicular (E_r) and parallel (E_l) to the scattering plane, which holds incident and scattered beams. Then it follows:

$$E_r^s = -E_\phi^s, \; E_l^s = E_\theta^s. \tag{2.87}$$

Also we have for the incident electric vector

$$E_r^i = e^{-ikz} \sin\phi, \; E_l^i = e^{-ikz} \cos\phi. \tag{2.88}$$

Therefore, it follows instead of Eqs. (2.82):

$$\begin{pmatrix} E_l^s \\ E_r^s \end{pmatrix} = \frac{e^{ik(z-r)}\hat{S}(\theta)}{ikr} \begin{pmatrix} E_l^i \\ E_r^i \end{pmatrix}, \tag{2.89}$$

where

$$\hat{S} = \begin{pmatrix} S_2 & 0 \\ 0 & S_1 \end{pmatrix} \tag{2.90}$$

for homogeneous isotropic spherical particles. The elements of the matrix $\hat{S}$ are $S_{11}(\theta) = S_2(\theta)$, $S_{22} = S_1(\theta)$, $S_{12}(\theta) = S_{21}(\theta) = 0$. The amplitude matrix $\hat{S}$ plays a paramount role in the light scattering theory. Elements of this matrix can be used to calculate all relevant light scattering and extinction characteristics. This is considered in next sections in more detail.

2.3.2 Cross Sections

2.3.2.1 Scattering and absorption cross sections

The values of E_l^s, E_r^s and, therefore, elements of the amplitude scattering matrix can be directly measured using microwave techniques at large wavelengths. This is not the case for optical frequencies ($\lambda \to 0$). Then only parameters quadratic with respect to the field can be measured (e.g., scattered and absorbed energy). This issue is considered in detail by Rozenberg (1973). It is known from the electromagnetic theory that the intensity of energy flow at any point in the field is given by the absolute value of the Poynting's vector

$$\vec{P} = \left[\vec{\mathrm{E}} \times \vec{\mathrm{H}}\right]. \tag{2.91}$$

This can be shown as follows. Let us multiply Eq. (2.1) by $\vec{\mathrm{H}}$ and Eq. (2.2) by $\vec{\mathrm{E}}$ and subtract resulting equations from one another. Then it follows:

$$\vec{\mathrm{H}}\vec{\nabla} \times \vec{\mathrm{E}} - \vec{\mathrm{E}}\vec{\nabla} \times \vec{\mathrm{H}} = -\frac{1}{c}\left[\vec{\mathrm{H}}\frac{\partial \vec{B}}{\partial t} + \vec{\mathrm{E}}\frac{\partial \vec{D}}{\partial t} + 4\pi\vec{\mathrm{E}}\vec{j}\right] \tag{2.92}$$

or accounting for the vector identity

$$\vec{\mathrm{H}}\vec{\nabla} \times \vec{\mathrm{E}} - \vec{\mathrm{E}}\vec{\nabla} \times \vec{\mathrm{H}} = \nabla\left[\vec{\mathrm{E}} \times \vec{\mathrm{H}}\right] \tag{2.93}$$

we have:

$$\vec{\nabla}\left[\vec{\mathrm{E}} \times \vec{\mathrm{H}}\right] + \frac{4\pi}{c}\vec{\mathrm{E}}\vec{j} = -\frac{\partial \omega}{\partial t}, \tag{2.94}$$

where we introduced the derivative

$$\frac{\partial \omega}{\partial t} = \frac{1}{c}\left\{\vec{\mathrm{H}}\frac{\partial \vec{B}}{\partial t} + \vec{\mathrm{E}}\frac{\partial \vec{D}}{\partial t}\right\}. \tag{2.95}$$

This derivative can be interpreted as the rate of decrease of electric and magnetic energy stored in the volume V (Stratton, 1941).

Let us integrate Eq. (2.94) over a volume V bounded by a surface S. Then it follows:

$$\int_S \vec{P}\vec{n}ds + \frac{4\pi}{c}\int_V \vec{\mathrm{E}}\vec{j}dV = -\frac{\partial W}{\partial t}, \tag{2.96}$$

where

$$W = \int_V w dV \tag{2.97}$$

and we used Eq. (2.91) and the Stokes theorem:

$$\int_V \vec{\nabla} \times \vec{P}dV = \int_S \vec{P}d\vec{S}, \tag{2.98}$$

where $d\vec{S} = \vec{n}ds$ is the unit area vector directed along the outer normal to the closed surface S. Equation (2.96) states the decrease of energy in the volume is due to two processes. The first one is given by the flux A_{sca} of the vector $\vec{P}$ through the surface S:

$$A_{sca} = \int_S \vec{P}\vec{n}ds. \tag{2.99}$$

The second integral in Eq. (2.96) can be interpreted as the electromagnetic power A_{abs} absorbed in the medium enclosed inside S per unit time. So we have:

$$\frac{\partial W}{\partial t} = -A_{sca} - A_{abs} \tag{2.100}$$

and A_{sca} can be interpreted as the total scattered power (per unit time). Optical detectors are characterized by a time constant τ, which is very large in comparison with the period of oscillations $T \sim 10^{-15}c$ of optical waves. So only time-averaged values of A_{abs} and A_{sca}can be measured. In particular, we have:

$$\langle A_{sca}\rangle = \int_S \langle\vec{P}\rangle\vec{n}ds, \tag{2.101}$$

$$\langle A_{abs}\rangle = \frac{4\pi}{c}\int_V \langle\vec{\mathrm{E}}\vec{j}\rangle dV, \tag{2.102}$$

where

$$\langle P\rangle = \frac{1}{\tau}\int_0^{\tau} \vec{P}(t)dt, \tag{2.103}$$

$$\langle \vec{\mathrm{E}}\vec{j}\rangle = \frac{1}{\tau}\int_0^{\tau} \vec{\mathrm{E}}(t)\vec{j}(t)dt. \tag{2.104}$$

Let us use the following representations: $\vec{\mathrm{E}} = \vec{E}e^{i\omega t}$ and $\vec{\mathrm{H}} = \vec{H}e^{i\omega t}$. Then it follows

$$\vec{P}(t)=[\mathrm{Re}\{\vec{\mathrm{E}}\}\times\mathrm{Re}\{\vec{\mathrm{H}}\}]=\frac{1}{4}[\vec{E}e^{-i\omega t}+\vec{E}^*e^{i\omega t}][\vec{H}e^{-i\omega t}+H^*e^{i\omega t}] \tag{2.105}$$

or

$$\vec{P}(t)=\frac{1}{4}\{[\vec{E}\times\vec{H}^*]+[\vec{E}^*\times\vec{H}]+[\vec{E}^*\times\vec{H}^*]e^{2i\omega t}+[\vec{E}\times\vec{H}]e^{-2i\omega t}\}, \tag{2.106}$$

where we accounted for the fact that only real parts of electric and magnetic complex vectors contribute to the measured signal. Taking into account that the last two terms do not contribute to the measured signal due to their high frequency oscillations, we obtain:

$$\langle A_{sca}\rangle = \frac{1}{2}\int_S \mathrm{Re}[\vec{E}\times\vec{H}^*]\vec{n}ds, \tag{2.107}$$

where we accounted for the fact that

$$\mathrm{Re}[\vec{E}\times\vec{H}^*] = \frac{1}{2}\{[\vec{E}\times\vec{H}^*]+[\vec{E}^*\times\vec{H}]\}. \tag{2.108}$$

The intensity I is defined as the energy for unit time through the unit surface. So we have:

$$I = \frac{1}{2}\mathrm{Re}[\vec{E}\times\vec{H}^*]\vec{n}, \tag{2.109}$$

where high oscillations not relevant to optical measurements are removed. We may write for a scattered part of the field using Eq. (2.89):

$$\vec{E} = E_l^s\vec{e_l} + E_r^s\vec{e_r}, \tag{2.110}$$

$$\vec{H} = -E_r^s\vec{e_l} - E_l^s\vec{e_r}, \tag{2.111}$$

where the vector $\vec{e_l}$ is in the parallel to the scattering plane and $\vec{e_r}$ is perpendicular to this plane (see Eq. (2.89)). Equations (2.88), (2.89), (2.109)–(2.111) lead us to the following expression for the scattered light intensity:

$$I_s = \frac{i_1\sin^2\phi + i_2\cos^2\phi}{k^2r^2}I_0, \tag{2.112}$$

where I_0 is the intensity of incident light and

$$i_1 = |S_2|^2 \equiv |S_{11}(\theta)|^2, \tag{2.113}$$

$$i_2 = |S_1| \equiv |S_{22}(\theta)|^2. \tag{2.114}$$

Substitute $I_0 = I_{l0} = I_{r0}$ by $\langle \cos^2 \phi \rangle = \langle \sin^2 \phi \rangle = 1/2$

$$I_s = \frac{i_1 + i_2}{2k^2 r^2} I_0. \tag{2.115}$$

This important equation allows to study the measurable quantity, namely, the scattered light intensity in a given direction. The total scattered power is given by Eq. (2.107). Taking into account that $ds = r^2 \sin\theta d\theta d\phi$ in spherical coordinates, we obtain:

$$\langle A_{sca} \rangle = \frac{I_0}{2k^2} \int_0^{2\pi} d\phi \int_0^{\pi} d\theta (i_1 + i_2), \tag{2.116}$$

where we used Eqs. (2.107), (2.109) and (2.115). Functions i_1 and i_2 depend only on the scattering angle θ for spherical particles. Therefore, it follows

$$\langle A_{sca} \rangle = \frac{\pi I_0 \langle i \rangle}{k^2}, \tag{2.117}$$

where

$$\langle i \rangle = \int_0^{\pi} [i_1(\theta) + i_2(\theta)] \sin\theta d\theta. \tag{2.118}$$

The parameter

$$C_{sca} = \frac{\pi \langle i \rangle}{k^2}, \tag{2.119}$$

where $\langle i \rangle = \langle i_1 \rangle + \langle i_2 \rangle$, has the dimension of the area and is called the scattering cross section. It gives the characteristic area for a scattering process. The value of C_{sca} allows to determine $\langle A_{sca} \rangle$ for a given I_0. It is defined as

$$C_{sca} = \frac{\langle A_{sca} \rangle}{I_0} \tag{2.120}$$

for a particle of an arbitrary shape. One can also introduce the differential scattering cross section

$$C_{sca}(\theta) = \frac{r^2 I_s(\theta)}{I_0} \tag{2.121}$$

or

$$C_{sca}(\theta) = \frac{i_1 + i_2}{2k^2} \tag{2.122}$$

such that

$$C_{sca} = \int_0^{2\pi} d\phi \int_0^{\pi} \sin\theta d\theta C_{sca}(\theta). \tag{2.123}$$

The results for scattering patterns are usually presented in the form of plots $C_{sca}(\theta)$ because then a constant factor r^2/I_0 is eliminated. Those working in the field of radiative transfer prefer to use the normalized differential cross section

$$p(\theta) = \frac{4\pi C_{sca}(\theta)}{C_{sca}}, \tag{2.124}$$

which is called the phase function. We have for the integral of this function with respect to the solid angle $d\Omega = \sin\theta d\theta d\phi$

$$\int_{4\pi} p(\theta)\frac{d\Omega}{4\pi} = 1, \tag{2.125}$$

where we used Eqs. (2.122) and (2.123). Clearly [see Eq. (2.124)], the function $p(\theta)$ gives the conditional probability of photon scattering in a given direction. It follows from Eq. (2.125) in the case of an equal probability of light scattering in all directions: $p(\theta) = 1$. For Mie scattering considered here, we have:

$$p(\theta) = \frac{2\pi(i_1 + i_2)}{k^2 C_{sca}}. \tag{2.126}$$

This follows from Eqs. (2.123)–(2.125). The phase function is often expressed in terms of the following expansions:

$$p(\theta) = \sum_{n=1}^{\infty} x_n P_n(\cos\theta), \tag{2.127}$$

where

$$x_n = \frac{2n+1}{2} \int_0^{\pi} p(\theta) P_n(\cos\theta) \sin\theta d\theta \tag{2.128}$$

and $P_n(\cos\theta)$ are Legendre polynomials. The discrete representation of the phase function in terms of coefficients x_n is largely equivalent to the continuous representation given by Eq. (2. 126). However, Eq. (2.127) is very convenient in studies of multiple light scattering. The coefficient $x_1 = 3g$, where [see Eq. (2.128)]

$$g = \frac{1}{2} \int_0^{\pi} p(\theta) \cos\theta \sin\theta d\theta \tag{2.129}$$

is of a special importance for radiative transfer problems (van de Hulst, 1980). Integrals $\langle i\rangle = \langle i_1\rangle + \langle i_2\rangle$ and g can be calculated analytically for spheres. In particular, we have:

$$\langle i_1\rangle = \int_0^{\pi} |S_1|^2 \sin\theta d\theta = \sum_{n,m=1}^{\infty} \frac{(2m+1)(2n+1)}{(m+1)n(n+1)} \left(a_n a_m^* \langle \pi_n \pi_m\rangle + b_n b_m^* \langle \tau_n \tau_m\rangle \right.$$
$$\left. + a_n b_m^* \langle \pi_n \tau_m\rangle + b_n a_m^* \langle \tau_n \pi_m\rangle\right), \tag{2.130}$$

where $\langle\ \rangle$ represents the integral operator having the property

$$\langle f(\theta)\rangle = \int_0^{\pi} f(\theta) \sin\theta d\theta \tag{2.131}$$

for any function $f(\theta)$. Similarly, we have:

$$\langle i_2\rangle = \sum_{n,m=1}^{\infty} \frac{(2m+1)(2n+1)}{m(m+1)n(n+1)} (b_n b_m^* \langle \pi_n \pi_m\rangle + a_n a_m^* \langle \tau_n \tau_m\rangle + b_n a_m^* \langle \pi_n \tau_m\rangle$$
$$+ a_n b_m^* \langle \tau_n \pi_m\rangle). \tag{2.132}$$

Therefore, it follows

$$\langle i\rangle = \sum_{n,m=1}^{\infty} \frac{(2m+1)(2n+1)}{m(m+1)n(n+1)} \left\{\left(a_n a_m^* + b_n b_m^*\right) \mu_{nm} + \left(a_n b_m^* + b_n a_m^*\right) \nu_{nm}\right\}, \tag{2.133}$$

where integrals

$$\mu_{nm} = \langle \pi_n \pi_m\rangle + \langle \tau_n \tau_m\rangle \tag{2.134}$$

and

$$\nu_{nm} = \langle \pi_n \tau_m\rangle + \langle \pi_n \tau_m\rangle \tag{2.135}$$

can be calculated analytically. Namely, we have (Debye, 1909):

$$\mu_{nm} = \frac{2n^2(n+1)^2}{2n+1}\delta_{nm},\ \nu_{nm} = 0, \tag{2.136}$$

where $\delta_{nm} = 1$ at $n = m$ and $\delta_{nm} = 0$ at $n \neq m$. Therefore, one can derive the following important relationships:

$$\langle i\rangle = 2\sum_{n=1}^{\infty} (2n+1)\left(|a_n|^2 + |b_n|^2\right), \tag{2.137}$$

$$C_{sca} = \frac{2\pi}{k^2}\sum_{n=0}^{\infty} (2n+1)\left(|a_n|^2 + |b_n|^2\right). \tag{2.138}$$

Sometimes the dimensionless scattering efficiency factor Q_{sca} is introduced:

$$Q_{sca} = \frac{C_{sca}}{\pi a^2}. \tag{2.139}$$

It follows for this factor from Eq. (2.138):

$$Q_{sca} = \frac{2}{x^2} \sum_{n=1}^{\infty} (2n+1) \left(|a_n|^2 + |b_n|^2 \right). \tag{2.140}$$

For a given refractive index, Q_{sca} depends only on the size parameter $x = ka$.

Let us find the analytical expression for g now. For this we substitute Eq. (2.126) in Eq. (2.129) and obtain:

$$g = \frac{2\pi \langle j \rangle}{k^2 C_{sca}}, \tag{2.141}$$

where

$$\langle j \rangle = \int_0^{\pi} (i_1 + i_2) \cos\theta \sin\theta d\theta \tag{2.142}$$

Eq. (2.142) is similar to Eq. (2.130). Taking the same steps as in the derivation of Eq. (2.133), we have:

$$\langle j \rangle = \sum_{n,m=1}^{\infty} \frac{(2n+1)(2m+1)}{mn(m+1)(n+1)} \left\{ \left(a_n a_m^* + b_n b_m^* \right) \bar{\mu}_{nm} + \left(a_n b_m^* + b_n a_m^* \right) \bar{\nu}_{nm} \right\}, \tag{2.143}$$

where

$$\bar{\mu}_{nm} = \overline{\pi_n \pi_m} + \overline{\tau_n \tau_m}, \tag{2.144}$$

$$\bar{\nu}_{nm} = \overline{\pi_n \tau_m} + \overline{\tau_n \pi_m}, \tag{2.145}$$

and the overbar represents the integral operator having the property

$$\bar{f}(\theta) = \int_0^{\pi} f(\theta) \sin\theta \cos\theta d\theta \tag{2.146}$$

for arbitrary $f(\theta)$. Integrals $\bar{\mu}_{nm}$ and $\bar{\nu}_{nm}$ can be found analytically. The answer is (Debye, 1909):

$$\bar{\mu}_{nm} = 2\frac{(n-1)^2 (n+1)^2}{(2n-1)(2n+1)} \delta_{m,n-1} + 2\frac{n^2 (n+1)(n+2)^2}{(2n+1)(2n+3)} \delta_{m,n+1}, \tag{2.147}$$

$$\bar{\nu}_{nm} = 2\frac{n(n+1)(n+2)^2}{2n+1} \delta_{mn}. \tag{2.148}$$

Therefore, it follows from Eq. (2.143):

$$j = 2\sum_{n,m=1}^{\infty} \frac{(2n+1)(2m+1)}{mn(m+1)(n+1)} \left\langle \left(a_n a_m^* + b_n b_m^*\right) \left\{ \frac{(n-1)^2 n (n+1)^2}{(2n-1)(2n+1)} \delta_{m,n-1} + \frac{n^2 (n+1)(n+2)^2}{(2n+1)(2n+3)} \delta_{m,n+1} \right\} + \frac{n(n+1)}{2n+1} \left(a_n b_m^* + b_n a_m^*\right) \delta_{nm} \right\rangle \tag{2.149}$$

or

$$j = 2\sum_{n=1}^{\infty} \left\{ \frac{2n+1}{n(n+1)} \left[a_n b_m^* + b_n a_m^*\right] + \frac{n^2-1}{n} \left[a_n a_{n-1}^* + b_n b_{n-1}^*\right] + \frac{n(n+2)}{n+1} \left[a_n a_{n+1}^* + b_n b_{n+1}^*\right] \right\}. \tag{2.150}$$

The sum

$$A = 2\sum_{n=2}^{\infty} \frac{n^2-1}{n} \left(a_n a_{m-1}^* + b_n b_{n-1}^*\right) \tag{2.151}$$

in Eq. (2.150) can be reduced to

$$A = 2\sum_{l=2}^{\infty} \frac{l(l+2)}{l+1} \left(a_{l+1} a_l^* + b_{l+1} b_l^*\right) \tag{2.152}$$

using the index $l = n - 1$. So we obtain:

$$j = 4\,\mathrm{Re} \sum_{n=1}^{\infty} \left\{ \frac{2n+1}{n(n+1)} a_n b_n^* + \frac{n(n+2)}{n+1} \left(a_n a_{n+1}^* + b_n b_{n+1}^*\right) \right\} \tag{2.153}$$

and therefore,

$$g = \frac{4\pi}{k^2 C_{sca}} \mathrm{Re} \sum_{n=1}^{\infty} \left\{ \frac{2n+1}{n(n+1)} a_n b_n^* + \frac{n(n+2)}{n+1} \left(a_n a_{n+1}^* + b_n b_{n+1}^*\right) \right\}, \tag{2.154}$$

where we used the property: $xy^* + x^*y = 2\,\mathrm{Re}(xy^*)$. Sometimes Eq. (2.154) is written in the form:

$$g = \frac{4}{x^2 Q_{sca}} \mathrm{Re} \sum_{n=1}^{\infty} \left\{ \frac{2n+1}{n(n+1)} a_n b_n^* + \frac{n(n+2)}{n+1} \left(a_n a_{n+1}^* + b_n b_{n+1}^*\right) \right\}, \tag{2.155}$$

where $x = ka$ and Eq. (2.139) was used. It follows that g depends only on the size parameter x and the complex refractive index of particles m.

Clearly, using the formula similar to Eq. (2.120), we can also introduce the absorption cross section C_{abs}, which describes the absorptive power of the particle (see Eq. (2.102)):

$$C_{abs} = \frac{\langle A_{abs} \rangle}{I_0} \tag{2.156}$$

or

$$C_{abs} = \frac{2\pi}{cI_0} \int_V \sigma \left|\vec{E}\right|^2 dV, \tag{2.157}$$

where we accounted for the relationship:

$$\vec{j} = \sigma \vec{E}, \tag{2.158}$$

and the property given by Eq. (2.106) as applied to the absorption term. It follows from Eq. (2.157):

$$C_{abs} = k \left|\vec{E}_0\right|^{-2} \int_V \varepsilon'' \left|\vec{E}\right|^2 dV. \tag{2.159}$$

where $\varepsilon'' = 4\pi\sigma/\omega$, $\omega = kc$.
The total losses can be expressed as the sum of C_{abs} and C_{sca}. Namely, we have:

$$C_{ext} = C_{abs} + C_{sca}. \tag{2.160}$$

The analytical expression for C_{abs} (and, therefore, for C_{ext}) can be obtained directly from Eq. (2.159) (Babenko et al., 2003). However, we prefer to calculate C_{ext} first. Then Eq. (2.160) can be used to find C_{abs} avoiding calculations of the volume integral (2.159). It should be pointed out that for numerical reasons, the calculation of C_{abs} for weakly absorbing particles is usually done with Eq. (2.159) because the calculation using the difference $C_{ext} - C_{sca}$ can have large errors because $C_{ext} \approx C_{sca}$ then.

2.3.2.2 *Extinction cross section*

The total electric $\vec{\mathrm{E}}$ and magnetic $\vec{\mathrm{H}}$ fields outside a particle can be written as

$$\vec{\mathrm{E}} = \vec{\mathrm{E}}_i + \vec{\mathrm{E}}_s, \vec{\mathrm{H}} = \vec{\mathrm{H}}_i + \vec{\mathrm{H}}_s, \tag{2.161}$$

where the superposition principle was used. The substitution of Eq. (2.161) in Eq. (2.109) gives:

$$I = \frac{1}{2}\,\mathrm{Re}\big[\vec{\mathrm{E}}_i\vec{\mathrm{H}}_i^* + \vec{\mathrm{E}}_s\vec{\mathrm{H}}_i^* + \vec{\mathrm{E}}_i\vec{\mathrm{H}}_s^*\big]\vec{n} = I_i + I_s + I_e, \tag{2.162}$$

where

$$I_i = \frac{1}{2}\,\mathrm{Re}\big[\vec{\mathrm{E}}_i\vec{\mathrm{H}}_i^*\big]\vec{n}, \tag{2.163}$$

$$I_s = \frac{1}{2}\,\mathrm{Re}\big[\vec{\mathrm{E}}_s\vec{\mathrm{H}}_s^*\big]\vec{n}, \tag{2.164}$$

The value of I_i gives the intensity of incident light. I_s is the intensity of scattered light. The term I_e arises due to the interference of incident and scattered waves. The surface integral of Eq. (2.162) with respect to the concentric spherical surface placed at a large distance from the particle gives the total flux of energy

$$A = A_i + A_s - A_e, \tag{2.166}$$

where A_s is given by Eq. (2.103) and

$$A_i = \int_s I_i ds, \tag{2.167}$$

$$A_e = -\int_s I_e ds, \tag{2.168}$$

$$A_s = \int_s I_s ds \tag{2.169}$$

Let us assume that both a particle and a surrounding medium do not absorb incident radiation. This means that the total flux passing through the surface in both directions (to and from a particle) vanishes. This is a direct consequence of the conservation of energy principle. The same applies to A_i. Then it follows: $A_e = A_s$. We see that fluxes A_e and A_s coincide then. Now we assume that a particle can absorb incident light.

Then the value of A is not equal to zero. It becomes negative because the escaping light flux at a given wavelength is smaller than the incident one. The value of $-A$ can be interpreted as the absorbed light flux. Therefore, it follows: $A_{abs} = A_s - A_e$, and, therefore, $A_e = A_{abs} + A_s$. This relation allows us to interpret A_e as the total scattered and absorbed power. So we obtain for the cross section $C_{ext} = A_e/I_0$ of the correspondent process:

$$C_{ext} = C_{abs} + C_{sca}, \tag{2.170}$$

which is the extinction cross section introduced above. We see that the extinction or total losses are solely due to the interference of incident and scattered waves. This property allows us to find the value of C_{ext} directly from the scattering amplitude avoiding calculations of energy absorbed by a particle. Let us show this. It follows for C_{ext} (see Eqs. (2.165), (2.168)):

$$C_{ext} = -\frac{1}{2I_0}\int_0^{2\pi} d\phi \int_0^{\pi} r^2 \sin\theta d\theta \mathrm{Re}\left(\left[\vec{E}_s \vec{H}_i^*\right]_p + \left[\vec{E}_i \vec{H}_s^*\right]_p\right) \tag{2.171}$$

where we used Eq. (2.165) and p means the projection of the correspondent vector on the direction specified by the vector $\vec{n}$ (see Eq. (2.165)). Taking into account that

$$[\vec{E}_s \times \vec{H}_i^*]_p = E_\theta^s H_\phi^{i*} - E_\phi^s H_\theta^{i*} \tag{2.172}$$

and

$$[\vec{E}_i \times \vec{H}_s^*]_p = E_\theta^i H_\phi^{s*} - E_\phi^i H_\theta^{s*}, \tag{2.173}$$

we obtain:

$$C_{ext} = -\frac{1}{2I_0}\mathrm{Re}\int_0^{2\pi} \mathbb{N}(\phi)d\phi, \tag{2.174}$$

where

$$\mathbb{N}(\phi) = \int_0^{\pi} F(\theta,\phi)\sin\theta d\theta \tag{2.175}$$

and

$$F(\theta,\phi) = r^2\left[E_\theta^s H_\phi^{i*} - E_\phi^s H_\theta^{i*} + E_\theta^i H_\phi^{s*} - E_\phi^i H_\theta^{s*}\right]. \tag{2.176}$$

The functions E_θ^s, E_ϕ^s, H_θ^s, H_ϕ^s are given in Table 2.2. Correspondent incident field components are presented in Table 2.2 as well. The substitution of these results valid as $r \to \infty$ in Eq. (2.176) allows us to perform the integration and find the convenient analytical expression for C_{ext}. For this we write field components in the following form:

$$E_\theta^s = H_\phi^s = -ip_n e^{-ikr}(a_n\tau_n + b_n\pi_n)\cos\phi, \tag{2.177}$$

$$E_\phi^s = -H_\theta^s = ip_n e^{-ikr}(a_n\pi_n + b_n\tau_n)\sin\phi, \tag{2.178}$$

$$E_\theta^i = H_\theta^i \cot\phi = (-i)^n p_n(\psi_n\pi_n + i\psi_n'\tau_n)\cos\phi, \tag{2.179}$$

$$E_\phi^i = -H_\phi^i \tan\phi = -(-i)^n p_n(\psi_n\tau_n + i\psi_n'\pi_n)\sin\phi, \tag{2.180}$$

where arguments and the summation with respect to the index n are omitted for simplicity. The coefficients p_n are given by:

$$p_n = \frac{E_0}{kr}\frac{2n+1}{n(n+1)} \tag{2.181}$$

and $\psi_n = \sin(kr - \pi n/2)$, $\psi_n' = \cos(kr - \pi n/2)$ for the case $kr \to \infty$ considered here. First of all, we note that the angle ϕ in Eqs. (2.177)–(2.180) occurs in combinations $\cos^2\phi$, $\sin^2\phi$ only. This allows to perform integration with respect to ϕ immediately, taking into account that

$$\int_0^{2\pi} \sin^2\phi d\phi = \int_0^{2\pi} \cos^2\phi d\phi = \pi. \tag{2.182}$$

Let us introduce azimuthally averaged light field characteristics

$$D_1 = \frac{r^2}{2\pi}\int_0^{2\pi} (E_\phi^s H_\theta^{i*} - E_\theta^s H_\phi^{i*})d\phi \tag{2.183}$$

and

$$D_2 = \frac{r^2}{2\pi} \int_0^{2\pi} \left(E^i_\phi H^{s*}_\theta - E^i_\theta H^{s*}_\phi \right) d\phi. \tag{2.184}$$

Then the extinction cross section can be written as

$$C_{ext} = \frac{\pi}{\Phi_0} \mathrm{Re} \int_0^{\pi} (D_1 + D_2) \sin\theta d\theta \tag{2.185}$$

with

$$\begin{aligned} D_1 &= 2^{-1} i^{m+1} r^2 p_n p_m e^{-ikr} \left[(a_n \tau_n + b_n \pi_n) \left(\psi_m \tau_m - i \psi'_m \pi_m \right) \right. \\ &\quad \left. + (a_n \pi_n + b_n \tau_n) \left(\psi_m \pi_m - i \psi'_m \tau_m \right) \right], \\ D_2 &= 2^{-1} (-i)^{m+1} r^2 p_n p_m e^{-ikr} \left[\left(a_n^* \tau_n + b_n^* \pi_n \right) \left(\psi_m \pi_m + i \psi'_m \tau_m \right) \right. \\ &\quad \left. + \left(a_n^* \pi_n + b_n^* \tau_n \right) \left(\psi_m \tau_m + i \psi'_m \pi_m \right) \right], \end{aligned} \tag{2.186}$$

where the summation with respect to repeating indices is understood. Performing multiplications in brackets and substituting the result in Eq (2.185), we obtain;

$$\begin{aligned} C_{ext} = \frac{\pi}{E_0^2} \mathrm{Re} &\left[i^{n+1} r^2 p_n^2 \mu_{nn} e^{-ikr} \left(a_n \psi_n - i b_n \psi'_n \right) \right. \\ &\left. + (-i)^{n+1} r^2 p_n^2 \mu_{nn} e^{ikr} \left(b_n^* \psi_n + i a_n^* \psi'_n \right) \right], \end{aligned} \tag{2.187}$$

where we used integrals and the equality $I_0 = E_0^2/2$. This can be written as

$$C_{ext} = \frac{\pi(2n+1)}{k^2} \mathrm{Re} \{ Q_1 + Q_2 \}, \tag{2.188}$$

where

$$Q_1 = 2 i^{n+1} e^{-ikr} (a_n \psi_n - i b_n \psi'_n), \tag{2.189}$$

$$Q_2 = 2(-i)^{n+1} e^{ikr} (b_n^* \psi_n + i a_n^* \psi'_n) \tag{2.190}$$

and we used the substitution (see Eqs. (2.136), (2.181)):

$$\frac{p_n^2 \mu_{nn}}{E_0^2} = \frac{2(2n+1)}{k^2 r^2}. \tag{2.191}$$

It follows that functions Q_1 and Q_2 depend on the distance from the particle r. This dependence should disappear for the combination $\mathrm{Re}(Q_1 + Q_2)$. Indeed, both scattered light intensity and attenuated light flux for a single particle placed in vacuum should not depend on kr as $kr \to \infty$. Let us show that this is the case. We take into account that

$$\bar{\psi}_n = \frac{e^{i(y-\alpha)} - e^{-i(y-\alpha)}}{2i} \tag{2.192}$$

and

$$\psi_n' = \frac{e^{i(y-\alpha)} + e^{-i(y-\alpha)}}{2i} \tag{2.193}$$

with $y = kr$, $\alpha = \pi n/2$ in the asymptotic case considered here. Then it follows:

$$Q_1 = i^{n+1} e^{-ikr} \left(-\sqrt{i a_n} \left[(-i)^n e^{iy}\right] - i b_n \left[(-i)^n e^{iy} + i e^{-iy}\right]\right) \tag{2.194}$$

and

$$Q_2 = (-i)^{n+1} e^{iy} \left(-i b_n^* \left[(-i)^n e^{iy} + (i)^n e^{-iy}\right] - i a_n^* \left[(-i)^n e^{iy} + i^n e^{-iy}\right]\right), \tag{2.195}$$

where we accounted for the equality $\exp(\pm i\alpha) = (\pm i)^n$, where $\alpha = \pi n/2$. Performing multiplications, we obtain:

$$Q_1 = a_n(1 - (-1)^n e^{-2iy}) + b_n(1 + (-i)^n(-1)^n e^{-2iy}), \tag{2.196}$$
$$Q_2 = a_n^*(1 + (-i)^n e^{2iy}) + b_n^*(1 - (-i)^n e^{2iy}). \tag{2.197}$$

Therefore, it follows

$$Q_1 + Q_2 = 2\text{Re}\,(a_n + b_n) + 2i\ \text{Im}\,(f_n)\,, \tag{2.198}$$

where

$$f_n = (-1)^n (b_n - a_n) e^{-2iy}. \tag{2.199}$$

So we have for the extinction cross section:

$$C_{ext} = \frac{2\pi}{k^2} \text{Re} \sum_{n=1}^{\infty} (2n+1)(a_n + b_n), \tag{2.200}$$

where the previously ignored the sign of the summation is written explicitly now. This simple result for the extinction cross section was obtained by Mie (1908) and has proved to be useful in a broad range of applications since then. Taking into account Eq. (2.86), we obtain:

$$C_{ext} = \frac{4\pi}{k^2} \text{Re}\,[S(0)]\,. \tag{2.201}$$

This important relationship is called the optical theorem. It manifests itself in a broad range of wave scattering phenomena.

The absorption cross section

$$C_{abs} = C_{ext} - C_{sca} \tag{2.202}$$

or

$$C_{abs} = \frac{2\pi}{k^2} \sum_{n=1}^{\infty} (2n+1) \left(a_n - |a_n|^2 + b_n - |b_n|^2 \right) \tag{2.203}$$

should vanish for nonabsorbing particles. This means that

$$a_n = |a_n|^2, \; b_n = |b_n|^2 \tag{2.204}$$

in this case. Therefore, real phase angles α_n and β_n can be introduced:

$$a_n = \frac{1}{2}(1 - e^{-2i\alpha_n}), \; b_n = \frac{1}{2}(1 - e^{-2i\beta_n}). \tag{2.205}$$

They insure that properties (2.204) are satisfied. Also we have then:

$$C_{ext} = \frac{2\pi}{k^2} \sum_{n=1}^{\infty} (2n+1) \left(\sin^2 \alpha_n + \sin^2 \beta_n \right), \tag{2.206}$$

where real angles are

$$\alpha_n = -\frac{1}{2i} \ln(1 - 2a_n), \; \beta_n = -\frac{1}{2i} \ln(1 - 2b_n). \tag{2.207}$$

Using the property

$$\sin^2 z = \frac{1}{2}(1 - \cos z), \tag{2.208}$$

we obtain:

$$C_{ext} = \frac{2\pi}{k^2} \sum_{n=1}^{\infty} (2n+1)(1-s), \tag{2.209}$$

where

$$s = \frac{1}{2}(\cos 2\alpha_n + \cos 2\beta_n). \tag{2.210}$$

The investigation of series for C_{ext} shows that the main contribution to C_{ext} comes from values of $n \leq ka$. It appears that s is a highly oscillating function for large values of x. The summation of these oscillations results in no contribution to C_{ext} as the size parameter $x \to \infty$. So we can write approximately:

$$C_{ext} = \frac{2\pi}{k^2} \sum_{n=1}^{x} (2n+1) \approx \frac{2\pi}{k^2} \int_0^x (2n+1)dn \approx 2\pi a^2, \tag{2.211}$$

where we accounted for the fact that $x = ka \gg 1$. Then summation can be substituted by integration. Our derivations are confirmed by direct calculations as shown in Fig. 2.1 for the ratio $Q_{ext} = C_{ext}/\pi a^2$. Also they are easily generalized to account for light absorption by a particle. Then phase angles become

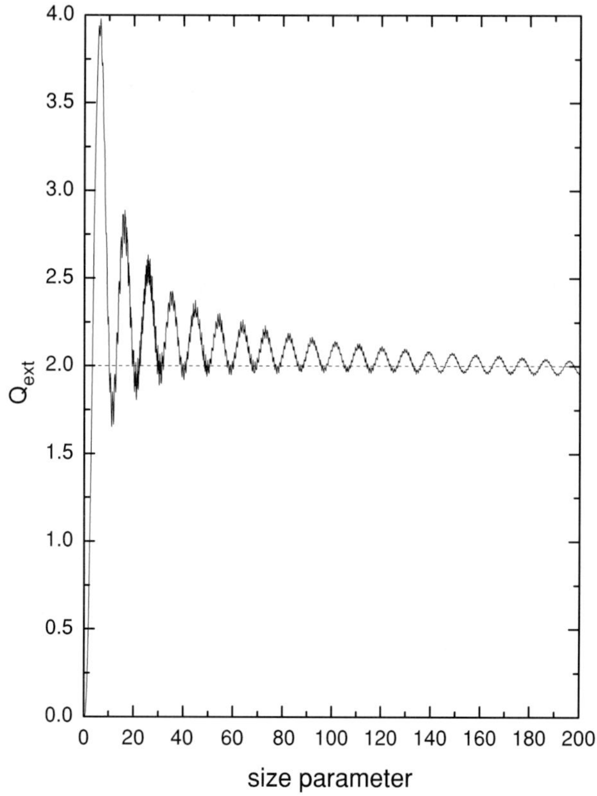

Fig. 2.1. Dependence of the extinction efficiency factor on the size parameter at $n = 1.34$.

complex numbers. It follows that $Q_{ext} = 2$ independent of the refractive index of a particle and the wavelength as $x \to \infty$. This simple relationship appears very useful in cloud optics studies. In particular, it explains the cloud colour (or the whiteness of clouds due to white light illumination conditions). It should be emphasized that the values of Q_{abs} and Q_{sca} also saturate to a constant value as $x \to \infty$. This is shown in Fig. 2.2 for absorbing particles. Clearly, it follows: $Q_{abs} = 1 - \omega$, $Q_{sca} = 1 + \omega$ as $x \to \infty$. So the asymptotic result $Q_{ext} = Q_{abs} + Q_{sca} = 2$ is satisfied. The value of ω depends on the refractive index only. Interestingly, all asymptotic efficiency factors are smaller than their values for finite x (see Fig. 2.2). This is not the case for the asymmetry parameter (see Fig. 2.3).

2.3.3 Mueller Matrix

The amplitude scattering matrix $\hat{S}(\theta)$ allows us to find the electric vector $\vec{E}^s$ of scattered light for a given electric vector $\vec{E}^i$ of an incident wave. In particular,

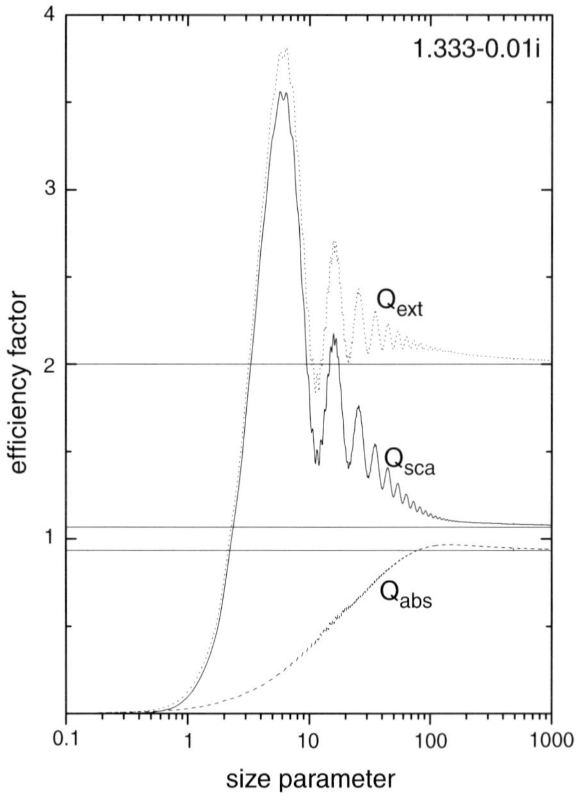

Fig. 2.2. Dependence of the extinction efficiency factor Q_{ext}, the scattering efficiency factor Q_{ext}, the absorption efficiency factor Q_{ext} on the size parameter at $n = 1.333 - 0.01i$.

we have:

$$\vec{E}^s = \frac{\alpha}{kr} \hat{S}(\theta) \vec{E}^i, \tag{2.212}$$

where $\alpha = e^{ik(r-z)-\pi i/2}$. Due to high-frequency oscillations of optical waves, the direct measurements of electric fields is difficult to achieve in the optical range. Instead quadratic values (e.g., $\vec{E}\vec{E}^*$) are measured.

Let us consider a completely polarized wave. Then it follows:

$$\vec{E} = E_l \vec{e}_1 + E_r \vec{e}_2, \tag{2.213}$$

where

$$E_l = a \exp(i\varphi), \; E_r = b \exp(i\psi). \tag{2.214}$$

A completely polarized electromagnetic wave is characterized by four real numbers:

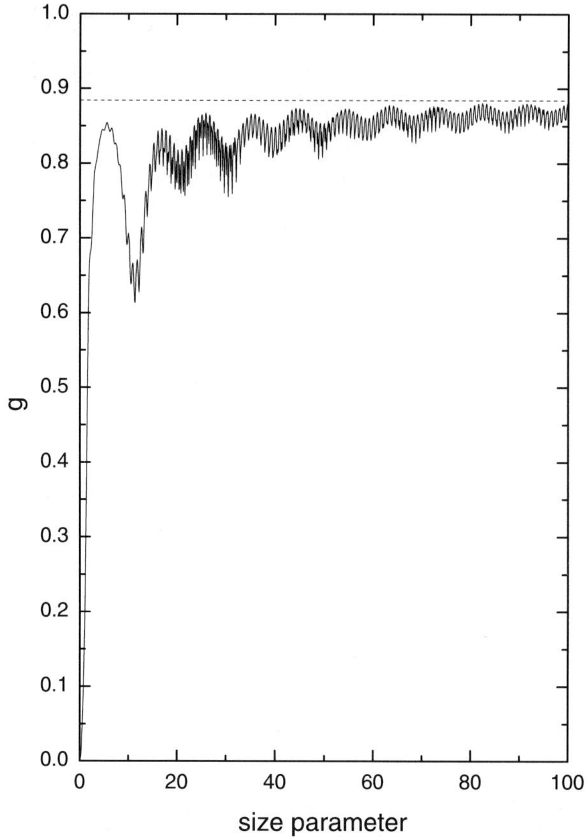

Fig. 2.3. Dependence of the asymmetry parameter g on the size parameter at $n = 1.34$.

usually introduced:

$$I = E_l E_l^* + E_r E_r^*, \tag{2.215}$$

$$Q = E_l E_l^* - E_r E_r^*, \tag{2.216}$$

$$U = E_l E_r^* + E_r E_l^*, \tag{2.217}$$

$$V = i(E_l E_r^* - E_r E_l^*) \tag{2.218}$$

or

$$I = a^2 + b^2, \tag{2.219}$$

$$Q = a^2 - b^2, \tag{2.220}$$

$$U = 2ab\cos\delta, \tag{2.221}$$

$$V = 2ab\sin\delta, \tag{2.222}$$

where $\delta = \varphi - \psi$ is the phase difference and the common multiplier is omitted. Note that these parameters depend only on the difference $\delta = \varphi - \psi$ and not separately on φ and ψ. This underlines the fact that only parameters a, b and δ can be measured by quadratic detectors (and not φ, ψ separately). Parameters I, Q, U and V can be easily measured by optical instruments. In particular, the value of I is proportional to the intensity of light. It can be measured, e.g., using a standard multiplier. Components Q, U and V are measured using retarders and polarizers. A retarder is a phase-shifting optical element, which has the property that the phase of the x – component (E_1) is advanced by $\varphi/2$ and the phase of the y component (E_2) is retarded by $\varphi/2$, written as $-\varphi/2$. Then components E_1', E_2' emerging from the phase-shifting element are

$$E_1' = E_1 e^{i\varphi/2}, \; E_2' = E_2 e^{-i\varphi/2}. \tag{2.223}$$

A polarizer has the property that the optical field can only pass along an axis known as the transmission axis. If one places a polarizer after a retarder, the component of E_1' along the transmission axis becomes $E_1' \cos\theta$, where θ is the angle between the axis specified by the vector $\vec{e}_1$ and the transmission axis of a polarizer. Similarly, the component of E_2' becomes $E_2' \sin\theta$. Clearly, the total field emerging from the system retarder-polarizer is

$$E = E_1' \cos\theta + E_2' \sin\theta \tag{2.224}$$

or

$$E = E_1 e^{i\varphi/2} \cos\theta + E_2 e^{-i\varphi/2} \sin\theta. \tag{2.225}$$

The intensity of the beam is defined by $J = EE^*$. So we have:

$$\begin{aligned} J(\theta, \varphi) = {} & E_1 E_1^* \cos^2\theta + E_2 E_2^* \sin^2\theta + \frac{1}{2} E_1^* E_2 \sin 2\theta e^{-i\varphi} \\ & + \frac{1}{2} E_2^* E_1 \sin 2\theta e^{i\varphi} \end{aligned} \tag{2.226}$$

or after simple transformations using relationships: $E_1 E_1^* = (I + Q)/2$, $E_1 E_1^* = (I + Q)/2$, $E_2 E_2^* = (I - Q)/2$, $E_1 E_2^* = (U - iV)/2$, and $E_2 E_1^* = (U + iV)/2$, it follows:

$$J(\theta, \varphi) = \frac{1}{2}(I + Q\cos 2\theta + U \cos\varphi \sin 2\theta + V \sin\varphi \sin 2\theta). \tag{2.227}$$

This means that measuring the intensity at four combinations (θ, φ) we obtain parameters I, Q, U and V. In particular, it follows:

$$I = J(0^\circ, 0^\circ) + J(90^\circ, 0^\circ), \tag{2.228}$$

$$Q = J(0^\circ, 0^\circ) - J(90^\circ, 0^\circ), \tag{2.229}$$

$$U = 2J(45^\circ, 0^\circ) - J(0^\circ, 0^\circ) - J(90^\circ, 0^\circ), \tag{2.230}$$

$$V = 2J(45^\circ, 90^\circ) - J(0^\circ, 0^\circ) - J(90^\circ, 0^\circ). \tag{2.231}$$

One obtains for a completely polarized wave that

$$I^2 = Q^2 + U^2 + V^2. \tag{2.232}$$

Therefore, it is enough to measure only three parameters (e.g., Q, U and V) in this case. The fourth parameter can be easily found (e.g., $I = \sqrt{Q^2 + U^2 + V^2}$). In practice, however, light fields are usually only partially polarized and

$$Q^2 + U^2 + V^2 \leq I^2. \tag{2.233}$$

We can introduce, a degree of polarization

$$P = \frac{\sqrt{Q^2 + U^2 + V^2}}{I}, \tag{2.234}$$

which changes from 0 (for unpolarized light) to 1 for completely polarized light. Therefore, measuring I, Q, U and V we can determine

– the intensity of light field I
– the degree of polarization P
– the phase shift $\delta = \mathrm{a}\tan(V/U)$
– the amplitudes a and b.

Independently, we can give a geometric representation of a simple electromagnetic wave with the most general (elliptic) polarization:

$$\vec{E} = c\cos\beta\sin(\omega t - kz + \alpha)\vec{h}_1 + c\sin\beta\cos(\omega t - kz + \alpha)\vec{h}_2, \tag{2.235}$$

where $\vec{h}_1$ and $\vec{h}_2$ are unit vectors along the long and short axes of the polarization ellipse. Simple algebraic calculations lead us to the following equations:

$$I = c^2, \tag{2.236}$$

$$Q = c^2\cos 2\beta\cos 2\chi, \tag{2.237}$$

$$U = c^2\cos 2\beta\sin 2\chi, \tag{2.238}$$

$$V = c^2\sin 2\beta, \tag{2.239}$$

where the orientation of the ellipse is given by the angle χ. It follows that values of χ and β can easily be obtained from measured parameters I, Q, U and V. In particular, we have:

$$\chi = \frac{1}{2}\arctan\left(\frac{U}{Q}\right) \tag{2.240}$$

and

$$\beta = \frac{1}{2}\arcsin\left(\frac{V}{I}\right). \tag{2.241}$$

The value of β specifies the ellipticity of radiation. In particular, we have at $V = \pm I$: $\beta = \pm 1$ and radiation is circularly polarized. The positive sign means a right-hand polarized light. In particular, an observer looking in the direction from which light propagates will see the rotation of an electric vector in the clockwise direction. The opposite is true for left-hand polarized light.

We see that the description of radiation in terms (I, P, χ, β) and (I, Q, U, V) is largely equivalent. Intuitively, the description in terms (I, P, χ, β) is easily understood using well-known conceptions of intensity, degree of polarization, the polarization ellipse orientation and ellipticity. However, (unlike I, P, χ, β) parameters I, Q, U and V have the same dimensions and can be directly measured (Stokes, 1852). Therefore, in our analysis below we prefer to use them. Physically based characteristics P, χ and β can easily be calculated using rules given above.

The question arises—how we can calculate Stokes parameters of scattered light knowing the amplitude scattering matrix $\hat{S}$. The problem can be easily solved using Eq. (2.89). Let us introduce the density matrix $\hat{\rho}$ as follows:

$$\hat{\rho} = \vec{E} \otimes \vec{E}^{+}, \tag{2.242}$$

where $\otimes$ means the direct product and $+$ gives the simultaneous operation of conjugation and transportation. This can be written also in the matrix form:

$$\hat{\rho} = \begin{pmatrix} E_l E_l^* & E_l E_r^* \\ E_r E_l^* & E_r E_r^* \end{pmatrix}. \tag{2.243}$$

The relationship between matrices $\hat{\rho}$ of scattered beam and the matrix $\hat{\rho}^i$ of the incident beam can be easily established now. In particular, we have (see Eq. (2.89)):

$$\hat{\rho} \equiv \vec{E} \otimes \vec{E}^{+} = \frac{1}{k^2 r^2} \hat{S}\vec{E}_i \otimes \left(\hat{S}\vec{E}_i\right)^{+} \tag{2.244}$$

or

$$\hat{\rho} = \frac{1}{k^2 r^2} \hat{S} \hat{\rho}_i \hat{S}^{+}. \tag{2.245}$$

This gives us the law of transformation of the density matrix $\hat{\rho}$. This matrix is related to the Stokes vector by the following equation:

$$\hat{\rho} = \frac{1}{2}(I\hat{\sigma}_1 + Q\hat{\sigma}_2 + U\hat{\sigma}_3 + V\hat{\sigma}_4), \tag{2.246}$$

where

$$\hat{\sigma}_1 = \begin{pmatrix} 1 & 0 \\ 0 & 1 \end{pmatrix}, \hat{\sigma}_2 = \begin{pmatrix} 1 & 0 \\ 0 & -1 \end{pmatrix}, \hat{\sigma}_3 = \begin{pmatrix} 0 & 1 \\ 1 & 0 \end{pmatrix}, \hat{\sigma}_4 = \begin{pmatrix} 0 & -i \\ i & 0 \end{pmatrix}. \tag{2.247}$$

Therefore, we have

$$I_j = Tr(\hat{\sigma}_j \hat{\rho}), \tag{2.248}$$

where $I_1 = I$, $I_2 = Q$, $I_3 = U$, $I_4 = V$ and we used the property $Tr(\hat{\sigma}_q \hat{\sigma}_P) = 2\delta_{qp}$. Here Tr means the trace operation. Finally, we obtain:

$$I_j = \frac{1}{k^2 r^2} Tr(\hat{\sigma}_j \hat{S} \hat{\rho}_0 \hat{S}^+). \tag{2.249}$$

Also it follows:

$$\rho_0 = \frac{1}{2} \sum_{k=1}^{4} I_{0k} \hat{\sigma}_k \tag{2.250}$$

and, therefore,

$$I_j = \frac{1}{k^2 r^2} M_{jk} I_{0k}, \tag{2.251}$$

where

$$M_{jk} = \frac{1}{2} Tr(\hat{\sigma}_j \hat{S} \hat{\sigma}_k \hat{S}^+) \tag{2.252}$$

are elements of 4*4 Mueller matrix $\hat{M}$. This establishes the law of transformation of the Stokes vector of the incident light due to the scattering process. We see that the dimensionless 4×4 transformation matrix M_{jk} is determined solely by the 2*2 amplitude scattering matrix $\hat{S}$. Simple calculations give for nonzero elements of this matrix for spheres:

$$M_{11} = M_{22} = \frac{1}{2}(i_1 + i_2), \tag{2.253}$$

$$M_{12} = M_{21} = \frac{1}{2}(i_1 - i_2), \tag{2.254}$$

$$M_{33} = M_{44} = \mathrm{Re}(S_{11} S_{22}^*), \tag{2.255}$$

$$M_{34} = -M_{43} = \mathrm{Im}(S_{11} S_{22}^*), \tag{2.256}$$

where $i_1 = S_{11} S_{11}^*$, $i_2 = S_{22} S_{22}^*$. Note that we have for randomly oriented nonspherical particles: $M_{22} \neq M_{11}$, $M_{33} \neq M_{44}$. It is useful to introduce the normalized Mueller matrix having elements $m_{ij} = M_{ij}/M_{11}$. We can write for spheres:

$$m_{11} = m_{22} = 1, \tag{2.257}$$

$$m_{12} = m_{21} = \frac{i_1 - i_2}{i_1 + i_2}, \tag{2.258}$$

$$m_{22} = m_{33} = \frac{\mathrm{Re}(S_1 S_2^*)}{i_1 + i_2}, \tag{2.259}$$

$$m_{34} = -m_{43} = \frac{\mathrm{Im}(S_1 S_2^*)}{i_1 + i_2} \tag{2.260}$$

with all other elements equal to zero. These elements have a simple physical sense. Namely m_{12} is equal to the degree of polarization P of scattered light under unpolarized (e.g., solar) light illumination conditions. The element m_{44} gives the degree of circular polarization $P_c = V/I$ of scattered light assuming that incident light is right-hand circularly polarized. The element m_{34} gives the value of P_c for the case of linearly polarized incident light with the azimuth $-45°$. To establish this meaning, the following experiment can be performed. Let us illuminate a spherical particle by light with the Stokes vector with components I_0, Q_0, U_0, V_0. Then it follows:

$$I = M_{11}I_0 + M_{12}Q_0 \quad Q = M_{12}I_0 + M_{11}Q_0, \tag{2.261}$$

$$U = M_{44}U_0 + M_{34}V_0, \quad V = -M_{34}U_0 + M_{44}V_0, \tag{2.262}$$

where we neglected a common multiplier $(kr)^{-2}$.

The physical sense of ratios $m_{12} = M_{12}/M_{11}, m_{34} = M_{34}/M_{11}, m_{44} = M_{44}/M_{11}$ can be obtained using various assumptions on the polarization state of an incident beam. For instance, assuming that incident light is unpolarized $(Q_0 = U_0 = V_0)$, we have:

$$I = M_{11}I_0, Q = M_{12}\mathrm{I}_0, U = V = 0. \tag{2.263}$$

This means that a spherical droplet polarizes incident unpolarized light (e.g., from the Sun). The scattered light becomes linearly polarized with the degree of polarization

$$P_l = -\frac{Q}{I} \tag{2.264}$$

or

$$P_l = -m_{12}. \tag{2.265}$$

Therefore, the value of m_{12} gives (with the opposite sign) the degree of polarization of scattered light for the unpolarized light illumination conditions. The sign is chosen in such a way that the degree of polarization becomes positive as $|mka| \to 0$ (for Rayleigh scattering). Then light is predominantly polarized in the plane perpendicular to the scattering plane.

Let us assume now that a droplet is illuminated by a right-hand circularly polarized beam. Then we have: $I_0 = V_0$ and $U_0 = Q_0 = 0$. This means that

$$I = M_{11}I_0, Q = M_{12}I_0, U = M_{34}V_0, V = M_{44}V_0. \tag{2.266}$$

Scattered light becomes elliptically polarized with the total degree of polarization

$$P = \frac{\sqrt{Q^2 + U^2 + V^2}}{I} \tag{2.267}$$

or

$$P = \frac{\sqrt{M_{12}^2 + M_{34}^2 + M_{44}^2}}{M_{11}}. \tag{2.268}$$

Note that the spherical particle does not depolarize the totally polarized scattering beam. It means that $P \equiv 1$ in this case. The equality

$$M_{12}^2 + M_{34}^2 + M_{42}^2 = M_{11}^2 \tag{2.269}$$

indeed holds for monodispersed spheres. This can be easily checked using exact results given above. It also means that a spherical particle does not produce entropy during the scattering process (Kokhanovsky, 2003a).

The total degree of polarization P can be written as

$$P = \sqrt{P_l^2 + P_c^2}, \tag{2.270}$$

where

$$P_l = \frac{\sqrt{Q^2 + U^2}}{I} \tag{2.271}$$

is the degree of linear polarization of scattered light and

$$P_c = \frac{V}{I} \tag{2.272}$$

is the degree of circular polarization of scattered light. Note that $m_{44} \equiv P_c$ for illumination of a particle by the circularly polarized beam ($I_0 = V_0 = 1,\ Q_0 = U_0 = 0$). This gives the physical sense of the element m_{44}. In particular, this elements describes the reduction ($m_{44} \leq 1$) of the degree of circular polarization of the completely circularly polarized incident light beam due to a scattering event.

Assuming that incident light is linearly polarized with the azimuth -45° ($Q_0 = 0$, $U_0 = -I_0$, $V_0 = 0$), we obtain:

$$I = M_{11}I_0,\ Q = M_{12}I_0,\ U = M_{44}U_0,\ V = -M_{34}U_0. \tag{2.273}$$

Again we have for the degree of total polarization:

$$P = \frac{\sqrt{M_{12}^2 + M_{34}^2 + M_{44}^2}}{M_{11}} \equiv 1 \tag{2.274}$$

and $P_c = m_{34}$. Therefore, the element m_{34} describes the degree of linear to circular polarization mode conversion. It coincides with the degree of circular polarization of a scattered light beam, assuming that illuminated beam is linearly polarized (with the azimuth -45°). This element is usually small for cloud droplets. So we conclude that such a transformation is not an effective process for cloudy media.

Summing up, we may write the single scattering law in the following form:

$$\vec{I} = \alpha p(\theta) \begin{pmatrix} 1 & m_{12} & 0 & 0 \\ m_{12} & 1 & 0 & 0 \\ 0 & 0 & m_{44} & m_{34} \\ 0 & 0 & -m_{34} & m_{44} \end{pmatrix} \vec{I}_0, \quad (2.275)$$

or

$$\vec{I} = \alpha p(\theta)\hat{m}(\theta)\vec{I}_0, \quad (2.276)$$

where $p(\theta) = 4\pi M_{11}/k^2 C_{sca}$ is the phase function and $\alpha = d\Omega_s/4\pi$ is the relative solid angle associated with scattering:

$$d\Omega_s = \frac{C_{sca}}{r^2}. \quad (2.277)$$

The element of the Mueller matrix M_{11} gives the transformation law for the unpolarized incident light due to a scattering event:

$$I = \frac{M_{11}}{k^2 r^2} I_0. \quad (2.278)$$

It is useful to introduce the phase matrix

$$\hat{P}(\theta) = p(\theta)\hat{m}(\theta). \quad (2.279)$$

This matrix is of paramount importance for vector radiative transfer studies. Note that it follows for the elements of the normalized phase matrix: $p_{ij} \equiv P_{ij}/P_{11} = m_{ij}$.

To conclude, we underline that Mie theory can be used to calculate not only intensity but also other Stokes vector components of the scattered light beam.

2.3.4 Spherical Polydispersions

Water droplets in clouds have different sizes. Usually only average characteristics of a cloud are of interest. In particular, the average extinction cross section for a given particle size distribution $f(a)$ can be presented as:

$$\bar{C}_{ext} = \int_0^\infty C_{ext}(a) f(a) da. \quad (2.280)$$

In a similar way other light scattering and absorption characteristics can be calculated. For instance, it follows for the average scattering cross section:

$$\bar{C}_{sca} = \int_0^\infty C_{sca}(a) f(a) da, \quad (2.281)$$

and for the average phase function:

$$p(\theta) = \frac{4\pi \bar{i}(\theta)}{k^2 \bar{C}_{sca}}, \tag{2.282}$$

where

$$\bar{i}(\theta) = \frac{1}{2}\int_0^\infty (i_1(a) + i_2(a)) f(a) da. \tag{2.283}$$

Also we have for the average phase matrix elements:

$$P_{ij}(\theta) = \frac{4\pi \bar{M}_{ij}}{k^2 \bar{C}_{sca}}, \tag{2.284}$$

where

$$\bar{M}_{ij} = \int_0^\infty M_{ij}(a) f(a) da. \tag{2.285}$$

The average asymmetry parameter is given by:

$$\bar{g} = \frac{1}{2}\int_0^\pi \bar{p}(\theta) \sin\theta \cos\theta d\theta \tag{2.286}$$

or

$$\bar{g} = \frac{\int_0^\infty g(a) C_{sca}(a) f(a) da}{\int_0^\infty C_{sca}(a) f(a) da}. \tag{2.287}$$

Extinction σ_{ext}, absorption σ_{abs}, and scattering σ_{sca} coefficients are determined by the correspondent average cross sections, $\bar{C}_{ext}$, $\bar{C}_{abs} = \bar{C}_{ext} - \bar{C}_{sca}$, and $\bar{C}_{sca}$, respectively, as follows: $\sigma_{ext} = N\bar{C}_{ext}$, $\sigma_{abs} = N\bar{C}_{abs}$, $\sigma_{sca} = N\bar{C}_{sca}$, where $N = C_v \bar{V}^{-1}$, C_v is the volumetric concentration of particles, and $\bar{V}$ is the average volume of particles. Furthermore, one can introduce the single scattering albedo $\omega_0 = \sigma_{sca}/\sigma_{ext}$ and the probability of photon absorption $\beta = 1 - \omega_0 = \sigma_{abs}/\sigma_{ext}$.

In Section 2.3.5 we show examples of functions given above calculated from the Mie theory for the gamma droplet size distribution

$$f(a) = Aa^6 e^{-1.5a/a_{ef}}, \tag{2.288}$$

where and a_{ef} is the effective radius of droplets and A is the PSD normalization constant.

But first of all we would like to demonstrate how closely theoretical results obtained from Mie theory follow experimental data. This is shown in Fig. 2.4. Experimental data given in Fig. 2.4 were obtained by Volten (2001) for spherical polydispersions of water droplets at the wavelengths 441.6 and 632.8 nm. It follows

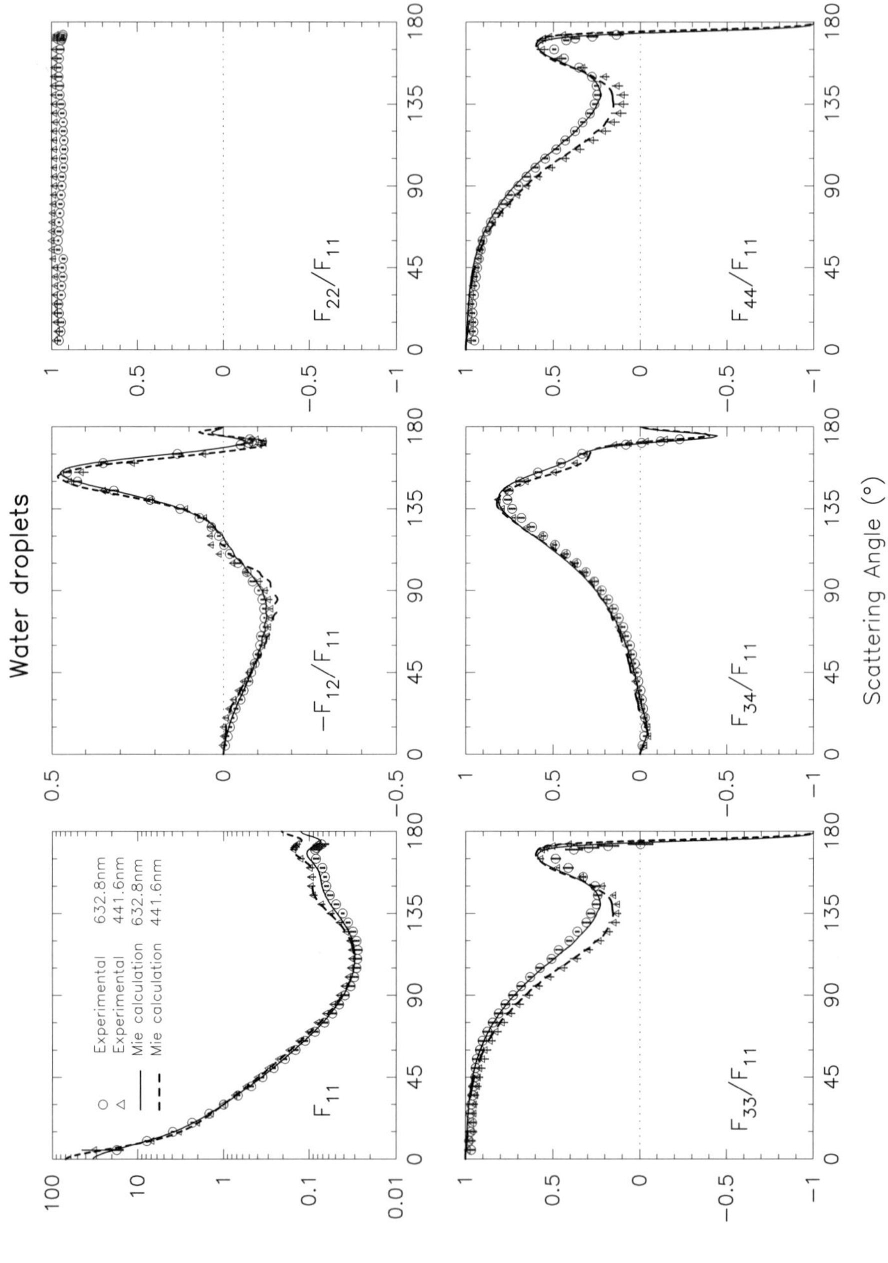

Fig. 2.4. The phase matrix of a light scattering medium composed of water droplets. Measurements and calculations are performed at wavelengths 441.6 and 632.8 nm (Volten, 2001).

that experimental data can be explained by the theory developed in this chapter using the lognormal size distribution with the effective radius 1.1 μm and the effective variance equal to 0.5 and the refractive index $m = 1.33$. Mie theory and experiment agree in every detail.

The following conclusions can be drawn from Fig. 2.3:

- Light scattering properties of water droplets little change in the visible.
- The phase function is highly extended in the forward scattering direction as opposite to the Rayleigh scattering, where $p(0) = p(\pi)$.
- The probability of photon scattering reaches a minimum not at $\theta = 90°$ (as it is the case for the molecular scattering) but rather at somewhat larger scattering angle $\theta \approx 110°$.
- The dependence of the phase function on the wavelength is more pronounced in the directions close to forward and backward scattering.
- The degree of polarization P of scattered initially unpolarized light changes its sign as opposite to the case of Rayleigh scattering, where it is positive for all scattering angles. For droplets, P is negative (oscillations are predominantly in the scattering plane) in the forward scattering hemisphere. The scattered light in the backward hemisphere is generally characterized by positive values of P but the sign of the degree of polarization changes at $\theta \approx \pi$.
- For most scattering angles, the direction of rotation of incident circularly polarized light is preserved (see the element P_{44}). However, the absolute value of the degree of circular polarization is substantially reduced due to the scattering process and can vanish at the angle θ close to 170°.
- The linear-to-circular polarization light conversion (see the element P_{34}) is not very effective in the forward direction. However, such a conversion could be of importance in the backward hemisphere (e.g., at $\theta \approx 135°$).

Although these conclusions are derived from a particular case presented in Fig. 2.4, most of them remain valid for arbitrary $f(a)$ in the range of radii characteristic for water clouds.

Having shown that Mie theory satisfies experimental data in every detail, we can investigate light scattering by water droplets theoretically not referring to experimental measurements. This is presented in the next section.

2.3.5 Local Optical Characteristics of Clouds

The phase function of a water cloud calculated using Mie theory and PSD (2.288) at $a_{ef} = 6$μm is shown in Fig 2.5 at wavelengths 0.532, 1.064, 1.64, and 2.13 6μm. These wavelengths are often used in cloud remote sensing from ground and space. We see that the rainbow feature around 145° is more pronounced in the

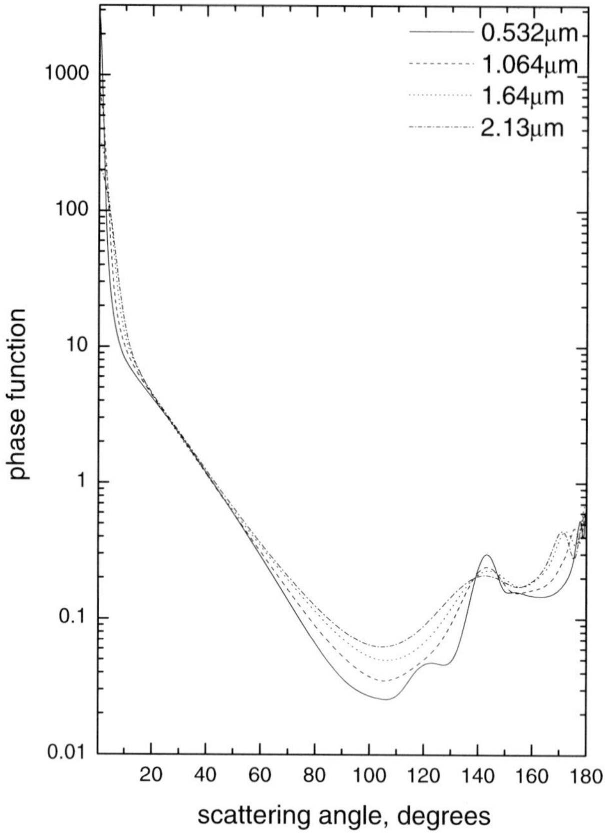

Fig. 2.5. The phase function of water clouds at wavelengths 0.532, 1.064, 1.64, and 2.13 μm for the cloud C1 PSD.

visible. There is a large asymmetry in forward—backward scattering. Interestingly, the phase function around 30° is almost insensitive to the wavelength. The value of the phase function is close to 2 at this angle. The peak in the forward direction is more pronounced in the visible.

One can find even more differences in the degree of light polarization presented in Fig. 2.6 at several wavelengths for solar light illumination conditions. We see that the degree of polarization generally decreases with the wavelength. The largest polarization is in the region of the rainbow. The degree of polarization is low at small scattering angles. But it is quite large at $\theta \approx \pi$, where the degree of polarization changes its sign in the vicinity of backward scattering.

The normalized scattering matrix elements p_{34} and p_{44} are shown in Figs. 2.7 and 2.8. The element p_{34} coincides with the degree of circular polarization for linearly polarized light illumination conditions (with the azimuth –45°). We see that

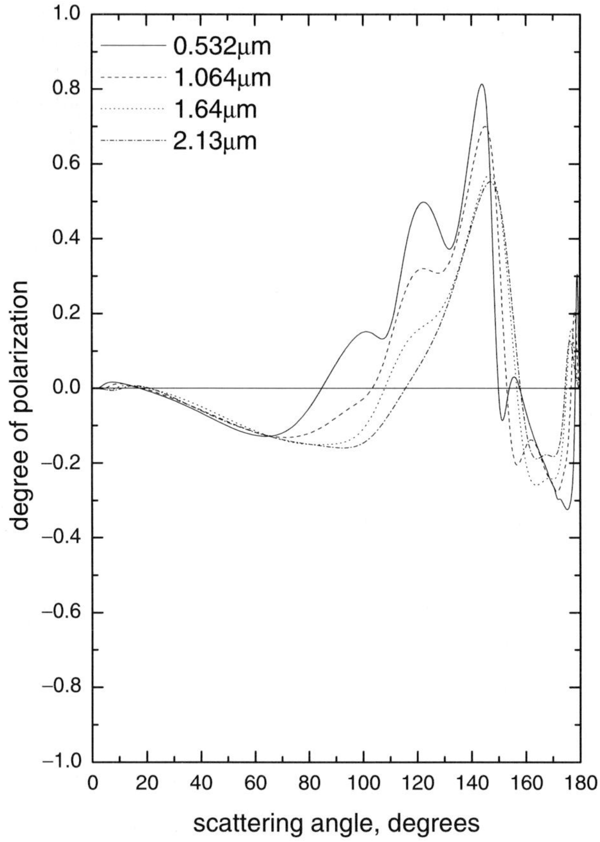

Fig. 2.6. The same as in Fig. 2.5 except for the degree of polarization.

there is a strong linear-to-circular polarization mode conversion at the scattering angle close to 130°. The element p_{44} also gives the degree of circular polarization (DCP) but for the right-hand circularly polarized incident light conditions. Generally, DCP decreases with the scattering angle in the forward hemisphere. Then it oscillates and reaches the value equal to –1 in the exact backward direction. We see that the degree of circular polarization (DCP) of the scattered light is 100% both for forward and backward scattering directions. However, the scattering at $\theta = \pi$ changes the direction of rotation and DCP is equal to -1 then (see Fig. (2.8)).

The dependence of the phase function and normalized phase matrix elements on the PSD halfwidth parameter is shown in Fig. 2.9. It follows that the dependence of the phase function and, generally, the phase matrix on this parameter is rather weak with exception of scattering close to 150° in polarized light. This makes it difficult to obtain the halfwidth parameter from light scattering experiments

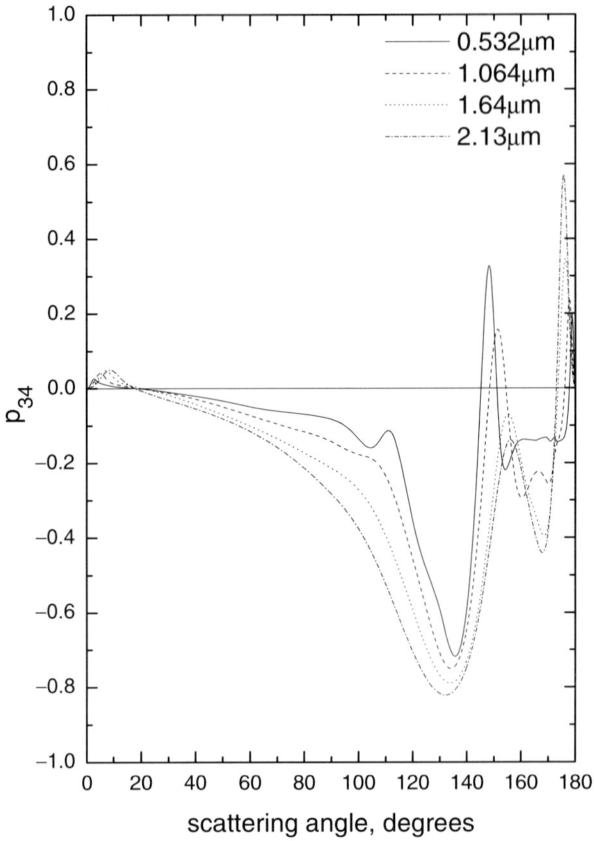

Fig. 2.7. The same as in Fig. 2.5 except for the normalized phase matrix element p_{34}.

performed in cloudy media. Generally, only the value of the effective radius can be retrieved. The dependence of the same characteristics on the imaginary part of the refractive index is shown in Fig. 2.10. We see that absorbing additives to clouds (e.g., soot) can change the phase function and phase matrix elements considerably.

The spectral dependence of the cloud extinction coefficient is shown in Fig. 2.11a. It is in the range 0.26–0.38 m^{-1} at $a_{ef} = 6\,\mu\mathrm{m}$, the volumetric concentration of droplets $C_v = 10^{-6}$ and λ smaller than 5μm. Usually measurements give values of σ_{ext} around 0.03 m^{-1}. This means that C_v is at least one order of magnitude smaller ($\sim 10^{-7}$) in natural clouds. The spectral absorption coefficient, the probability of photon absorption, and the co-asymmetry parameter $G = 1 - g$ of water clouds are shown in Figs. 2.11b–d. It follows that absorption generally increases with the wavelength. The asymmetry parameter does not change considerably in the visible. However, it increases at the water absorption band close to

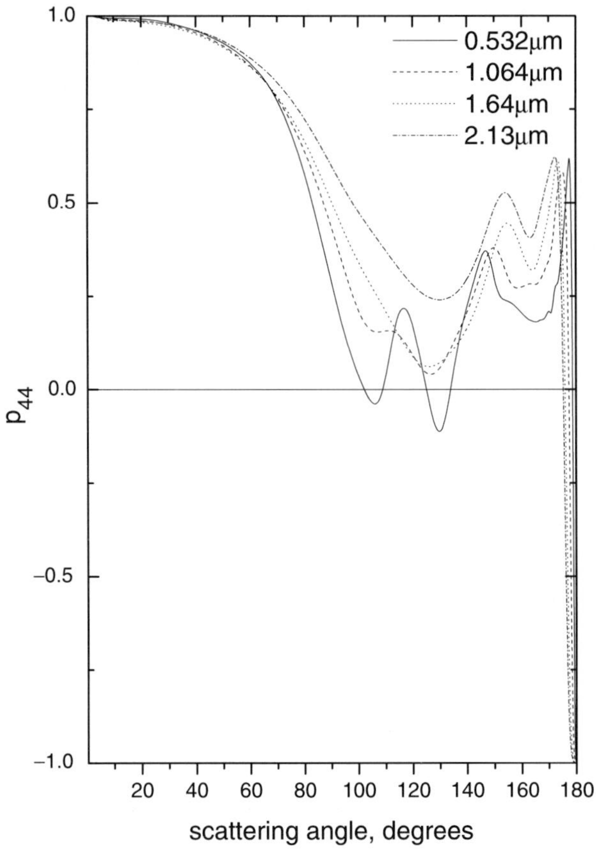

Fig. 2.8. The same as in Fig. 2.5 except for the normalized phase matrix element p_{44}.

3 μm, where extinction is at a minimum. The increase of the droplet size leads to a larger absorption as demonstrated in Fig. 2.12. Therefore, one can perform droplet optical sizing measuring the absorption (or ratio absorption/extinction).

2.4 Geometrical Optics

2.4.1 Water Droplets

2.4.1.1 Scattered light intensity

A typical radius of a cloud droplet a is 10μm. This is approximately 20 times larger than the wavelength of visible light. Sizes of ice crystals are even larger (typically, 100 μm). It means that the geometrical optics techniques valid

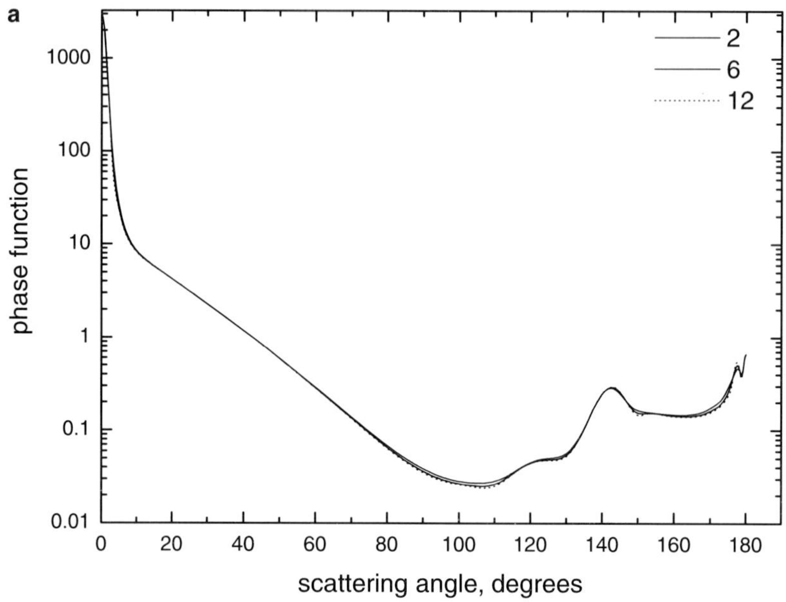

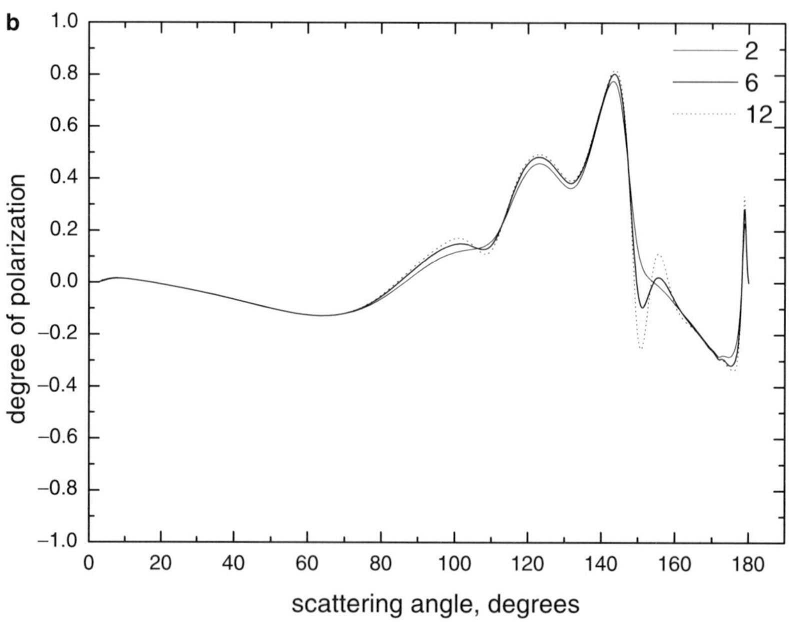

Fig. 2.9. The phase function (a), the degree of polarization (b), the element p_{34} (c), and the element p_{44} (d) of water clouds calculated using Mie theory for the gamma PSD with $\mu = 2, 6, 12$ at $a_{\mathrm{ef}} = 6\ \mu\mathrm{m}$ and $\lambda = 0.55\ \mu\mathrm{m}$.

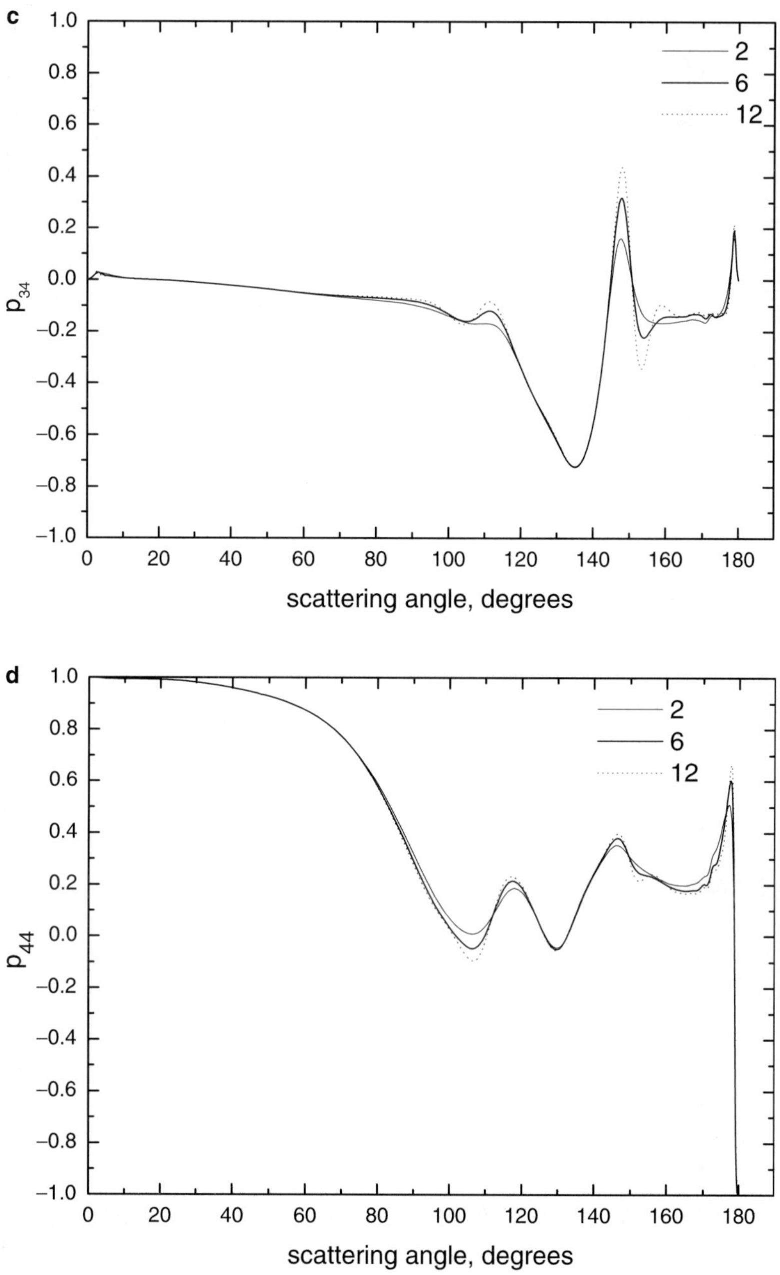

Fig. 2.9. *Continued.*

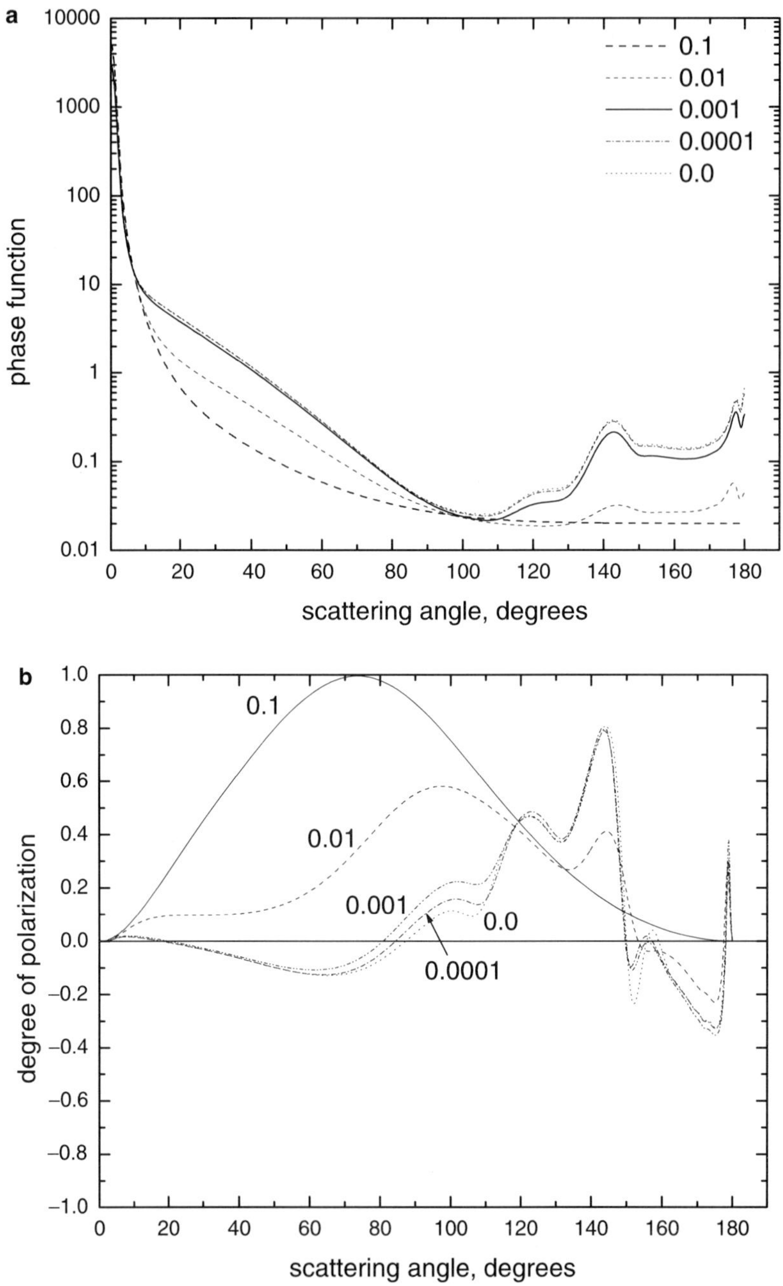

Fig. 2.10. The phase function (a), the degree of polarization (b), the element p_{34} (c), and the element p_{44} (d) of light scattering media with water droplets calculated using Mie theory for the gamma PSD with $\mu = 6$, $a_{ef} = 6$ μm and $\lambda = 0.55$ μm. The imaginary part of the refractive index is varied is in the range 0.0–0.1 as indicated in the Figure. The real part of the refractive index n is equal to 1.333.

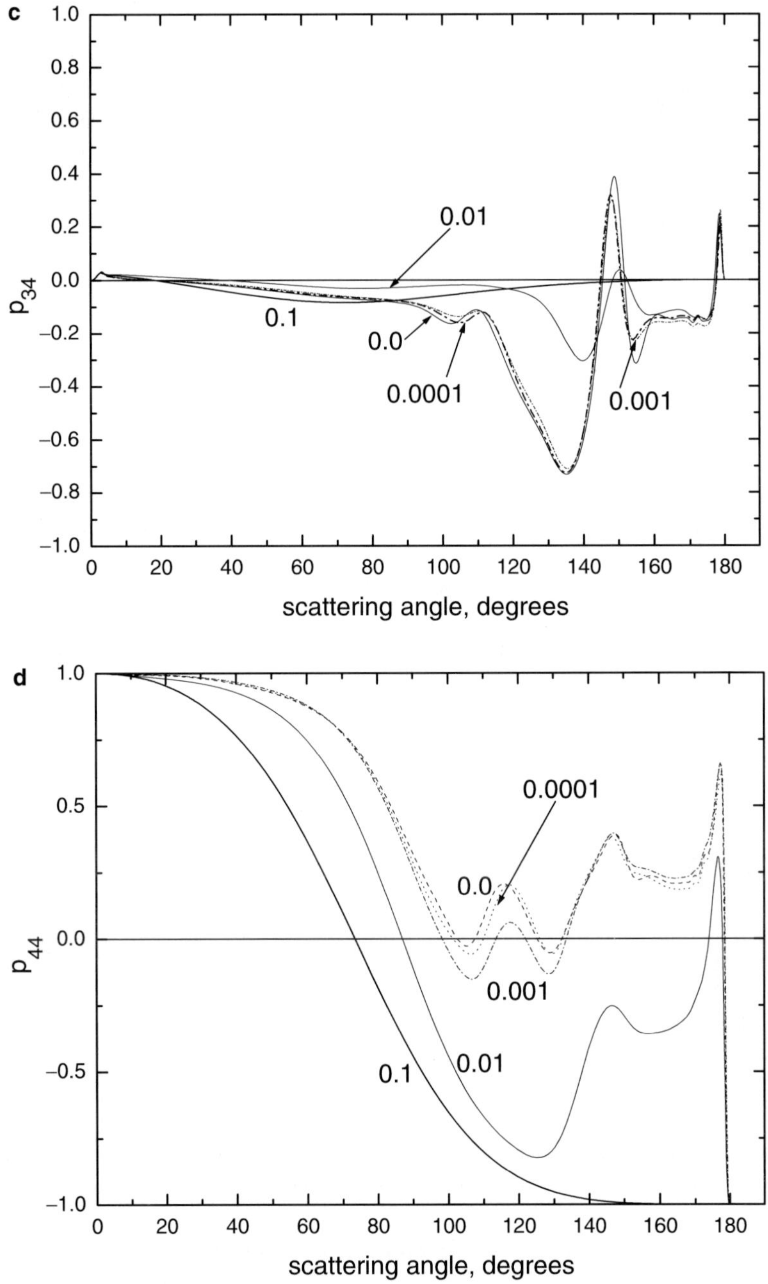

Fig. 2.10. *Continued*

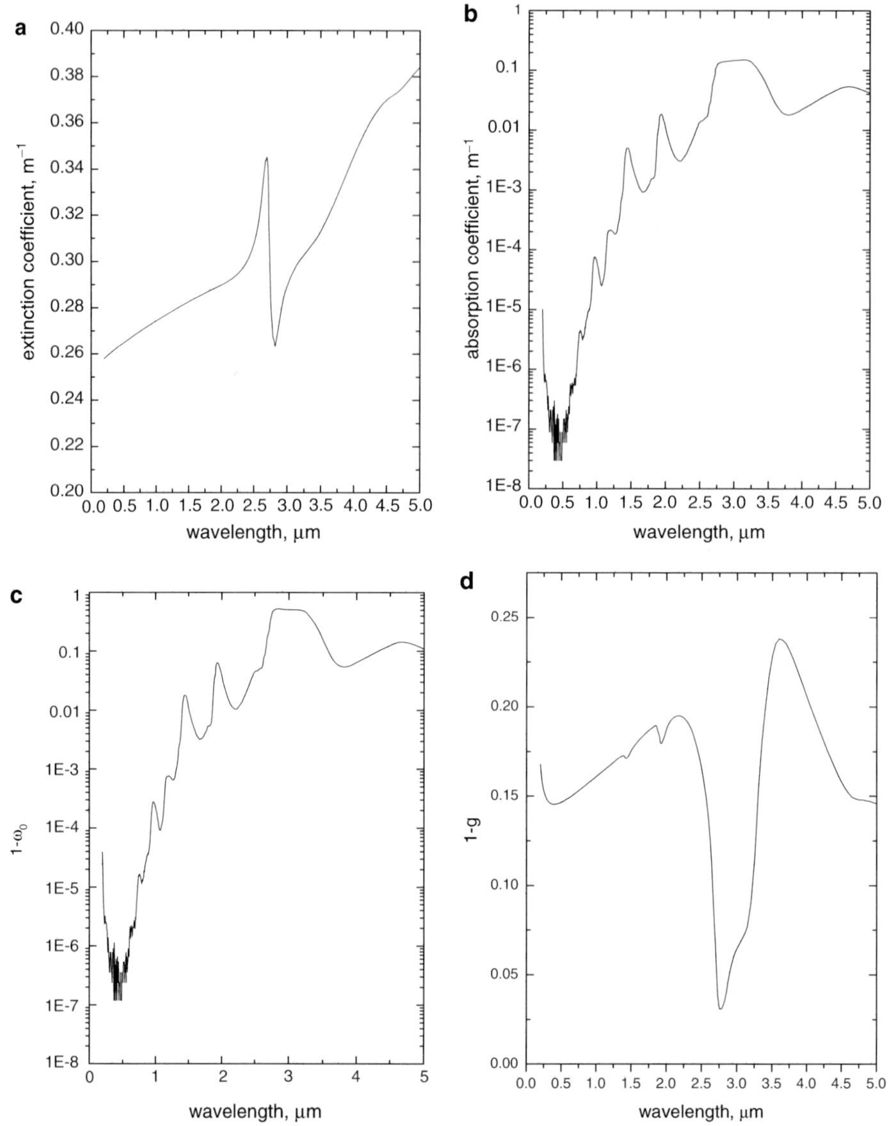

Fig. 2.11. Extinction coefficient (a), absorption coefficient (b), probability of photon absorption (c), and $1-g$ (d) as functions of the wavelength for the cloud model C1.

as $\lambda/a \to 0$ can be applied to cloud optics studies. Of course, geometrical optics results provide only an approximate answer to the problem at hand. A full wave solution (see, e.g., Mie theory above) must be used whenever it is possible. The importance of geometrical optics is mainly due to the fact that the geometrical

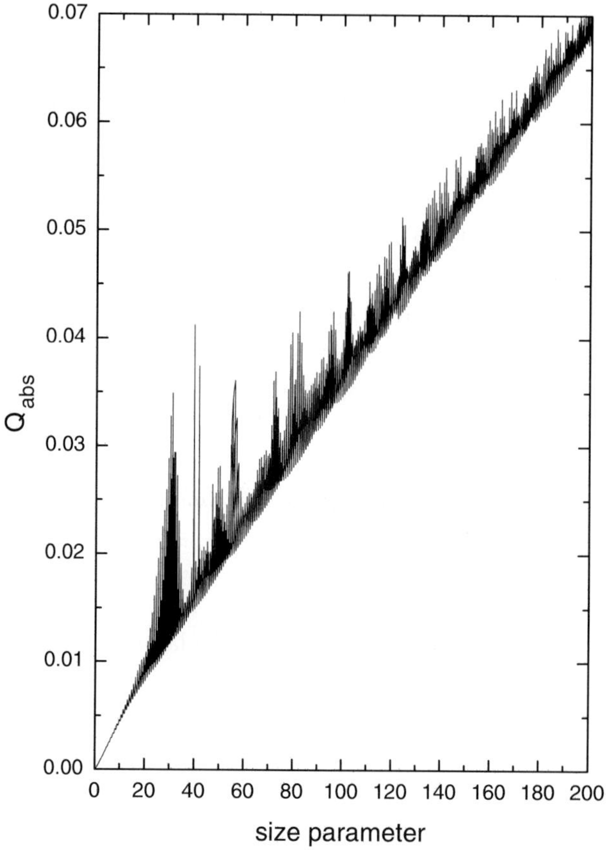

Fig. 2.12. Dependence of absorption efficiency factor on the size parameter $x = ka$ at $n = 1.333$, $\chi = 0.0001$.

optics approach offers the possibility of physical insights and explanations of the behaviour of various light scattering characteristics as functions of the wavelength, size, and refractive index. For instance, it follows from Figs. 2.1 and 2.2 that $Q_{ext} \to 2$ and $Q_{abs} \to 1 - w$ as $\lambda/a \to 0$ but what is the reason for this? Also we see that $Q_{abs} \sim a$ as $\lambda/a \to 0$ and $\alpha a \to 0$, where $\alpha = 4\pi\chi/\lambda$ (see Fig. 2.12). What is the underlying physics behind these and many other effects? To answer this question the geometrical optics approach can be used. Geometrical optics also allows us to obtain simple analytical relationships between light scattering and microphysical characteristics, which are used for rapid estimations of various light scattering characteristics, physical parameterizations and the solution of inverse problems. The geometrical optics approach is of special importance for optics of nonspherical ice crystals. Then Mie theory can not be applied.

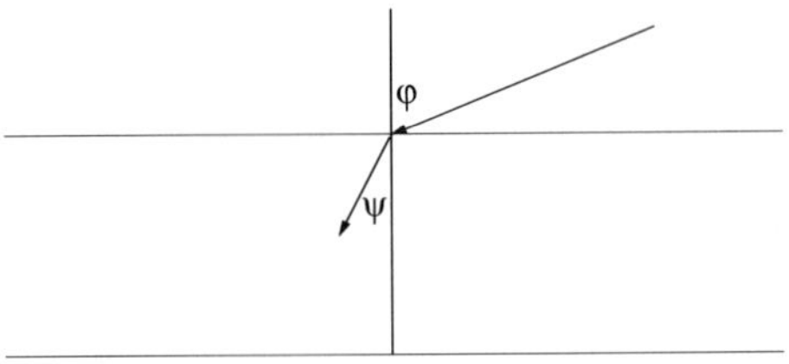

Fig. 2.13. The refraction of a light beam in a homogeneous plane parallel slab.

The background of geometrical optics is quite simple. It is based on the Fresnel theory. In particular, it was shown by Fresnel (1866) that the intensity of light reflected from a plane surface can be presented as:

$$I_l = |R_l|^2 I_0 \tag{2.289}$$

for incident light with the intensity I_0 polarized in the incidence plane, which holds incident wave and normal to the layer, and

$$I_r = |R_r|^2 I_0 \tag{2.290}$$

for incident light polarized in the plane perpendicular to the incidence plane. The Fresnel coefficients R_l and R_r are given by the following simple equations:

$$R_l = \frac{\cos\varphi - m\cos\psi}{\cos\varphi + m\cos\psi}, \tag{2.291}$$

$$R_r = \frac{m\cos\varphi - \cos\psi}{m\cos\varphi + \cos\psi}, \tag{2.292}$$

where $m = n - i\chi$ is the relative refractive index of the medium (e.g., $m = m_2/m_1$, where m_1 is the refractive index of a medium from which light comes and m_2 is the refractive index of a medium, where light propagates after the refraction event). The incidence φ and refraction ψ angles (see Fig. 2.13) are related by the following equation:

$$\sin\varphi = m\sin\psi, \tag{2.293}$$

which is called the Snellius law. Therefore, the refracted wave propagates in the direction specified by the angle

$$\psi = \arcsin\left(\frac{\sin\varphi}{m}\right) \tag{2.294}$$

in the incidence plane. The reflected light beam is in the same plane as an incident light beam and makes the angle with the normal equal to the incidence angle φ (see Fig. 2.12).

These simple laws allow the study a lot of practically important optical problems. For instance, we can obtain for the reflectivity $\widehat{R}$ in the case of a plane-parallel homogeneous layer illuminated by unpolarized light (Stokes, 1862):

$$\widehat{R} = R + \frac{c^2 R(1-R)^2}{1-c^2R^2}, \tag{2.295}$$

where

$$R = \frac{1}{2}(|R_l|^2 + |R_r|^2) \tag{2.296}$$

and

$$c = \exp(-\alpha L/\cos\psi) \tag{2.297}$$

where $\alpha = 4\pi\chi/\lambda$ and L is the thickness of the layer. We have for the transmittivity T (Stokes, 1862):

$$T = \frac{c(1-R)^2}{1-c^2R^2}. \tag{2.298}$$

It follows for a nonabsorbing layer ($c = 1$) that $\widehat{R} + T = 1$. However, one obtains for an absorbing layer: $T + \widehat{R} = 1 - A$, where

$$A = \frac{(1-R)(1-c)}{1-cR} \tag{2.299}$$

is the absorptivity of the layer. Similar but somewhat more complex results can be obtained for a spherical scatterer. The difference of a particle from a slab is due to the fact that a particle has finite dimensions and the scattered light energy is composed of both reflected and refracted beams.

A finite pencil of light is characterized by the azimuth range $d\phi$ and the range of incidence angles $d\varphi$. The flux of energy F_0 in this pencil is

$$F_0 = I_0 dS, \tag{2.300}$$

where

$$dS = a^2 \cos\varphi \sin\varphi d\varphi d\phi \tag{2.301}$$

is the illuminated surface element for a spherical particle with the radius a. The reflected fractions of energy are

$$R_1 = |R_l|^2 \tag{2.302}$$

and

$$R_2 = |R_r|^2 \tag{2.303}$$

depending on the polarization of incident light. Here we assume that the spherical surface element can be substituted by the plane surface element at $a \gg \lambda$.

Therefore, we have for the reflected flux F^r:

$$F_j^r = R_j F_0 = R_j I_0 dS. \tag{2.304}$$

To find the intensity of reflected light we need to divide this expression by the area $d\Sigma$, where the scattered beam is distributed. It is easy to show that this area is equal to $r^2 \sin\theta d\theta d\phi$, which is the surface element in the spherical coordinate system. Here θ is the scattering angle and r is the distance to the observation point. Therefore, we have for the intensity:

$$I_j^r = R_j I_0 \frac{dS}{d\Sigma} \tag{2.305}$$

or

$$I_j^r = \frac{R_j a^2 \cos\varphi \sin\varphi I_0}{r^2 \sin\theta N}, \tag{2.306}$$

where

$$N = \left|\frac{d\theta}{d\varphi}\right|. \tag{2.307}$$

We take the absolute value of the derivative because $I_j^r \geq 0$. Usually the value I_j^r is represented in the following form:

$$I_j^r = \frac{a^2}{r^2} R_j I_0 D, \tag{2.308}$$

where

$$D = \frac{\sin\varphi \cos\varphi}{N \sin\theta} \tag{2.309}$$

is the so-called divergence.

Let us consider the application of this formula to the calculation of light intensity reflected in the direction specified by the scattering angle θ. First of all, we have:

$$\theta = \pi - 2\varphi. \tag{2.310}$$

for the reflected light. This means that

$$\varphi = \frac{\pi - \theta}{2}, N = 2, D = \frac{1}{4} \tag{2.311}$$

and

$$\psi = \arcsin\left(\frac{1}{m}\cos\left(\frac{\theta}{2}\right)\right). \tag{2.312}$$

Therefore, it follows:

where

$$R_1 = \left| \frac{\sin\frac{\theta}{2} - \sqrt{m^2 - \cos^2\frac{\theta}{2}}}{\sin\frac{\theta}{2} + \sqrt{m^2 - \cos^2\frac{\theta}{2}}} \right|^2, \tag{2.314}$$

$$R_2 = \left| \frac{m\sin\frac{\theta}{2} - \sqrt{1 - m^{-2}\cos^2\frac{\theta}{2}}}{m\sin\frac{\theta}{2} + \sqrt{1 - m^{-2}\cos^2\frac{\theta}{2}}} \right|^2. \tag{2.315}$$

In particular, we have for incident unpolarized light:

$$I^r = \frac{i_1 + i_2}{2k^2r^2} I_0, \tag{2.316}$$

where

$$i_1 = \frac{x^2}{4} R_1, i_2 = \frac{x^2}{4} R_2. \tag{2.317}$$

This allows us to introduce a simple approximate way to calculate Mie dimensionless intensities [see Eq. (2.114)] using geometrical optics results. Clearly, the accuracy of such an approximation depends on the contribution of reflected light in the general light scattered intensity.

Generally speaking, we need to consider the contribution of transmitted light as well. The only change in Eq. (2.306) is the substitution of R_j by (van de Hulst, 1981)

$$\varepsilon_j^p = (1 - R_j)^2 R_j^{2(p-1)} e^{-\bar{c}p}, \tag{2.318}$$

where $\bar{c} = c\cos\psi$, $c = 2\alpha a$. In particular, we have at $p = 1$:

$$\varepsilon_j^1 = (1 - R_j)^2 \exp(-\bar{c}). \tag{2.319}$$

This term has a simple interpretation as a product of three processes:

– transmission of light through a host medium-particle interface $(1 - R_j)$;
– wave attenuation on the length $\Lambda = d\cos\psi$, where $d = 2a$;
– transmission through the particle-host medium interface back in outer space.

It follows for the scattering angle for rays with $p = 1$:

$$\theta = 2(\varphi - \psi) \tag{2.320}$$

and, therefore,

$$N = 2 - \frac{2\cos\varphi}{m\cos\psi}, \tag{2.321}$$

$$I_j^1 = \frac{a^2}{r^2}(1 - R_j)^2 e^{-\bar{c}} \frac{\sin\varphi\cos\varphi}{2\left(1 - \frac{\cos\varphi}{m\cos\psi}\right)\sin\theta} I_0, \tag{2.322}$$

It means that

$$i_j^t = \frac{x^2}{4}(1 - R_j)^2 e^{-\bar{c}} \Phi(\theta) \tag{2.323}$$

for transmitted light at $p = 1$. Here,

$$\Phi(\theta) = \frac{m \sin\varphi \cos\varphi \cos\psi}{(m \cos\psi - \cos\varphi) \sin\theta}. \tag{2.324}$$

The next step is to express the pair (φ, ψ) via θ. Clearly, we have (see above):

$$\cos\left(\frac{\theta}{2}\right) = \cos(\varphi - \psi) = \cos\varphi \cos\psi + \sin\varphi \sin\psi \tag{2.325}$$

or

$$\sin\varphi = n\sqrt{\frac{1 - q^2}{1 - 2qn + n^2}} \tag{2.326}$$

and

$$\sin\psi = \frac{\sin\varphi}{n} = \sqrt{\frac{1 - q^2}{1 - 2qn + n^2}}, \tag{2.327}$$

where $q = \cos(\theta/2)$. Also it follows:

$$\cos\varphi = \sqrt{1 - \sin^2\varphi}, \ \cos\psi = \sqrt{1 - n^{-2}\sin^2\varphi}. \tag{2.328}$$

Using these equations, we derive:

$$I_j^1 = \frac{a^2 e^{-\bar{c}}}{4r^2 b_j}\left(\frac{2n}{n^2 - 1}\right)^4 \frac{\{nq - 1\}^3 \{n - q\}^3}{q(n^2 + 1 - 2nq)^2}, \tag{2.329}$$

where $b_1 = 1$, $b_2 = q^4$.

In particular, one obtains for incident unpolarized light

$$I^t = \frac{i_1^t + i_2}{2k^2 r^2} I_0, \tag{2.330}$$

where

$$i_1^t = \frac{x^2 e^{-\bar{c}}}{4q}\left(\frac{2n}{n^2 - 1}\right)^4 \frac{(nq - 1)^3 (n - q)^3}{(1 + n^2 - 2nq)^2}, \tag{2.331}$$

and

$$i_2 = \frac{x^2 e^{-\bar{c}}}{4q^5}\left(\frac{2n}{n^2 - 1}\right)^4 \frac{(nq - 1)^3 (n - q)^3}{(1 + n^2 - 2nq)^2}. \tag{2.332}$$

Accounting for both reflected and transmitted ($p = 1$) light, we have:

$$I^{r,t} = \frac{i_1 + i_2}{2k^2r^2} I_0, \tag{2.333}$$

where

$$i_1 = \frac{x^2}{4}\left(R_1 + \frac{e^{-\bar{c}}}{q}\left(\frac{2n}{n^2-1}\right)^4 \frac{(nq-1)^3(n-q)^3}{(1+n^2-2nq)^2}\right), \tag{2.334}$$

$$i_2 = \frac{x^2}{4}\left(R_2 + \frac{e^{-\bar{c}}}{q^5}\left(\frac{2n}{n^2-1}\right)^4 \frac{(nq-1)^3(n-q)^3}{(1+n^2-2nq)^2}\right) \tag{2.335}$$

and

$$R_1 = \left|\frac{\sqrt{1-q^2} - \sqrt{m^2-q^2}}{\sqrt{1-q^2} + \sqrt{m^2-q^2}}\right|^2 \tag{2.336}$$

$$R_2 = \left|\frac{m^2\sqrt{1-q^2} - \sqrt{m^2-q^2}}{m^2\sqrt{1-q^2} + \sqrt{m^2-q^2}}\right|^2 \tag{2.337}$$

Similarly, we can account for rays with $p \geq 2$. Then we need to add additional terms in expressions for i_1, i_2. In particular, it follows:

$$i_j = \sum_{p=0}^{\infty} i_{jp}. \tag{2.338}$$

Terms i_{j0}, i_{j1} are given above. It is difficult to derive explicit expressions for i_{jp} at $p \geq 2$.

Therefore, to find the total scattered intensity

$$I = \frac{I_0}{2k^2r^2} \sum_{p=0}^{\infty} (i_{1p} + i_{2p}) \tag{2.339}$$

the following procedure is used. First of all for a given incidence angle φ, the refraction angle $\psi = \arcsin(\sin\varphi/n)$ and the scattering angle

$$\theta = (p-2)\pi + 2(\varphi - (p-1)\psi) \tag{2.340}$$

are found. Then the intensity i_{jp} is calculated:

$$i_{jp} = x^2 \varepsilon_j^p D \tag{2.341}$$

for all values of p at fixed angle φ. Repeating this procedure for different values of φ (and, therefore, θ) we can derive the dependence $I(\theta)$.

One can also introduce the gain function

$$G_j(\theta) = \frac{4r^2 I_j(\theta)}{a^2 I_0}. \tag{2.342}$$

This function does not depend on r and I_0. It is defined as the ratio of the scattered intensity to the intensity that would be found in any direction if the drop scattered the entire incident light isotropically ($\varepsilon_j^p \equiv 1$). The gain function for water droplets is tabulated by van de Hulst (1981).

It should be underlined that we have completely ignored the wave phenomena in our derivations. In particular, the diffraction of light by an obstacle is of importance at small scattering angles. To account for this effect we should add the diffraction to the geometrical component I^G. So the full intensity I_{sca} is given by:

$$I_{sca} = I^G + I^D. \tag{2.343}$$

Let us find the approximate equation for I^D from Mie theory results. Namely, we have as $\theta \to 0$:

$$I^D = \frac{i(\theta)}{k^2 r^2}, \tag{2.344}$$

where we assumed that $I_0 = 1$ and

$$i(\theta) = \frac{i_1(\theta) + i_2(\theta)}{2}, \tag{2.345}$$

$$i_1 = \left| \sum_{n=1}^{\infty} \frac{2n+1}{n(n+1)} (a_n \tau_n + b_n \pi_n) \right|^2, \quad i_2 = \left| \sum_{n=1}^{\infty} \frac{2n+1}{n(n+1)} (a_n \pi_n + b_n \tau_n) \right|^2. \tag{2.346}$$

Let us represent amplitude coefficients as:

$$a_n = \frac{1}{2}\{1 - \exp(-2i\alpha_n)\}, \quad b_n = \frac{1}{2}\{1 - \exp(-2i\beta_n)\} \tag{2.347}$$

and use the fact that

$$\pi_n(\theta) \approx \frac{n(n+1)}{2} \left\{ J_0\left(\left[n + \frac{1}{2}\right]\theta\right) + J_2\left(\left[n + \frac{1}{2}\right]\theta\right) \right\}, \tag{2.348}$$

$$\tau_n(\theta) \approx \frac{n(n+1)}{2} \left\{ J_0\left(\left[n + \frac{1}{2}\right]\theta\right) - J_2\left(\left[n + \frac{1}{2}\right]\theta\right) \right\} \tag{2.349}$$

as $\theta \to 0$. Then one can write:

$$i_1(\theta) \approx i_2(\theta) \approx \left| \sum_{n=1}^{\infty} \left(n + \frac{1}{2}\right) J_0\left(\left(n + \frac{1}{2}\right)\theta\right) \right|^2, \tag{2.350}$$

where we have neglected small contributions from exponential terms. The summation can be performed analytically as $n \to \infty$. Namely, it holds approximately:

$$\sum_{n=1}^{\infty}\left(n+\frac{1}{2}\right)J_0\left(\left(n+\frac{1}{2}\right)\theta\right) \approx \int_0^x yJ_0(y\theta)dy, \tag{2.351}$$

where we have taken into account that the number of terms in the sum is close to $x = ka$. Therefore, using the representation:

$$J_1(x\theta) = \theta x^{-1}\int_0^x J_0(y\theta)ydy, \tag{2.352}$$

we easily obtain:

$$S(\theta) = \frac{x^2}{2}\varsigma(\theta) \tag{2.353}$$

and

$$i(\theta) = \frac{x^4}{4}\varsigma^2(x\theta), \tag{2.354}$$

where

$$\varsigma(x\theta) = \frac{2J_1(x\theta)}{x\theta}. \tag{2.355}$$

Note that we ignore effects of interference of diffracted and geometrical optics parts of a general light scattered field. Therefore, it follows: $I^D = x^4\varsigma^2(x\theta)/4k^2r^2$.

The function $F(x\theta) = \varsigma^2(x\theta)$ is shown in Fig. 2.14. We clearly see that its contribution to the overall scattered light is of importance only for small scattering angles ($\theta \le 3/x$).

2.4.1.2 Cross sections

Let us calculate now the scattering cross section defined as

$$C_{sca} = \frac{1}{I_0}\int_0^{2\pi} d\phi \int_0^{\pi} I_{sca}(\theta)R^2 \sin\theta d\theta. \tag{2.356}$$

Then we obtain

$$C_{sca} = C_{sca}^D + C_{sca}^G, \tag{2.357}$$

where

$$C_{sca}^D = \pi a^2 \tag{2.358}$$

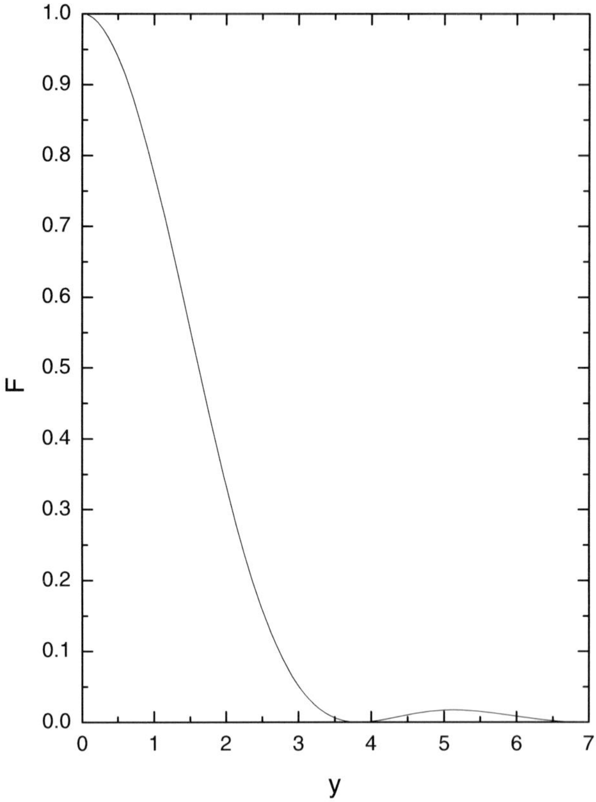

Fig. 2.14. The dependence of F on $y = x\theta$.

and

$$C^G_{sca} = \frac{1}{2}\pi a^2 \sum_{j=1}^{2} \int_0^{\pi/2} (R_j + t_j) \sin\varphi \cos\varphi d\varphi, \tag{2.359}$$

where

$$t_j = \sum_{p=1}^{\infty} \varepsilon_j^p. \tag{2.360}$$

In particular, we obtain for nonabsorbing particles:

$$t_j = (1 - R_j)^2 \{1 + R_j - R_j^2 + \dots\} = \frac{(1 - R_j)^2}{1 - R_j} \tag{2.361}$$

and

$$R_j + t_j = 1. \tag{2.362}$$

Therefore, it follows in this specific case that

$$C_{sca} = C_{ext} = 2\pi a^2. \tag{2.363}$$

This explains the asymptotic limit $Q_{ext} \equiv C_{ext}/\pi a^2 = 2$ in Fig. 2.1. It is interesting to note that although C_{sca}^G is smaller than πa^2 for absorbing particles, the value C_{ext} is equal to $2\pi a^2$ for absorbing particles as well in the approximation under study. This also means that $C_{abs} \equiv C_{ext} - C_{sca}$ is given by the following integral

$$C_{abs} = \frac{1}{2}\pi a^2 \sum_{j=1}^{2} \int_0^{\pi/2} (1 - R_j - t_j) \sin\varphi \cos\varphi d\varphi. \tag{2.364}$$

The result for C_{ext} can be easily derived from the optical theorem, which states that

$$C_{ext} = \frac{4\pi}{k^2} \mathrm{Re}(S(0)). \tag{2.365}$$

Because in the forward direction ($\theta = 0$) the diffracted light dominates, we have:

$$C_{ext} = \frac{4\pi}{k^2} \cdot \frac{x^2}{2} = 2\pi a^2 \tag{2.366}$$

at any absorption. The value t_j can be presented in the following analytical form:

$$t_j = \sum_{p=1}^{\infty} (1 - R_j)^2 (-R_j)^{2(p-1)} e^{-\bar{c}p} = \frac{(1 - R_j)^2 e^{-\bar{c}}}{1 - R_j e^{-\bar{c}}} \tag{2.367}$$

and, therefore,

$$C_{sca} = \pi a^2 (1 + W), \tag{2.368}$$

where

$$W = \frac{1}{2} \sum_{j=1}^{2} \int_0^{\pi/2} \left(R_j + \frac{(1 - R_j)^2 e^{-\bar{c}}}{1 - R_j e^{-\bar{c}}} \right) \sin\varphi \cos\varphi d\varphi. \tag{2.369}$$

It means that

$$C_{abs} = \pi a^2 (1 - W) \tag{2.370}$$

or

$$C_{abs} = \frac{\pi a^2}{2} \sum_{j=1}^{2} \int_0^{\pi/2} \frac{(1 - R_j)(1 - e^{-\bar{c}})}{1 - R_j e^{-\bar{c}}} \sin\varphi \cos\varphi d\varphi. \tag{2.371}$$

Table 2.3. Parameters W_∞, g_0, g_∞ and y for different values of the refractive index n.

n	W_∞	g_0	g_∞	y
1.1	0.0252	0.9731	0.9946	0.5180
1.2	0.0443	0.9341	0.9856	0.6528
1.25	0.0529	0.9147	0.9806	0.6948
1.3	0.0611	0.8961	0.9751	0.7280
1.333	0.0664	0.8843	0.9714	0.7468
1.34	0.0675	0.8818	0.9706	0.7505
1.35	0.0691	0.8783	0.9695	0.7555
1.4	0.0768	0.8613	0.9638	0.7785
1.45	0.0844	0.8542	0.9579	0.7985
1.5	0.0918	0.8299	0.9520	0.8160
1.55	0.0991	0.8154	0.9460	0.8315
1.60	0.1063	0.8015	0.9400	0.8453
1.65	0.1133	0.7884	0.9340	0.8580
1.70	0.1203	0.7759	0.9280	0.8695
1.90	0.1475	0.7314	0.9046	0.9080
2.00	0.1606	0.7121	0.8933	0.9340
2.10	0.1734	0.6945	0.8823	0.9383

Note that C_{sca} decreases with c (from $2\pi a^2$ at $c = 0$ to $(1 + W_\infty)\pi a^2$ as $c \to \infty$). Here we introduced the factor $W_\infty \equiv W(c \to \infty)$ given by the following expression:

$$W_\infty = \frac{1}{2}\sum_{j=1}^{2}\int_0^{\pi/2} R_j \sin\varphi \cos\varphi d\varphi. \tag{2.372}$$

This integral can be evaluated analytically (see also Table 2.3). It follows:

$$W_\infty = \Im_1 \ln n + \Im_2 \ln\left(\frac{n-1}{n+1}\right) + \Im_3, \tag{2.373}$$

$$\Im_1 = \frac{8n^4(n^4+1)}{(n^4-1)^2(n^2+1)}, \Im_2 = \frac{n^2(n^2-1)^2}{(n^2+1)^3}, \Im_3 = \frac{\sum_{j=0}^{7} A_j n^j}{3(n^4-1)(n^2+1)(n+1)}, \tag{2.374}$$

where $A_j = (-1, -1, -3, 7, -9, -13, -7, 3)$ and it was assumed that $n > 1$ and $\chi/n \ll 1$.

The same parameter W_∞ determines the absorption cross section as $c \to \infty$:

$$C_{abs} = \pi a^2(1 - W_\infty). \tag{2.375}$$

It follows as $c \to \infty$:

$$C_{abs} = f(n)\alpha V, \tag{2.376}$$

where

$$f(n) = 1 - (1 - n^{-2})^{3/2}, \tag{2.377}$$

$$V = \frac{4\pi a^3}{3}, \alpha = \frac{4\pi \chi}{\lambda}. \tag{2.378}$$

Therefore, we see that $C_{abs} \sim V$ as $c \to 0$. This explains the dependence $Q_{abs} = C_{abs}/\pi a^2 \sim a$ shown in Fig. 2.12. We underline that the absorption cross section of large weakly absorbing particles is proportional to their volume. This general result is valid for nonspherical particles with the geometrical cross section Σ as well. For strongly absorbing particles ($c \to \infty$), we have:

$$C_{abs} = \Sigma(1 - W_\infty), \tag{2.379}$$

where the integral W_∞ can be calculated analytically as it was shown above. The curve $Q_{abs}(c)$ can be approximated as:

$$Q_{abs} = 1 - M(c) - W(1 - e^{-cb})^2, \tag{2.380}$$

where

$$M = \frac{2n^2}{c^2}(e^{-bc}(1 + bc) - e^{-c}(1 + c)), b = \sqrt{1 - n^2}. \tag{2.381}$$

2.4.1.3 *van de Hulst approximation*

It was shown above that the extinction cross section of a large spherical particle is equal to $2\pi a^2$ as $x \to \infty$ if the interference effects can be neglected. These effects are of no importance for large phase shifts ρ on the diameter of the sphere d: $\rho = 2x|m - 1|$. However, the interference plays an important role at small values ρ, which is the case for some selected wavelength regions both for water and ice in air. Then the diffraction pattern will be modified by effects of the interference and Mie theory should preferably be used for calculations.

We derive here the analytical equation for C_{ext} valid as $x \to \infty$ and arbitrary phase shifts ρ assuming that $|m - 1| \to 0$ and using the optical theorem. Then we find:

$$S(0) = \sum_{n=1}^{\infty} \left(n + \frac{1}{2}\right) \left(1 - \frac{1}{2} \left[\exp(-2i\alpha_n) + \exp(-2i\beta_n)\right]\right), \qquad (2.382)$$

where we used Eqs. (2.83) and (2.205) and accounted for the fact that

$$\pi_n(\theta = 0) = \tau_n(\theta = 0) = \frac{n(n+1)}{2}. \qquad (2.383)$$

If we neglect exponentials, then we arrive at the following diffraction optics result

$$S(0) = x^2/2 \qquad (2.384)$$

as it was discussed above. So we will account for exponents but in an approximate way. Taking into account Eqs. (2.205), (2.73) and (2.78), we obtain for $tg\alpha_n$, $tg\beta_n$:

$$tg\alpha_n = \frac{m\psi_n(y)\psi_n'(x) - \psi_n'(y)\psi_n(x)}{\psi_n'(y)\chi_n(x) - m\psi_n(y)\chi_n'(x)}, \qquad (2.385)$$

$$tg\beta_n = \frac{\psi_n(y)\psi_n'(x) - m\psi_n'(y)\psi_n(x)}{m\psi_n'(y)\chi_n(x) - \psi_n(y)\chi_n'(x)}. \qquad (2.386)$$

Let us take into account that $m \to 1$. Then the dominator D in both expressions is given approximately by:

$$D = \psi_n'(x)\chi_n(x) - \psi_n(x)\chi_n'(x). \qquad (2.387)$$

One can show that $D = 1$ using the well-known properties of spherical Bessel functions. Putting $m = 1$, we obtain that $tg\alpha_n = tg\beta_n$ and

$$tg\alpha_n = \psi_n(y)\psi_n'(x) - \psi_n'(y)\psi_n(x). \qquad (2.388)$$

Now we take into account that $x \to \infty$. Then it follows:

$$\psi_n(x) = \frac{\cos\left(xf' - \frac{\pi}{4}\right)}{\sqrt{\sin\tau'}}, \ \psi_n'(x) = -\sqrt{\sin\tau'}\sin\left(xf' - \frac{\pi}{4}\right), \qquad (2.389)$$

$$\psi_n(y) = \frac{\cos\left(yf' - \frac{\pi}{4}\right)}{\sqrt{\sin\tau'}}, \ \psi_n'(y) = -\sqrt{\sin\tau'}\sin\left(yf' - \frac{\pi}{4}\right), \qquad (2.390)$$

where $x\cos\tau = n + 1/2$, $f = \sin\tau - \tau\cos\tau$, $y{=}mx, f' = \sin\tau' - \tau'\cos\tau'$. After simple algebraic calculations we get:

$$tg\alpha_n = \sin\left(yf' - \frac{\pi}{4}\right)\cos\left(xf - \frac{\pi}{4}\right) - \sin\left(xf - \frac{\pi}{4}\right)\cos\left(yf' - \frac{\pi}{4}\right) \qquad (2.391)$$

or

$$tg\alpha_n = \sin\Upsilon, \tag{2.392}$$

where

$$\Upsilon = yf' - xf. \tag{2.393}$$

We also can write:

$$\Upsilon = mx\sin\tau' - m\tau'\cos\tau' + \tau'\cos\tau' - x\sin\tau \tag{2.394}$$

and

$$\Upsilon = (m-1)x\sin\tau, \tag{2.395}$$

where we accounted for the fact that $\tau \approx \tau'$ and $\cos\tau \approx \cos\tau' \approx n + 1/2$. Because $m \to 1$, we can write: $\alpha_n \approx \Upsilon$. Therefore, transforming the sum to the integral, it follows:

$$S(0) = x^2 \int_0^x (1 - \exp(-i\rho\sin\tau)\cos\tau d(x\cos\tau) \tag{2.396}$$

or

$$S(0) = x^2 K(i\rho), \tag{2.397}$$

where

$$K(\upsilon) = \frac{1}{2} + \frac{e^{-\upsilon}}{\upsilon} + \frac{e^{-\upsilon} - 1}{\upsilon^2}. \tag{2.398}$$

Finally, we have using the optical theorem:

$$C_{ext} = 4\pi a^2 \mathrm{Re}\left[K(i\rho)\right]. \tag{2.399}$$

In particular, it follows that

$$C_{ext} = 2\pi a^2\{1 - 2\rho^{-1}\sin\rho + 2\rho^{-2}(1 - \cos\rho)\} \tag{2.400}$$

for nonabsorbing large spheres as $m \to 1$. The accuracy of this equation as applied to the calculation of the extinction efficiency factor $Q_{ext} = C_{ext}/\pi a^2$ is shown in Fig. 2.15a for nonabsorbing droplets. We see that results at the refractive index 1.01 closely correspond to the van de Hulst approximation shown by line. This is not the case for the water refractive index. The adding edge term contribution ($Q_{ext} \to Q_{ext} + 2(ka)^{-2/3}$) considerably improves the accuracy of van de Hulst approximation as applied to water clouds (Kokhanovsky, 2004a). The dependence of Q_{ext} on the attenuation parameter $c = 4\chi ka$ is shown in Fig. 2.15b at various values of n. It follows that the asymptotical value of the extinction efficiency factor is achieved more quickly for particles having larger refractive indices.

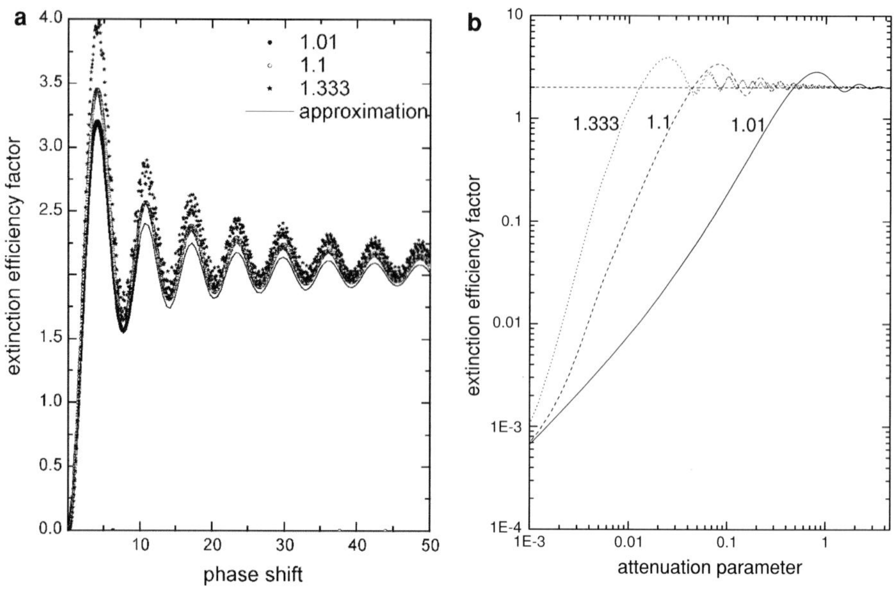

Fig. 2.15. (a) Dependence of extinction efficiency factor on the phase shift at $n = 1.01$, 1.1, and 1.333. The solid line shows the van de Hulst's approximation. Results of Mie calculations for the case of $n = 1.01$ are shown only for values of ρ smaller than 10. (b) Dependence of extinction efficiency factor on the attenuation parameter at $n = 1.01$, 1.1, and 1.333.

The absorption efficiency factor $Q_{abs} = C_{abs}/\pi a^2$can be obtained in the approximation of straight rays ($m \to 1$) as

$$Q_{abs} = 1 + \frac{2e^{-c}}{c} + \frac{2(e^{-c} - 1)}{c^2}. \tag{2.401}$$

These formulae were first derived by van de Hulst using the amplitude-phase screen model and the approximation of straight rays. Results presented here can be applied only for spherical particles. However, they can be easily generalized to account for the nonsphericity and inhomogeneity of scatterers (Kokhanovsky, 2004a). This proved to be an useful approach in the optics of crystalline clouds.

2.4.1.4 The asymmetry parameter

The asymmetry of light scattering in respect to the incident light beam is often described by the average cosine of the scattering angle or the asymmetry parameter g:

$$g = \frac{1}{2}\int_0^{\pi} p(\theta)\sin\theta\cos\theta d\theta, \tag{2.402}$$

where $p(\theta)$ is the phase function. Let us derive the geometrical optical asymptotic value for this important light scattering characteristic now.

The phase function is given by:

$$p(\theta) = \frac{2\pi(i_1 + i_2)}{k^2 C_{sca}} \tag{2.403}$$

or we can write in the case under consideration neglecting the interference effects:

$$p^D(\theta) = \frac{C_{sca}^D p^D(\theta) + C_{sca}^G p^G(\theta)}{C_{sca}^D + C_{sca}^G}, \tag{2.404}$$

where taking into account that $i^D = x^4 F(\theta)/4$ [see Eq. (2.354] and $C_{sca}^D = \pi a^2$

$$p^D(\theta) = \frac{x^2 F(\theta)}{2} \tag{2.405}$$

and

$$p^G(\theta) = \frac{2\pi(i_1^G + i_2^G)}{k^2 C_{sca}^G}. \tag{2.406}$$

The normalization conditions are:

$$\frac{1}{2}\int_0^\pi p^D(\theta)\sin\theta d\theta = 1. \tag{2.407}$$

$$\frac{1}{2}\int_0^\pi p^G(\theta)\sin\theta d\theta = 1. \tag{2.408}$$

Clearly, we can obtain for the asymmetry parameter:

$$g = \frac{C_{sca}^D g^D + C_{sca}^G g^G}{C_{sca}^D + C_{sca}^G}, \tag{2.409}$$

where

$$g^D = \frac{1}{2}\int_0^\pi p^D(\theta)\sin\theta\cos\theta d\theta, \tag{2.410}$$

$$g^G = \frac{1}{2}\int_0^\pi p^G(\theta)\sin\theta\cos\theta d\theta. \tag{2.411}$$

Let us take into account that

$$C_{sca}^D = \pi a^2, \tag{2.412}$$

$$C_{sca}^G = W\pi a^2, \quad g^D \approx 1. \tag{2.413}$$

Then we have:

$$g = \frac{1 + W g^G}{1 + W}. \tag{2.414}$$

Table 2.4. The phase function calculated in the framework of the geometrical optics approximation (the diffraction is ignored) at the refractive index $m = 1.333$.

θ, deg	$p^G(\theta)$	θ, deg	$p^G(\theta)$
0.0000	0.164E+02	44.0000	0.181E+01
1.0000	0.163E+02	45.0000	0.166E+01
2.0000	0.161E+02	46.0000	0.153E+01
3.0000	0.160E+02	47.0000	0.141E+01
4.0000	0.158E+02	48.0000	0.131E+01
5.0000	0.154E+02	49.0000	0.120E+01
6.0000	0.151E+02	50.0000	0.111E+01
7.0000	0.148E+02	51.0000	0.104E+01
8.0000	0.144E+02	52.0000	0.941E+00
9.0000	0.139E+02	53.0000	0.849E+00
10.0000	0.135E+02	54.0000	0.791E+00
11.0000	0.130E+02	55.0000	0.727E+00
12.0000	0.125E+02	56.0000	0.653E+00
13.0000	0.120E+02	57.0000	0.600E+00
14.0000	0.115E+02	58.0000	0.538E+00
15.0000	0.110E+02	59.0000	0.487E+00
16.0000	0.105E+02	60.0000	0.438E+00
17.0000	0.998E+01	61.0000	0.404E+00
18.0000	0.953E+01	62.0000	0.360E+00
19.0000	0.898E+01	63.0000	0.319E+00
20.0000	0.856E+01	64.0000	0.283E+00
21.0000	0.807E+01	65.0000	0.248E+00
22.0000	0.762E+01	66.0000	0.217E+00
23.0000	0.723E+01	67.0000	0.189E+00
24.0000	0.676E+01	68.0000	0.169E+00
25.0000	0.640E+01	69.0000	0.147E+00
26.0000	0.599E+01	70.0000	0.125E+00
27.0000	0.566E+01	71.0000	0.106E+00
28.0000	0.529E+01	72.0000	0.936E–01
29.0000	0.502E+01	73.0000	0.821E–01
30.0000	0.467E+01	74.0000	0.670E–01
31.0000	0.437E+01	75.0000	0.606E–01
32.0000	0.411E+01	76.0000	0.522E–01
33.0000	0.384E+01	77.0000	0.467E–01
34.0000	0.360E+01	78.0000	0.407E–01
35.0000	0.336E+01	79.0000	0.380E–01
36.0000	0.316E+01	80.0000	0.351E–01
37.0000	0.294E+01	81.0000	0.340E–01
38.0000	0.277E+01	82.0000	0.329E–01
39.0000	0.258E+01	83.0000	0.322E–01
40.0000	0.240E+01	84.0000	0.317E–01
41.0000	0.229E+01	85.0000	0.311E–01
42.0000	0.211E+01	86.0000	0.306E–01
43.0000	0.191E+01	87.0000	0.300E–01
		88.0000	0.297E–01

Table 2.4. *Continued*

θ, deg	$p^G(\theta)$	θ, deg	$p^G(\theta)$
89.0000	0.294E–01	135.0000	0.215E–01
90.0000	0.295E–01	136.0000	0.211E–01
91.0000	0.291E–01	137.0000	0.210E–01
92.0000	0.280E–01	138.0000	0.125E+01
93.0000	0.276E–01	139.0000	0.843E+00
94.0000	0.273E–01	140.0000	0.616E+00
95.0000	0.270E–01	141.0000	0.518E+00
96.0000	0.267E–01	142.0000	0.462E+00
97.0000	0.265E–01	143.0000	0.418E+00
98.0000	0.262E–01	144.0000	0.392E+00
99.0000	0.260E–01	145.0000	0.364E+00
100.0000	0.258E–01	146.0000	0.337E+00
101.0000	0.256E–01	147.0000	0.328E+00
102.0000	0.256E–01	148.0000	0.309E+00
103.0000	0.276E–01	149.0000	0.273E+00
104.0000	0.265E–01	150.0000	0.265E+00
105.0000	0.258E–01	151.0000	0.240E+00
106.0000	0.253E–01	152.0000	0.230E+00
107.0000	0.252E–01	153.0000	0.210E+00
108.0000	0.251E–01	154.0000	0.198E+00
109.0000	0.251E–01	155.0000	0.181E+00
110.0000	0.252E–01	156.0000	0.169E+00
111.0000	0.252E–01	157.0000	0.152E+00
112.0000	0.253E–01	158.0000	0.147E+00
113.0000	0.255E–01	159.0000	0.131E+00
114.0000	0.266E–01	160.0000	0.125E+00
115.0000	0.275E–01	161.0000	0.115E+00
116.0000	0.295E–01	162.0000	0.108E+00
117.0000	0.322E–01	163.0000	0.105E+00
118.0000	0.359E–01	164.0000	0.102E+00
119.0000	0.391E–01	165.0000	0.101E+00
120.0000	0.429E–01	166.0000	0.101E+00
121.0000	0.499E–01	167.0000	0.100E+00
122.0000	0.547E–01	168.0000	0.100E+00
123.0000	0.601E–01	169.0000	0.998E–01
124.0000	0.696E–01	170.0000	0.997E–01
125.0000	0.787E–01	171.0000	0.100E+00
126.0000	0.925E–01	172.0000	0.103E+00
127.0000	0.111E+00	173.0000	0.106E+00
128.0000	0.162E+00	174.0000	0.111E+00
129.0000	0.246E+00	175.0000	0.100E+00
130.0000	0.284E–01	176.0000	0.987E–01
131.0000	0.263E–01	177.0000	0.985E–01
132.0000	0.246E–01	178.0000	0.990E–01
133.0000	0.231E–01	179.0000	0.986E–01
134.0000	0.222E–01	180.0000	1.012E–01

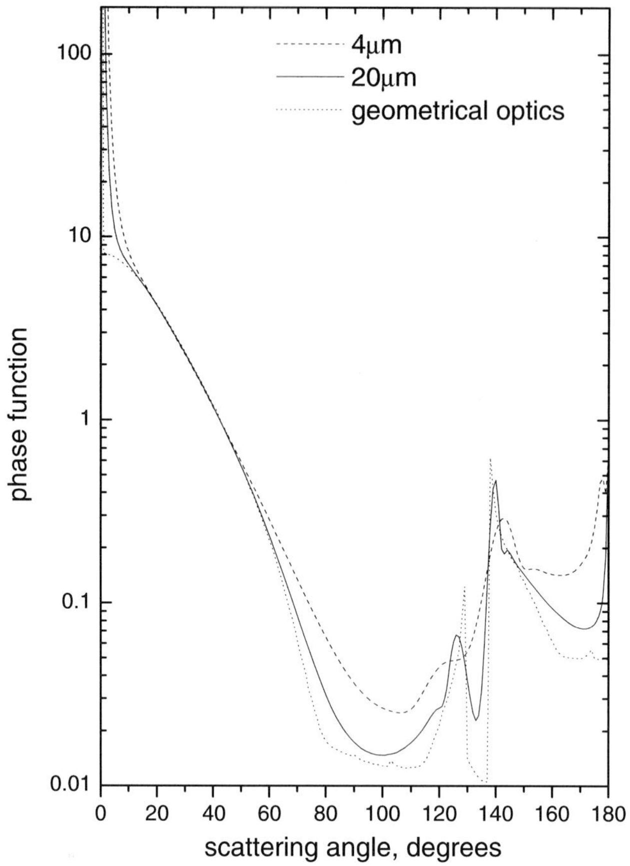

Fig. 2.16. The phase function of water droplets having the gamma PSD with $\mu = 6$, $a_{\mathrm{ef}} = 4$ and 20 μm and $\lambda = 0.55$ μm. Results of geometrical optics calculations are also shown.

Therefore, to find the value of g we need to calculate the geometrical optics contribution g^G and also W [see Eq. (2.369)]. This can be done using precalculated functions $p^G(\theta)$ (see Table 2.4 and Fig. 2.16). As an alternative, one can introduce geometrical optics expressions for intensities in Eq. (2.406) and then obtain from Eq. (2.411):

We account for the fact that

$$dB = p^G(\theta)\sin\theta\cos\theta d\theta = \frac{4\pi(i_1^G + i_2^G)}{k^2 C_{sca}^G}$$

$$= \frac{4\pi a^2 \cos\varphi \sin\varphi d\varphi}{k^2 C_{sca}^G \cos\theta d\theta} \sum_{j=1}^{2} \sum_{p=0}^{\infty} \varepsilon_{pj} \sin\theta \cos\theta d\theta \qquad (2.415)$$

or

$$g^G = \frac{1}{2W}\sum_{j=1}^{2}\int_0^{\pi/2} \frac{\mathbb{F}_j(\tau)\sin 2\tau d\tau}{1 - 2R_j e^{-c\xi}\cos(2\tau') + R_j^2 e^{-c\xi}} \tag{2.416}$$

where

$$\mathbb{F}_j(\tau) = e^{-c\xi}(1 - R_j)^2\cos 2(\tau - \tau') + R_j\cos 2\tau(1 - e^{-2c\xi}) \\ + 2R_j^2(e^{-c\xi} - \cos 2\tau')\cos 2\tau e^{-c\xi}, \tag{2.417}$$

$$\xi = \sqrt{1 - n^{-2}\cos^2\tau}, \tag{2.418}$$

$$n\cos\tau' = \cos\tau \tag{2.419}$$

Eq. (2.415) can be simplified for nonabsorbing media ($c = 0$) and also for strongly absorbing media ($c \to \infty$). Then it follows:

$$g_0^G = \frac{1}{2}\sum_{j=1}^{2}\int_0^{\pi/2} \frac{\left[(1 - R_j)^2\cos 2(\tau - \tau') + 2R_j^2(1 - \cos 2\tau')\cos 2\tau\right]\sin 2\tau d\tau}{1 - 2R_j\cos(2\tau') + R_j^2} \tag{2.420}$$

at $c = 0$ and

$$g_\infty^G = \frac{1}{2W}\sum_{j=1}^{2}\int_0^{\pi/2} R_j\cos 2\tau\sin 2\tau d\tau \tag{2.421}$$

as $c \to \infty$. This means that the asymmetry parameter for nonabsorbing particles can be derived from the following formula:

$$g_0 = \frac{1 + g_0^G}{2} \tag{2.422}$$

It follows for strongly absorbing media:

$$g_\infty = \frac{1 + W_\infty g_\infty^G}{1 + W_\infty}. \tag{2.423}$$

The analytical integration of Eq. (2.241) gives:

$$g_\infty^G = \frac{1}{W}\left\{1 + \Pi_1\ln n - \Pi_2\ln\frac{n+1}{n-1} + \Pi_3\right\}, \tag{2.424}$$

where

$$\Pi_1 = \frac{8n^4(n^6 - 3n^4 + n^2 - 1)}{(n^4 - 1)^2(n^2 + 1)^2}, \tag{2.425}$$

$$\Pi_2 = \frac{(n^2 - 1)^2(n^8 + 12n^6 - 54n^4 - 4n^2 + 1)}{16(n^2 + 1)^4}, \tag{2.426}$$

$$\Pi_3 = \frac{\mathfrak{M}}{24(n^2+1)^2(n^4-1)(n+1)} \tag{2.427}$$

$$\mathfrak{M} = \sum_{j=1}^{12} B_j n^j \tag{2.428}$$

$B_j = (-3, 13, -89, 151, 186, 138, -282, 22, 25, 25, 3, 3)$ and it was assumed that $n > 1$ and $\chi/n \ll 1$.

The value of g at given n changes from g_0 to g_∞ with c. This change can be approximately accounted for using the following approximation:

$$g = g_\infty - (g_\infty - g_0)e^{-cy}, \tag{2.429}$$

where y depends on n and is given in Table 2.3. Similar parameterizations can be derived for ice crystals as well.

2.4.1.5 Polarization characteristics

Geometrical optical technique allows us to find not only the intensity of scattered light but also its polarization. In particular, it follows for the degree of polarization of scattered light under unpolarized (e.g., solar) illumination conditions:

$$P_l = \frac{i_1^G - i_2^G}{i_1^G + i_2^G}. \tag{2.430}$$

The dependence $P_l(\theta)$ at $n = 1.33$ is given in Fig. 2.17. Mie results are also shown in this figure.

The Mueller matrix $\hat{M}$ allows to find the Stokes vector of scattered light at any illumination conditions:

$$\vec{I} = \frac{1}{k^2 r^2}\hat{M}\vec{I}_0. \tag{2.431}$$

We have for nonzero elements of this matrix for spheres:

$$\begin{aligned}
M_{11} &= M_{22} = \frac{1}{2}(|S_{11}|^2 + |S_{22}|^2), \\
M_{12} &= M_{21} = \frac{1}{2}(|S_{11}|^2 - |S_{22}|^2), \\
M_{33} &= M_{44} = \frac{1}{2}\mathrm{Re}\left(S_{11}S_{22}^*\right), \\
M_{34} &= -M_{43} = \mathrm{Im}(S_{11}S_{22}^*)
\end{aligned} \tag{2.432}$$

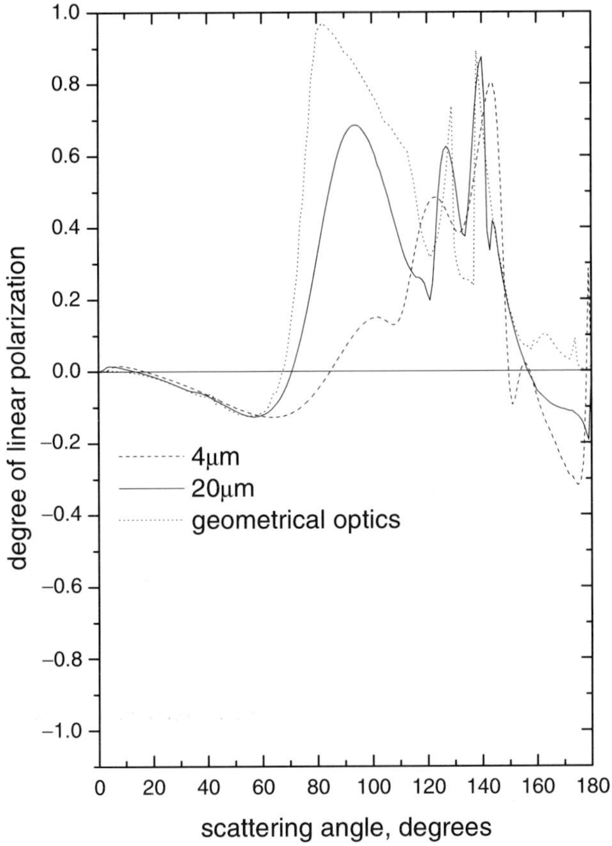

Fig. 2.17. The same as in Fig. 2.16 except for the degree of linear polarization.

Taking into account that

$$|S_{11}|^2 = i^D + i_2^G,\ |S_{22}|^2 = i^D + i_1^G,\ \mathrm{Re}(S_{11}S_{22}^*) = i^D + \sqrt{i_1^G i_2^G},\ \mathrm{Im}(S_{11}S_{22}^*) = 0, \tag{2.433}$$

where we neglected the effects of interference and phase shifts, we obtain:

$$\hat{M} = \begin{pmatrix} \dfrac{i_1^G + i_2^G}{2} & \dfrac{i_2^G - i_1^G}{2} & 0 & 0 \\ \dfrac{i_2^G - i_1^G}{2} & \dfrac{i_1^G + i_2^G}{2} & 0 & 0 \\ 0 & 0 & \sqrt{i_1^G i_2^G} & 0 \\ 0 & 0 & 0 & \sqrt{i_1^G i_2^G} \end{pmatrix} + i^D \hat{E}, \tag{2.434}$$

where $\hat{E}$ is the 4×4 unity matrix.

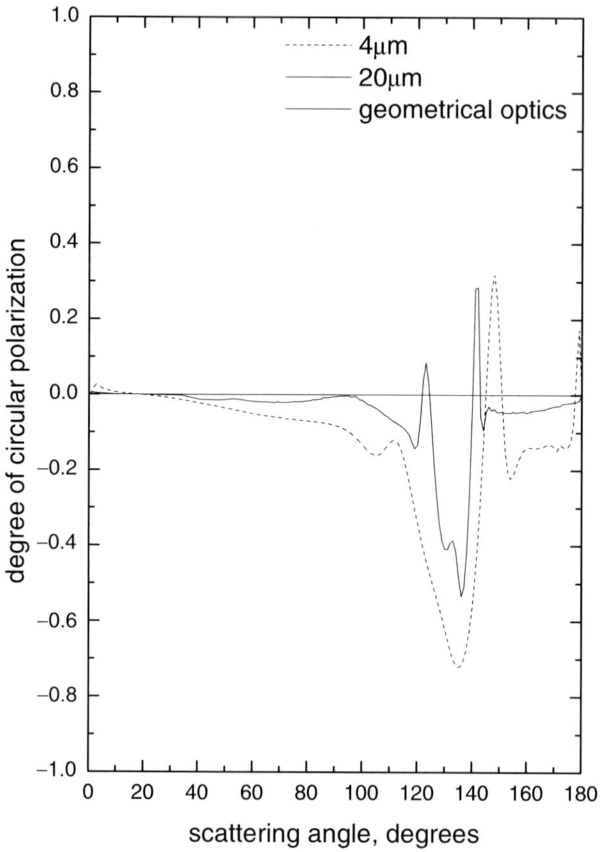

Fig. 2.18. The same as in Fig. 2.16 except for the degree of circular polarization (p_{34}).

We can also introduce the normalized matrix

$$\hat{m} = \frac{\hat{M}}{M_{11}} \begin{pmatrix} 1 & -P_l & 0 & 0 \\ -P_l & 1 & 0 & 0 \\ 0 & 0 & P_c & 0 \\ 0 & 0 & 0 & P_c \end{pmatrix} \tag{2.435}$$

where $P_c = (i_1^G - i_2^G)/(i_1^G + i_2^G)$, $P_c = 2\sqrt{i_1^G i_2^G}/(i_1^G + i_2^G)$ and we neglected the contribution of the diffraction, which is of importance only as $\theta \to 0$. The function P_c gives the reduction of the degree of circular polarization for incident right hand circularly polarized light as discussed above. $P_l(\theta)$ describes the degree of polarization of scattered light for incident unpolarized light beam. Functions $P_l(\theta)$

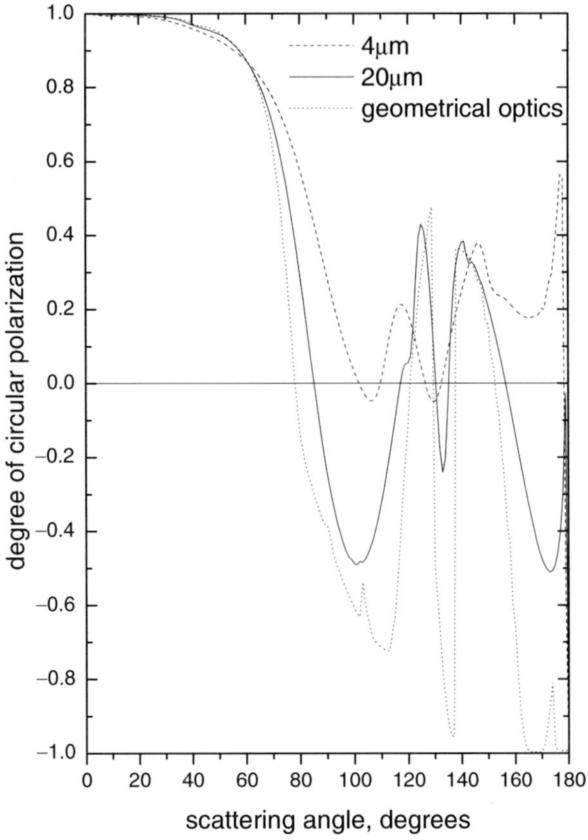

Fig. 2.19. The same as in Fig. 2.16 except for the degree of circular polarization (p_{44}).

and $P_c(\theta)$ are given in Figs. 2.18 and 2.19. The geometrical optics phase function of water droplets is shown in Fig. 2.20.

2.4.2 Ice Crystals

Geometrical optics techniques are of a special importance for studies of light scattering in ice clouds. This is due to the following reasons:

- the size of crystals is even larger than that of droplets (typically, 100μm);
- the Mie-type solutions for ice crystals do not exist or require (e.g., for hexagonal cylinders) large computational efforts.

An important problem is the variability of crystal shapes in ice clouds. It means that even if we derive scattering characteristics for a given fixed shape of crystals, we

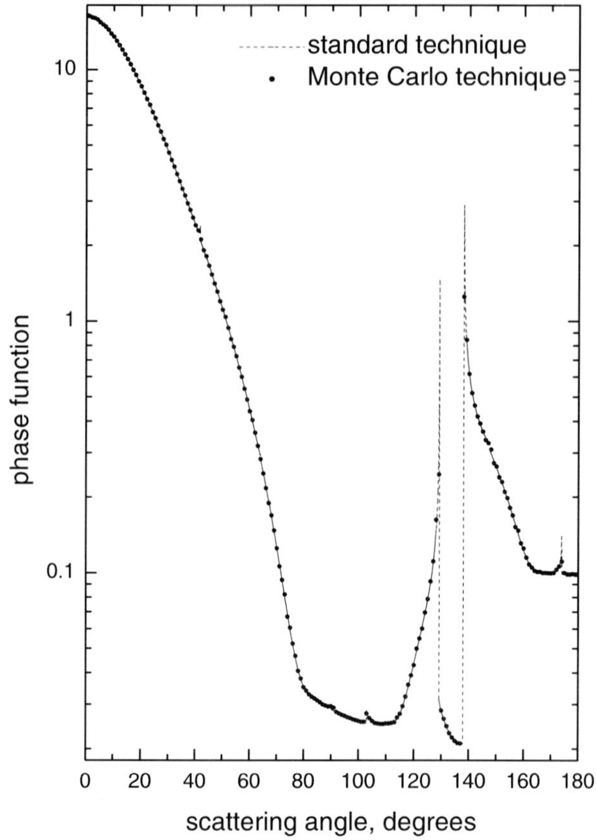

Fig. 2.20. Geometrical optics phase function of a water droplet calculated using the standard approach as described here (line) and the Monte-Carlo (Macke et al., 1996) technique (symbols) at $n = 1.333$.

are not sure that results of calculations represent in-situ measured optical characteristics (e.g., the phase function). Therefore, optical characteristics of ice clouds are usually represented by a linear mixture of correspondent characteristics for a finite number of shapes (e.g., hexagonal plates and columns, bullet rosettes, etc.). We call this a multi-particle approach (MPA). The problem with this technique is due to the fact that neither typical shapes of crystals nor the weighting factor for a given shape is known in advance for a crystalline cloud under study.

Yet another method is to represent the single scattering of light by an elementary volume of an ice cloud by the optical characteristics of single particles having an extremely complex shape. In particular, particles with random surfaces or fractals

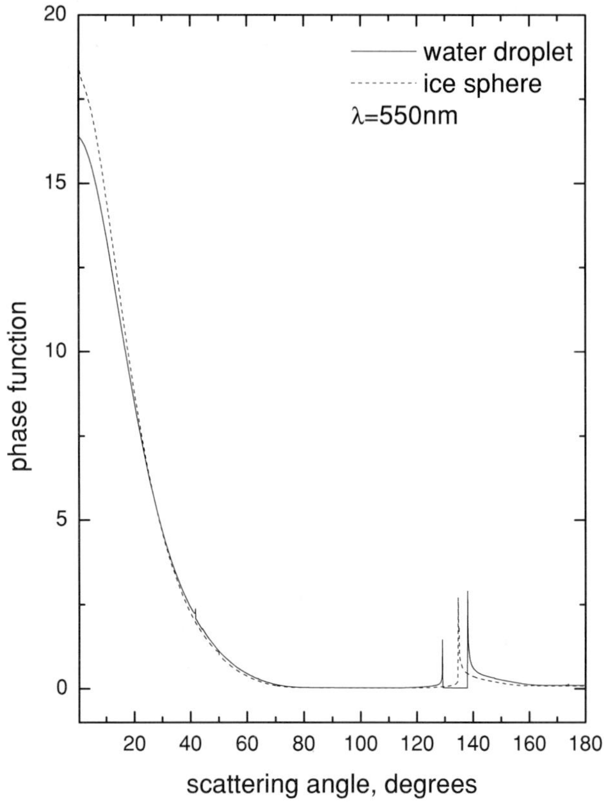

Fig. 2.21. Geometrical optics phase function of a water droplet (*solid line*, $n = 1.333$) and ice sphere (*broken line*, $n = 1.31$) calculated using the Monte-Carlo technique (Macke et al., 1996) at the wavelength 550 nm.

are used. Such particles do not necessarily represent geometrical characteristics of particles in clouds but they have scattering characteristics (e.g., phase functions) similar to those measured in natural ice clouds. We call this a single-particle approach (SPA).

The choice between these two techniques (MPA, SPA) is largely determined by the problem at hand. For instance, the MPA is routinely used in satellite ice cloud retrieval algorithms. Then certain shapes are prescribed and the aim is to find the size/concentration of particles in a given shape class. Retrieved sizes and concentrations depend on the *a priori* assumptions and pre-selected shape classes used in retrieval procedures. This constitutes a major problem of modern ice cloud retrieval techniques.

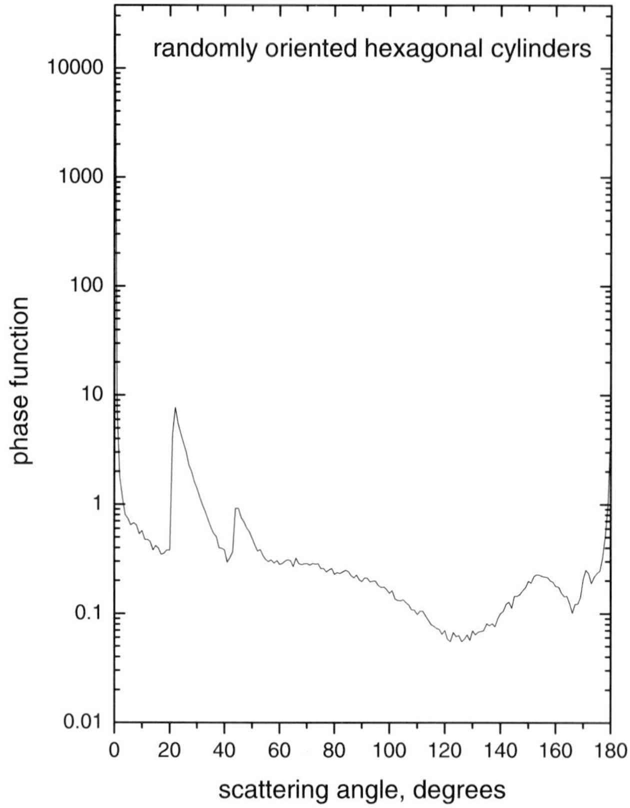

Fig. 2.22. Phase function of randomly oriented hexagonal ice crystals at $\lambda = 0.858\ \mu m$. The length of the cylinder is 400 μm. The ratio of the crystal length to the side of the hexagonal cross section is equal to 4.0.

In principle, geometrical optical characteristics of nonspherical ice crystals can be calculated using the same approach as outlined above for spheres. However, it is difficult to give an analytical result for the scattering angle as the function of the incidence angle. The problem is solved using the Monte Carlo technique (Macke et al., 1996). This technique is based on the following steps:

- an incident Stokes vector is selected;
- a shape of a particle is defined; an emission surface is defined; emission points (EP) on the emission surface are randomly determined;
- a light ray propagates from the EP in the direction of a particle;
- the check if the ray hits a particle is performed; if the ray does not hit a particle, another EP is selected;

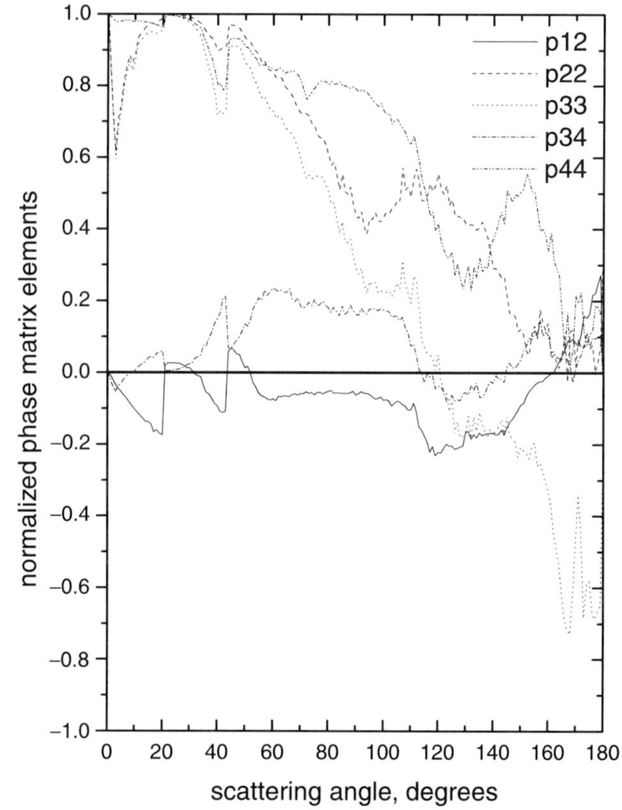

Fig. 2.23. The same as in Fig. 2.22 except for the normalized phase matrix elements.

– the reflection or refraction at the interaction point is randomly chosen. The photon is reflected if the reflection coefficient $r = (r_1 + r_2)/2$ is larger than the random number $q \in [0, 1]$ (otherwise, it is refracted);
– the incident Stokes vector is multiplied by a Mueller matrix for the reflection or refraction process, respectively;
– the path length L between an incident point and the next internal incident point is calculated using Snell's law and the specific shape of a crystal;
– the transmission factor $\gamma = \exp(-\alpha L)$, where $\alpha = 4\pi\chi/\lambda$ is the absorption coefficient, is calculated;
– if $\gamma \leq \exp(-\alpha La)$, where $La = -\ln(p/\alpha)$, $p \leq 1$, is the random number, then a photon is assumed to be absorbed; otherwise, a photon is transmitted; this procedure is repeated for all photons inside the crystal;
– the foregoing procedure is repeated until the photon escapes from the crystals;

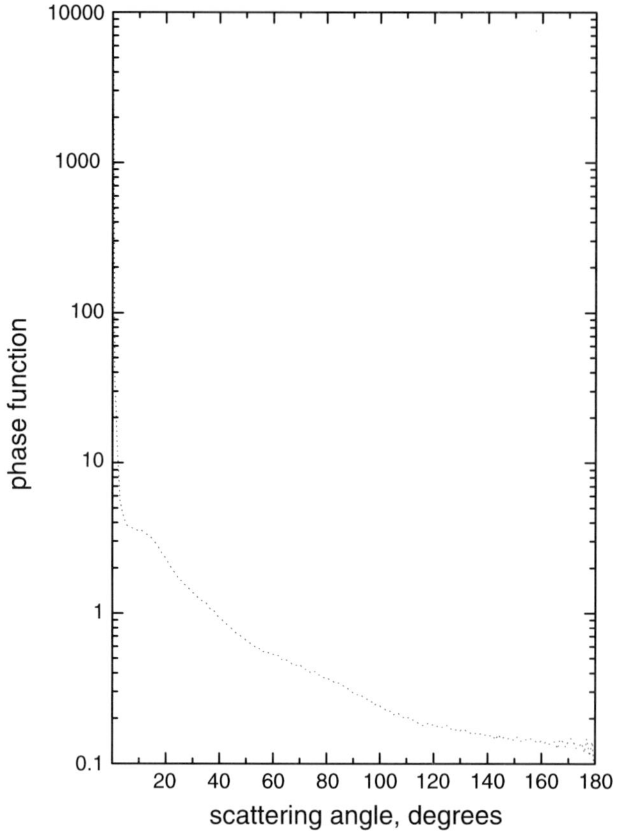

Fig. 2.24. Geometrical optics phase function of randomly oriented fractal particles (Macke et al., 1996) at $n = 1.31$.

- then a new EP is selected and the procedure is repeated till the convergence is reached (usually, $10^8 - 10^9$ photons are used);
- the single scattering albedo ω_0 is calculated as the ratio of the number of scattered photons to the number of incident photons;
- the extinction cross section is assumed to be equal to $2S$, where S is the surface area of the particle on the plane perpendicular to the direction of light propagation;
- the absorption cross section ($C_{abs} = (1 - \omega_0)C_{ext}$) and the asymmetry parameter are calculated.

The results of such calculations for spherical, hexagonal and fractal particles are given in Figs. 2.21–2.25. It follows from these figures that both intensity and

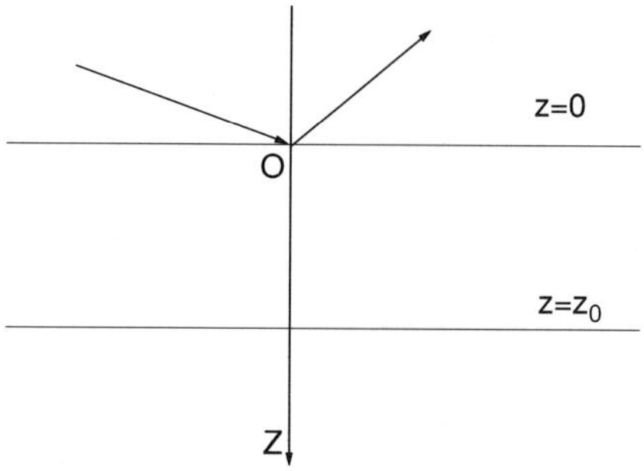

Fig. 3.1. Geometry of the problem.

is due to the fact that the matrix $\hat{\sigma}_{sca}$ in Eq. (3.5) is defined in the scattering plane and the vector $\vec{I}$ is defined with respect to the meridional plane. For spherical water droplets the matrix $\hat{\sigma}_{ext}$ is reduced to a scalar value and

$$\frac{d\vec{I}(\vec{\Omega})}{dl} = -\sigma_{ext}\vec{I}(\vec{\Omega}) + \int_{4\pi} \hat{L}_2\hat{\sigma}_{sca}\left(\vec{\Omega}, \vec{\Omega}'\right)\hat{L}_1\vec{I}(\vec{\Omega}')d\vec{\Omega}'. \tag{3.6}$$

However, for crystalline media, where particles can have a preferential orientation and be anisotropic, this is not the case. This fact adds an additional complexity in radiative transfer studies for the case of ice clouds.

We will be mostly concerned with solutions of Eq. (3.6) for a plane-parallel cloud layer illuminated by the Sun. The interaction of solar radiation with extended cloud fields is well approximated by the solution of this idealized problem. The geometry of the problem is given in Fig. 3.1. The solar light with the zenith angle ϑ_0 uniformly illuminates a plane-parallel scattering layer from above. We will assume that properties of the layer do not change in the horizontal direction. Then properties of light field depend only on the vertical coordinate Z (see Fig. 3.1) and the direction $\vec{\Omega}$, specified by the zenith angle ϑ and the azimuth φ. The main task is to calculate distributions $\vec{I}(\vartheta, \varphi, z)$. Usually only measurements of $\vec{I}(\vartheta, \varphi, 0)$ at the top of the cloud (reflected light) and $\vec{I}(\vartheta, \varphi, z_0)$ at the cloud base (transmitted light) are made (see Fig. 3.1). Therefore, we will be concerned mostly with these two angular distributions.

Equation (3.6) takes the following form for the case shown in Fig. 3.1:

$$\cos\vartheta\frac{d\vec{I}(\vartheta,\varphi)}{dz} = -\sigma_{ext}\vec{I}(\vartheta,\varphi) + \int_0^{2\pi} d\varphi' \int_0^{\pi} d\vartheta'\hat{L}_2\hat{\sigma}_{sca}\left(\vartheta',\varphi' \to \vartheta,\varphi\right) \times \hat{L}_1\vec{I}(\vartheta',\varphi') \tag{3.7}$$

The most simple case to study is that of idealized homogeneous clouds with values of σ_{ext} and $\hat{\sigma}_{sca}$ not dependent on the position inside the cloud. Then we have from Eq. (3.7):

$$\cos\vartheta \frac{d\vec{I}(\vartheta,\varphi)}{d\tau} = -\vec{I}(\vartheta,\varphi) + \frac{\omega_0}{4\pi}\int_0^{2\pi} d\varphi' \int_0^{\pi} d\vartheta' \hat{L}_2 \hat{P}(\vartheta',\varphi' \to \vartheta,\varphi) \times \hat{L}_1 \vec{I}(\vartheta',\varphi'), \tag{3.8}$$

where we introduced the optical depth

$$\tau = \sigma_{ext} l, \tag{3.9}$$

the phase matrix

$$\hat{P} = \frac{4\pi\hat{\sigma}_{sca}(\vartheta',\varphi' \to \vartheta,\varphi)}{\sigma_{sca}} \tag{3.10}$$

and the single-scattering albedo:

$$\omega_0 = \frac{\sigma_{sca}}{\sigma_{ext}}. \tag{3.11}$$

Studies of Eq. (3.8) open ways to treat more complex cases, including clouds with spatial varying microstructure characteristics and also cases of broken clouds and clouds having various complex shapes. Equation (3.8) is reduced to the following simpler form:

$$\cos\vartheta \frac{dI(\vartheta,\varphi)}{d\tau} = -I(\vartheta,\varphi) + \frac{\omega_0}{4\pi}\int_0^{2\pi} d\varphi' \int_0^{\pi} d\vartheta' p(\vartheta',\varphi' \to \vartheta,\varphi) I(\vartheta',\varphi'), \tag{3.12}$$

if the polarization effects are ignored. Here

$p(\vartheta',\varphi' \to \vartheta,\varphi) = \dfrac{4\pi\sigma_{sca}(\vartheta',\varphi' \to \vartheta,\varphi)}{\sigma_{sca}}$ is the phase function.

It is useful to distinguish between diffuse $\vec{I}_d$ and direct (or coherent) $\vec{I}_c = \vec{J}_0\delta(\mu-\mu_0)\delta(\varphi-\varphi_0)$ light in the general solution $\vec{I}(\vartheta,\varphi)$. Here $\vec{J}_0$ describes the Stokes vector of the incident attenuated flux. It is assumed that the layer is illuminated in the direction defined by the incidence zenith angle $\vartheta_0 = \arccos(\mu_0)$ and the azimuthal angle φ_0. The multiple scattered light is observed in the direction specified by the zenith observation angle $\vartheta = \arccos(\mu)$ and the azimuth φ. Namely, we write: $\vec{I}(\vartheta,\varphi) = \vec{I}_d(\vartheta,\varphi) + \vec{I}_c(\vartheta,\varphi)$. The substitution of this formula in Eq. (3.8) gives

$$\cos\vartheta \frac{d\vec{I}_d(\vartheta,\varphi)}{d\tau} = -\vec{I}_d(\vartheta,\varphi) + \frac{\omega_0}{4\pi}\int_0^{2\pi} d\varphi' \int_0^{\pi} d\vartheta' \hat{L}_2 \hat{P}(\vartheta',\varphi' \to \vartheta,\varphi)\hat{L}_1 \times \vec{I}_d(\vartheta',\varphi') + \frac{\omega_0}{4\pi}\hat{P}_*(\vartheta_0,\varphi_0 \to \vartheta,\varphi)\vec{F}\exp\left(-\frac{\tau}{\cos\vartheta_0}\right) \tag{3.13}$$

or

$$\cos\vartheta\frac{dI_d(\vartheta,\varphi)}{d\tau} = -I_d(\vartheta,\varphi) + \frac{\omega_0}{4\pi}\int_0^{2\pi} d\varphi' \int_0^{\pi} d\vartheta' p(\vartheta',\varphi' \to \vartheta,\varphi) I_d(\vartheta',\varphi')$$
$$+ \frac{\omega_0}{4\pi} p(\vartheta_0,\varphi_0 \to \vartheta,\varphi) F \exp\left(-\frac{\tau}{\cos\vartheta_0}\right) \quad (3.14)$$

for the scalar case. Here $\hat{P}_*$ is the correspondent scattering matrix defined with respect to the meridional plane, $\vec{F}(F, Q_0, U_0, V_0)$ is the Stokes vector of incident light flux at the top of a cloud. It follows for unpolarized solar light: $Q_0 = U_0 = V_0$. The solution of this equation under boundary conditions stating that there is no diffuse light entering the cloud from above and below, allows to find $\vec{I}_d(\vartheta,\varphi)$. Also it follows from Eqs. (3.8) and (3.13) that $\vec{I}_c(\vartheta,\varphi)$ is given simply by:

$$\vec{I}_c(\vartheta,\varphi) = \vec{F}(\vartheta,\varphi)\delta(\cos\vartheta - \cos\vartheta_0)\delta(\varphi-\varphi_0)\exp\left(-\frac{\tau}{\cos\vartheta_0}\right) \quad (3.15)$$

or

$$I_c(\vartheta,\varphi) = F(\vartheta,\varphi)\delta(\cos\vartheta - \cos\vartheta_0)\delta(\varphi-\varphi_0)\exp\left(-\frac{\tau}{\cos\vartheta_0}\right) \quad (3.16)$$

for the scalar case.

The solution of Eq. (3.13) is a more easy task than that of Eq. (3.8) because we avoid the necessity to deal with the divergence in the direction of incident light in the framework of Eq. (3.13).

3.2 Reflection and Transmission Functions

Reflectance and transmittance of light by cloud layers is usually defined in terms of reflection R and transmission T functions. They relate incident light intensity $I_0(\vartheta_0,\varphi_0)$ with reflected $I_R(\mu,\varphi)$ and transmitted $I_T(\mu,\varphi)$ light intensity. Namely, it follows by definition:

$$I_R(\mu,\varphi) = \frac{1}{\pi}\int_0^1 d\varphi' \int_0^{2\pi} R(\tau,\mu,\varphi,\mu',\varphi') I_0(\mu',\varphi')\mu' d\mu', \quad (3.17)$$

$$I_T(\mu,\varphi) = \frac{1}{\pi}\int_0^{2\pi} d\varphi' \int_0^{1} T(\tau,\mu,\varphi,\mu',\varphi') I_0(\mu',\varphi')\mu' d\mu'. \quad (3.18)$$

Reflection and transmission functions allow to find the intensity of reflected and transmitted light for arbitrary angular distributions of incident light with the intensity $I_0(\mu',\varphi')$.

If incident light is azimuthally independent, these formulae simplify:

$$I_R(\mu, \varphi) = 2\int_0^1 \bar{R}(\tau, \mu, \mu')I_0(\mu')\mu' d\mu', \tag{3.19}$$

$$I_T(\mu, \varphi) = 2\int_0^1 \bar{T}(\tau, \mu, \mu')I_0(\mu')\mu' d\mu', \tag{3.20}$$

where

$$\bar{R}(\tau, \mu, \mu') = \frac{1}{2\pi}\int_0^{2\pi} R(\tau, \mu, \mu'\varphi')d\varphi', \tag{3.21}$$

$$\bar{T}(\tau, \mu, \mu') = \frac{1}{2\pi}\int_0^{2\pi} T(\tau, \mu, \mu'\varphi')d\varphi'. \tag{3.22}$$

Note that relations (3.15) and (3.16) simplify for unidirectional illumination of a cloud by a wide beam (e.g., solar light). Then we can assume that

$$I_0(\mu', \varphi') = \delta(\mu' - \mu_0)\delta(\varphi' - \varphi_0)F, \tag{3.23}$$

where F is the solar flux density at the normal elementary area at the top of a cloud and $\delta(x)$ is the delta function, having the following property:

$$f(x_0) = \int_0^{\infty} \delta(x - x_0)f(x)dx. \tag{3.24}$$

Using this relation and equations for reflection and transmission functions given above, we arrive at the following results:

$$I_R(\mu, \varphi) = \frac{F\mu_0 R(\tau, \mu, \varphi, \mu_0, \varphi_0)}{\pi}, \tag{3.25}$$

$$I_T(\mu, \varphi) = \frac{F\mu_0 T(\tau, \mu, \varphi, \mu_0, \varphi_0)}{\pi}, \tag{3.26}$$

and, therefore,

$$R(\tau, \mu, \varphi, \mu_0, \varphi_0) = \frac{\pi I_R(\mu, \varphi)}{F\mu_0}, \tag{3.27}$$

$$T(\tau, \mu, \varphi, \mu_0, \varphi_0) = \frac{\pi I_T(\mu, \varphi)}{F\mu_0}. \tag{3.28}$$

These equations allow us to make the following interpretation of reflection and transmission functions. Indeed, we have for an absolutely white Lambertian surface by definition:

$$P_R^L(\mu, \varphi) = \int_{2\pi} I_R^L(\mu, \mu', \varphi, \varphi')\mu' d\Omega' = \int_0^{2\pi} d\varphi' \int_0^1 I_R^L(\mu, \mu', \varphi, \varphi')\mu' d\mu'$$
$$= \int_0^{2\pi} d\varphi' \int_0^1 C\mu_0\mu' d\mu' = \pi C\mu_0 \tag{3.29}$$

where $P_R^L(\vartheta, \varphi)$ is the total power scattered by a unit area of a Lambertian surface into the upper hemisphere and we used the fact that intensity of light reflected from a Lambertian surface is proportional to the cosine of the incidence angle $\mu_0 (I_R^L = C\mu_0)$. The constant C can be found from the condition that the scattered and incident powers are equal in the case of an absolute by white Lambertian surface by definition. We have for the incident power:

$$P_0 = \int_{2\pi} I_0(\mu', \mu_0, \varphi', \varphi_0)\mu' d\Omega' = \int_0^{2\pi} d\varphi \int_0^1 \delta(\mu' - \mu_0)\delta(\varphi' - \varphi_0) F\mu' d\mu' = F\mu_0 \tag{3.30}$$

and, therefore: $C = F/\pi$. It means that intensity of light reflected from an absolutely white Lambertian surface is given by:

$$I_R^L = \frac{F}{\pi}\mu_0, \tag{3.31}$$

and, therefore, $R(\tau, \mu, \varphi, \mu_0, \varphi_0)$ [see Eq. (3.25)] is equal to the ratio of light reflected from a given surface I_R to the value of I_R^L:

$$R = I_R / I_R^L. \tag{3.32}$$

Also it means that $R \equiv 1$ by definition for a Lambertian ideally white surface.

Accordingly, it follows:

$$T = I_T / I_R^L. \tag{3.33}$$

The results of calculations and measurements of cloud reflectance and transmittance will be mostly presented in terms of functions R and Tin this book. They do not depend on the intensity of incident light and characterize inherent properties of a cloud layer. The intensity of reflected and transmitted light can be calculated using Eqs. (3.17) and (3.18) for an arbitrary angular distribution of incident light. The integration of reflection and transmission functions with respect to angles allows us to find the cloud plane (r_d) and spherical (r) albedo, the cloud diffuse (t_d) and the global (t) transmittance, the cloud absorptance a_d and the global absorptance a (see Table 3.1 for definitions).

3.3 Polarization Characteristics

Clearly, Eqs. (3.15) and (3.16) should be modified if one would like to account for light polarization. Namely, it follows:

$$\vec{I}_R(\mu, \varphi) = \frac{1}{\pi}\int_0^{2\pi} d\varphi \int_0^1 \hat{R}(\tau, \mu, \varphi, \mu', \varphi')\vec{I}_0(\mu', \varphi')\mu' d\mu', \tag{3.34}$$

$$\vec{I}_T(\mu, \varphi) = \frac{1}{\pi}\int_0^{2\pi} d\varphi \int_0^1 \hat{T}(\tau, \mu, \varphi, \mu', \varphi')\vec{I}_0(\mu', \varphi')\mu' d\mu'. \tag{3.35}$$

Table 3.1. Radiative transfer characteristics $\left\{\bar{R} \equiv \frac{1}{2\pi}\int_0^{2\pi} R(\mu_0,\mu,\varphi)d\varphi,\ \bar{T} \equiv \frac{1}{2\pi}\int_0^{2\pi} T(\mu_0,\mu,\varphi)\mathrm{d}\varphi\right\}$.

Radiative characteristic	Symbol	Definition
Plane albedo	$r_d(\mu_0)$	$2\int_0^1 \bar{R}(\mu)\mu d\mu$
Spherical albedo	r	$2\int_0^1 r_d(\mu_0)\mu_0 d\mu_0$
Diffuse transmittance	$t_d(\mu_0)$	$2\int_0^1 \bar{T}(\mu_0,\mu)\mu d\mu$
Global transmittance	t	$2\int_0^1 t_d(\mu_0)\mu_0 d\mu_0$
Absorptance	$a_d(\mu_0)$	$1 - r_d(\mu_0) - t_d(\mu_0)$
Global absorptance	a	$1 - r - t$

Here $\vec{I}_R(\mu,\varphi)$ is the Stokes vector of reflected light, $\vec{I}_T(\mu,\varphi)$ is the Stokes vector of transmitted light, $\vec{I}_0(\mu,\varphi)$ is the Stokes vector of incident light, $\hat{R}$ and $\hat{T}$ are 4×4 reflection and transmission matrices, respectively, for illumination from above. Clearly, we have: $R \equiv R_{11}$, $T \equiv T_{11}$. The Stokes vector for unidirectional illumination of a cloud by a wide beam is

$$\vec{I}_0(\mu',\varphi') = \delta(\mu' - \mu_0)\delta(\varphi' - \varphi_0)\vec{F}. \tag{3.36}$$

The first element of $\vec{F}$ equals to the incident net flux F per unit perpendicular area at the top of a cloud. Other elements of this vector (Q_0, U_0, V_0) describe the polarization state of incident light. They equal to zero for incident unpolarized solar light. Obviously, we have in this case [see also Eqs. (3.25) and (3.26)]:

$$\vec{I}_R(\vartheta_0,\vartheta,\varphi) = \frac{\hat{R}\vec{F}}{\pi}\mu_0, \tag{3.37}$$

$$\vec{I}_T(\vartheta_0,\vartheta,\varphi) = \frac{\hat{T}\vec{F}}{\pi}\mu_0. \tag{3.38}$$

Let us assume that incident light is unpolarized. Then we have for the components of the Stokes vector of reflected light:

$$I_R(\vartheta_0,\vartheta,\varphi) = \frac{R_{11}F}{\pi}\mu_0, \quad Q_R(\vartheta_0,\vartheta,\varphi) = \frac{R_{21}F}{\pi}\mu_0, \tag{3.39}$$

$$U_R(\vartheta_0,\vartheta,\varphi) = \frac{R_{31}F}{\pi}\mu_0, \quad V_R(\vartheta_0,\vartheta,\varphi) = \frac{R_{41}F}{\pi}\mu_0, \tag{3.40}$$

and

$$I_T(\vartheta_0,\vartheta,\varphi) = \frac{T_{11}F}{\pi}\mu_0, \quad Q_T(\vartheta_0,\vartheta,\varphi) = \frac{T_{21}F}{\pi}\mu_0, \tag{3.41}$$

$$U_T(\vartheta_0,\vartheta,\varphi) = \frac{T_{31}F}{\pi}\mu_0, \quad V_T(\vartheta_0,\vartheta,\varphi) = \frac{T_{41}F}{\pi}\mu_0 \tag{3.42}$$

for transmitted light. The degree of linear polarization of reflected light is defined as

$$p_l^r \equiv -\frac{\sqrt{Q_R^2 + U_R^2}}{I_R}. \tag{3.43}$$

Therefore, it follows:

$$p_l^r \equiv -\frac{\sqrt{R_{21}^2 + R_{31}^2}}{R_{11}}. \tag{3.44}$$

It follows for the total degree of polarization:

$$p^r \equiv -\frac{\sqrt{R_{21}^2 + R_{31}^2 + R_{41}^2}}{R_{11}}. \tag{3.45}$$

The degree of circular polarization is given by:

$$p_c^r \equiv -\frac{R_{41}}{R_{11}}. \tag{3.46}$$

Note that the following relation holds for the total degree of polarization:

$$p^r \equiv \sqrt{p_l^r + p_c^r}. \tag{3.47}$$

Similar characteristics can be defined for the transmitted light:

$$p_l^t \equiv -\frac{\sqrt{T_{21}^2 + T_{31}^2}}{T_{11}}, \quad p_c^t \equiv \frac{T_{41}}{T_{11}}, \quad p^t \equiv -\frac{\sqrt{T_{21}^2 + T_{31}^2 + T_{41}^2}}{T_{11}}. \tag{3.48}$$

A number of simplifications arise for particular observation and incidence directions. For instance if incidence and observation directions coincide and normal to the scattering layer, then the principal plane is not defined and the reflection matrix for media with randomly oriented particles has the same structure as the scattering matrix for a local scattering volume. Therefore, light reflected in the nadir direction for the nadir illumination conditions is unpolarized. Also it follows that $p_c^r = p_c^t = 0$ and

$$p_l^r \equiv -\frac{R_{21}}{R_{11}}, \quad p_l^t \equiv -\frac{T_{21}}{T_{11}} \tag{3.49}$$

at $\mu_0 = 1$ and any observation conditions. Similarly, we have $p_c^r = p_c^t = 0$ at $\mu = 1$ for arbitrary illumination conditions. Equations (3.49) hold at $\mu = 1$ and $\varphi = \varphi_0$ as well.

3.4 Optically Thin Clouds

The radiative transfer in thin clouds (e.g., subvisual Cirrus) can be studied using the single-scattering approximation. Indeed multiple light scattering is of no importance in this case and the integral term in Eq. (3.13) can be omitted. Then it follows from Eq. (3.13)

$$\frac{d\vec{I}_d(x)}{dx} = -\vec{I}_d(x) + \frac{\omega_0(x)}{4\pi}\hat{P}(x)\vec{F}\exp(-sx), \tag{3.50}$$

where $x = \tau/\cos\vartheta$, $s = \cos\vartheta/\cos\vartheta_0$ and we omitted other arguments for simplicity. The matrix $\hat{P}$ in Eq. (3.50) is defined with respect to the meridional plane. This equation can be solved analytically. For this we multiply both sides of Eq. (3.50) by e^x. Then it follows:

$$\exp(x)\frac{d\vec{I}_d(x)}{dx} + \exp(x)\vec{I}_d(x) = \frac{\omega_0(x)}{4\pi}\hat{P}(x)\vec{F}\exp(-(s-1)x) \tag{3.51}$$

or

$$\frac{d\left[\vec{I}_d(x)e^x\right]}{dx} = \frac{\omega_0(x)}{4\pi}\hat{P}(x)\vec{F}\exp(-(s-1)x). \tag{3.52}$$

It means that

$$\vec{I}_d(x) = \frac{1}{4\pi}e^{-x}\int_a^x \omega_0(x')\hat{P}(x')\vec{F}\exp(-(s-1)x'). \tag{3.53}$$

Let us apply boundary conditions now:

$$\vec{I}_{d\uparrow}(x=0) = 0, \cos\vartheta > 0, \tag{3.54}$$

$$\vec{I}_{d\downarrow}(x=x_0) = 0, \cos\vartheta < 0. \tag{3.55}$$

They mean that there is no diffuse light entering the medium from above ($x = 0$) or below ($x = x_0 = (\tau_0/\cos\vartheta)$), τ_0 is the optical thickness of a scattering layer. So we have:

$$\vec{I}_{d\downarrow}(x) = \frac{1}{4\pi}e^{-x}\int_0^x \omega_0(x')\hat{P}(x')\vec{F}\exp(-(s-1)x'), \tag{3.56}$$

$$\vec{I}_{d\uparrow}(x) = \frac{1}{4\pi}e^{-x}\int_{x_0}^x \omega_0(x')\hat{P}(x')\vec{F}\exp(-(s-1)x') \tag{3.57}$$

and boundary conditions are satisfied automatically. A simple integration under the assumption of a homogeneous cloud layer gives:

$$\vec{I}_{d\downarrow}(x) = \frac{\omega_0\hat{P}\vec{F}}{4\pi(s-1)}\{e^{-x} - e^{-sx}\}, \tag{3.58}$$

$$\vec{I}_{d\uparrow}(x) = \frac{\omega_0\hat{P}\vec{F}}{4\pi(s-1)}\{e^{-x-(s-1)x_0} - e^{-sx}\} \tag{3.59}$$

for arbitrary x. It follows at the boundaries of a layer:

$$\vec{I}_{d\downarrow}(x) = \frac{\omega_0 \hat{P}\vec{F}}{4\pi(s-1)}\{e^{-x_0} - e^{-sx_0}\},\ \vec{I}_{d\uparrow}(x) = \frac{\omega_0 \hat{P}\vec{F}}{4\pi(s-1)}\{e^{-(s-1)x_0} - 1\} \tag{3.60}$$

So we have for the reflection and transmission vectors:

$$\vec{I}_r = \frac{\omega_0 \hat{P}\vec{h}}{4(\xi+\eta)}\left\{1 - \exp\left[-\left(\frac{1}{\xi}+\frac{1}{\eta}\right)\tau_0\right]\right\},$$
$$\vec{I}_T = \frac{\omega_0 \hat{P}\vec{h}}{4(\xi-\eta)}\left\{\exp\left[-\frac{\tau_0}{\xi}\right] - \exp\left[-\frac{\tau_0}{\eta}\right]\right\}, \tag{3.61}$$

where $\eta = |\cos\vartheta|$, $\xi = \cos\vartheta_0$, τ_0 is the optical thickness of the scattering layer, and $\vec{h} = \vec{F}/F_1$. We see that matrices $\hat{R}$ and $\hat{T}$ introduced above are proportional to $\hat{P}$ and have the same structure.

Obviously, it follows for unpolarized incident light

$$R = \frac{\omega_0 p(\theta)}{4(\xi+\eta)}\left\{1 - \exp\left[-\left(\frac{1}{\xi}+\frac{1}{\eta}\right)\tau_0\right]\right\}, \tag{3.62}$$

$$T = \frac{\omega_0 p(\theta)}{4(\xi-\eta)}\left\{\exp\left[-\frac{\tau_0}{\xi}\right] - \exp\left[-\frac{\tau_0}{\eta}\right]\right\}. \tag{3.63}$$

Equations given above are very useful in estimations of the influence of thin ($\tau_0 \to 0$) clouds on light fluxes in the terrestrial atmosphere.

3.5 Small-Angle Approximation

Let us consider now the case of a highly anisotropically light scattering layer (the asymmetry parameter $g \to 1$) illuminated along the normal. In this case most scattered photons propagate within the small-angle scattering region and there is a possibility to simplify the radiative transfer equation. Thus, we assume that $\cos\vartheta = 1$ in Eq. (3.12) and obtain:

$$\frac{dI(\tau,\mu)}{d\tau} = -I(\tau,\mu) + \frac{\omega_0}{2}\int_0^1 d\mu' I(\tau,\mu')p(\mu,\mu'), \tag{3.64}$$

where

$$p(\mu,\mu') = \frac{1}{2\pi}\int_{-1}^{2\pi} p(\mu,\mu',\phi)d\phi. \tag{3.65}$$

We used the fact that the intensity of scattered light field for layers with randomly oriented particles does not depend on the azimuth for the illumination of a layer along the normal. Note, that the value of $I(\tau,\mu)$ is the total intensity in the direction given by arccos(μ), not the diffuse intensity as in Eq. (3.14), and it includes the direct light. The phase function $p(\mu, \mu', \varphi)$ in Eq. (3.65) can be represented in the following form (Minin, 1988)

$$p(\mu, \mu', \phi) = p(\mu, \mu') + 2\sum_{m=1}^{\infty} \cos m(\phi - \phi') \sum_{i=m}^{\infty} x_i \frac{(i-m)!}{(i+m)!} P_i^m(\mu) P_i^m(\mu'), \tag{3.66}$$

where

$$p(\mu, \mu') = \sum_{i=0}^{\infty} x_i P_i(\mu) P_i(\mu') \tag{3.67}$$

and $P_i(\mu)$ and $P_i^m(\mu)$ are Legendre and associated Legendre polynomials, respectively. Note that it follows:

$$x_i = \frac{2i+1}{2} \int p(\theta) P_i(\theta) \sin\theta d\theta. \tag{3.68}$$

We seek the solution of Eq. (3.64) in the following form:

$$I(\tau, \mu) = \sum_{i=0}^{\infty} b_i(\tau) P_i(\mu). \tag{3.69}$$

Substituting Eqs. (3.67) and (3.69) in Eq. (3.64), one obtains:

$$\frac{db_i(\tau)}{d\tau} = -c_i b_i(\tau), \tag{3.70}$$

where

$$c_i = 1 - \omega_0 \frac{x_i}{2i+1} \tag{3.71}$$

and the orthogonality of Legendre polynomials was used. Thus, it follows:

$$b_i(\tau) = A_i \exp(-c_i \tau), \tag{3.72}$$

where $A_i = const.$

We see that it is possible to obtain the intensity of the transmitted light with the following formula:

$$I(\tau, \mu) = \sum_{i=0}^{\infty} A_i e^{-c_i \tau} P_i(\mu). \tag{3.73}$$

Values of A_i are found from initial conditions. We will assume that

$$I(0, \mu) = I_0 \delta(1 - \mu), \tag{3.74}$$

where $\delta(1-\mu)$ is the delta function, I_0 is the density of incident light flux. The delta function can be represented in the following form:

$$\delta(1-\mu) = \frac{1}{4\pi}\sum_{i=0}^{\infty}(2i+1)P_i(\mu). \tag{3.75}$$

Thus, it follows:

$$A_i = \frac{2i+1}{4\pi} \tag{3.76}$$

and, finally,

$$I(\tau,\mu) = B\sum_{i=0}^{\infty}\frac{2i+1}{2}e^{-c_i\tau}P_i(\mu), \tag{3.77}$$

where $B = I_0/2\pi$.

This is a solution of the problem under consideration. Equation (3.77) describes the angular distribution of the transmitted light at small angles for the normal incidence of a light beam. This important formula can be rewritten in the integral form. Indeed, the phase function $p(\theta)$ has a sharp peak in the forward-scattering direction ($\theta = 0$) for cloudy media and the main contribution to the integral Eq. (3.68) comes from the small-angle scattering region. Thus, it follows from Eq. (3.68) that

$$x_i = \frac{2i+1}{2}\int_0^{\infty} p(\theta)J_0\left(\theta\left(i+\frac{1}{2}\right)\right)\theta d\theta, \tag{3.78}$$

where the asymptotic relationship

$$\lim_{\theta\to 0} P_i(\cos\theta) = J_0\left(\theta\left(i+\frac{1}{2}\right)\right) \tag{3.79}$$

was used.

From Eqs. (3.78) and (3.79) and the sum formula

$$\sum_{i=0}^{\infty} f\left(i+\tfrac{1}{2}\right) \approx \int_0^{\infty} f(\sigma)d\sigma, \tag{3.80}$$

it follows:

$$I(\tau,\vartheta) = \frac{I_0}{2\pi}\int_0^{\infty} d\sigma\, J_0(\sigma\vartheta)\exp(-\tau(1-\omega_0 P(\sigma))), \tag{3.81}$$

where

$$P(\sigma) = \frac{1}{2}\int_0^{\infty} p(\theta)J_0(\theta\sigma)\theta d\theta. \tag{3.82}$$

Table 3.2. Phase functions $p(\theta)$ and their Fourier–Bessel transforms $P(\sigma)$ (Υ is the normalization constant, x is the size parameter).

$p(\theta)$	$P(\sigma)$
$\dfrac{2\Upsilon\exp(-\Upsilon\theta)}{\theta}$	$\dfrac{\Upsilon}{\sqrt{\Upsilon^2+\sigma^2}}$
$2\Upsilon^2\exp(-\Upsilon\theta)$	$\dfrac{\Upsilon^3}{(\Upsilon^2+\sigma^2)^{3/2}}$
$\dfrac{2}{\Upsilon^2}\exp\left(-\dfrac{\theta^2}{2\Upsilon^2}\right)$	$\exp\left(-\dfrac{\Upsilon^2\sigma^2}{2}\right)$
$\dfrac{4J_1^2(\theta x)}{\theta^2}$	$\begin{cases} \dfrac{2}{\pi}\left\{\arccos\left(\dfrac{\sigma}{2x}\right)-\dfrac{\sigma}{2x}\sqrt{1-\left(\dfrac{\sigma}{2x}\right)^2}\right\}, & \sigma\le 2x \\ 0, & \sigma>2x \end{cases}$

Equation (3.81) in many cases is more easy to handle than Eq. (3.77). For instance, Eq. (3.82) can be analytically integrated for special types of phase functions $p(\theta)$ (see Table 3.2).

It follows from Eq. (3.81) for the diffuse intensity $I_d(\tau,\vartheta)$:

$$I_d(\tau,\vartheta)=\frac{I_0}{2\pi}\int_0^\infty\left[e^{-\tau(1-\omega_0 P(\sigma))}-e^{-\tau}\right]J_0(\sigma\vartheta)\sigma\,d\sigma, \tag{3.83}$$

where we extracted the peak of light intensity exactly in the forward direction using the equality:

$$\frac{I_0}{2\pi}e^{-\tau}\int_0^\infty J_0(\sigma\vartheta)\sigma\,d\sigma=I_0e^{-\tau}\delta(\vartheta). \tag{3.84}$$

Equation (3.83) is often used for the solution of both direct and inverse problems of cloud optics. It is valid at $\tau\le 5$ and $\vartheta\to 0$. It follows that the phenomenon of multiple light scattering is responsible for broadening of the angular spectrum of the light transmitted by a scattering layer at small angles. The same effect is also observed in the single-scattering regime if the sizes of particles decrease. This remark is of importance for optical particle sizing in cloudy media.

The approximation, which was considered in this section, is called the small-angle approximation (SAA). Moments of different radiative characteristics in this approximation were studied by Lutomirski et al. (1995).

3.6 Optically thick clouds

3.6.1 Fundamental Relationships

Another case, where simple analytical results can be obtained, is that of thick ($\tau_0\to\infty$) clouds. Note that final equations derived here can be actually applied at

$\tau_0 \geq 10$ or at somewhat smaller values depending on the accuracy requirements. This means that this case covers most cloudy situations occurring in the terrestrial atmosphere. Before we proceed to the derivation of analytical equations for functions R and T at $\tau_0 \gg 1$, we need to establish some auxiliary asymptotic relations.

It is known that light intensity in deep layers of optically thick media is azimuthally independent. The azimuthally integrated radiative transfer equation [see Eq. (3.14)] can be written in the following form:

$$\eta \frac{dI(\tau, \eta)}{d\tau} = -I(\tau, \eta) + B(\tau, \eta) + B_0(\tau, \eta), \tag{3.85}$$

where

$$B(\tau, \eta) = \frac{\omega_0}{2} \int_{-1}^{1} p(\eta, \eta')I(\tau, \eta')d\eta', \tag{3.86}$$

$$B_0(\tau, \eta) = \frac{\omega_0 F}{4\pi} p(\eta, \xi)e^{-\tau/\xi}, \tag{3.87}$$

and

$$p(\eta, \xi) = \frac{1}{2\pi} \int_0^{2\pi} p(\eta, \xi, \varphi)d\varphi \tag{3.88}$$

is the azimuthally averaged phase function. We also neglect polarization effects.

Let us assume that $\tau \to \infty$. Then it follows: $B_0(\tau, \eta) \to 0$ and

$$I(\tau, \eta) = i(\eta)e^{-k\tau}. \tag{3.89}$$

The last equation corresponds to the so-called deep-layer regime, when parameters η and τ are decoupled. Then the overall light flux f decreases with the distance from the illuminated boundary preserving the scattered light angular pattern given by the function $i(\eta)$. The value of f decreases in e times on the optical depth $\tau_e = 1/k$. Both the function $i(\eta)$ and the diffusion exponent k, play an important role in the theory considered here. These characteristics of the deep-layer regime also define the intensity of transmitted and reflected light as will be shown below.

So we have from Eq. (3.85):

$$(1 - k\eta)i(\eta) = \frac{\omega_0}{2} \int_{-1}^{1} p(\eta, \eta')i(\eta')d\eta'. \tag{3.90}$$

This integral equation is usually solved numerically. Let us assume that $p = 1$. Then we have:

$$i(\eta) = \frac{\omega_0}{2(1 - k\eta)} \int_{-1}^{1} i(\eta')d\eta' \tag{3.91}$$

or

$$i(\eta) = \frac{D}{1 - k\eta}, \tag{3.92}$$

where

$$D = \frac{\omega_0}{2} \int_{-1}^{1} i(\eta')d\eta' \tag{3.93}$$

does not depend on the angle. Note that $i(\eta)$ satisfies the radiative transfer equation (3.90) for any constant D and, therefore,

$$i(\eta) = \frac{1}{1 - k\eta} \tag{3.94}$$

and

$$I(\tau, \eta) = \frac{\exp(-k\tau)}{1 - k\eta}, \tag{3.95}$$

where we used the normalization condition: $D = 1$. The diffusion constant k can be found as follows. Let us substitute (3.92) in (3.91) taking into account that $D = 1$. Then we have:

$$\frac{\omega_0}{2k} \ln\left(\frac{1+k}{1-k}\right) = 1. \tag{3.96}$$

This equation allows to find k at arbitrary ω_0 and $p(\theta) = 1$. Clearly, we have at $\omega_0 = 1 : k = 0$. The dependence $k(\omega_0)$ at $p(\theta) = 1$ is given in Fig. 3.2. For more complex phase functions a numerical solution is needed. The results of correspondent calculations for the Heney–Greenstein phase function

$$p(\theta) = \sum_{n=1}^{\infty} g^n P_n(\cos\theta) \tag{3.97}$$

at the asymmetry parameters $g = 0.75$ (ice clouds) and $g = 0.85$ (water clouds) are also given in Fig. 3.2. We see that $0 \leq k \leq 1$. This means that the decrease rate of the diffuse light in the deep-layer regime ($\exp(-k\tau)$) for absorbing media is smaller than that of the direct light ($\exp(-\tau)$).

It follows also:

$$(1 + k\eta)i(-\eta) = \frac{\omega_0}{2} \int_{-1}^{1} p(-\eta, \eta')i(\eta')d\eta' \tag{3.98}$$

or

$$(1 + k\eta)i(-\eta) = \frac{\omega_0}{2} \int_{-1}^{1} p(\eta, \eta')i(-\eta')d\eta', \tag{3.99}$$

where we used the property: $p(-\eta, -\eta') = p(\eta, \eta')$.

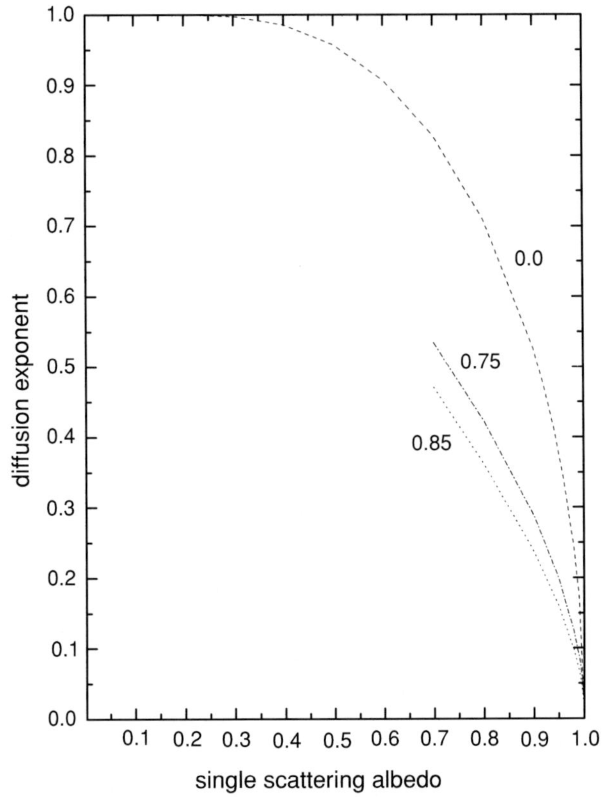

Fig. 3.2. Dependence of the diffusion exponent on the single-scattering albedo for isotropic scattering. Calculations of the diffusion exponent for single-scattering albedos typical for water (0.85) and ice (0.75) clouds are also shown. It follows that the value of k decreases with g for a given ω_0.

Let us establish now the relationship between the intensity $i^{\downarrow}(\eta)$ for light propagated downwards and the intensity $i^{\uparrow}(-\eta)$ for light propagated upwards. Arrows show the direction of light propagation. The value of $\eta = |\cos\vartheta|$ is positive by definition. Note that negative $\cos\vartheta$ corresponds to upwelling light flux. We consider a cut parallel to the upper boundary but at a large optical depth. The correspondent plane at $\tau \gg 1$ is illuminated not only by light coming from above and having the intensity i_a but also by light coming from below and reflected from the layer laying above the plane of cut. We denote this contribution to the total intensity as i_b. Clearly, we have:

$$i^{\downarrow}(\eta) = i_a(\eta) + i_b(\eta). \tag{3.100}$$

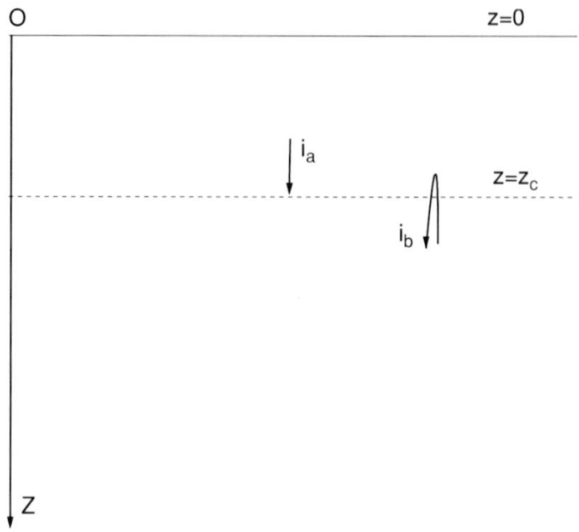

Fig. 3.3. Contributions to downward light flux.

So the function $i^{\downarrow}(\eta)$ can be presented as a sum of two terms (see Fig. 3.3). Clearly, $i_a(\eta)$ is proportional to the angular distribution $u(\eta)$ of light transmitted by the upper layer:

$$i_a(\eta) = Mu(\eta), \tag{3.101}$$

where M is the unknown proportionality constant. We will find this constant at later stages of our derivations. Also it follows from Eq. (3.19) for the intensity $i_b(\eta)$:

$$i_b(\eta) = 2\int_0^1 R(\eta, \eta')i(-\eta')\eta' d\eta', \tag{3.102}$$

where $R(\eta, \eta')$ is the azimuthally averaged reflection function of the upper layer under illumination from below ($\eta > 0, \eta' > 0$). This layer could be chosen to be arbitrary thick. So we will assume that $R(\eta, \eta')$ coincides with the azimuthally averaged reflection function of a semi-infinite layer $R_{\infty}(\eta, \eta')$.

Summing up, it follows:

$$i^{\downarrow}(\eta) = Mu(\eta) + 2\int_0^1 R_{\infty}(\eta, \xi)i(-\xi)\xi d\xi. \tag{3.103}$$

Let us find M. For this we multiply the last equation by $\eta i^{\downarrow}(\eta)$ and integrate it from 0 to 1 with respect to η. Then we have:

$$\int_0^1 \eta i^{\downarrow 2}(\eta) d\eta = M\int_0^1 u(\eta)i^{\downarrow}(\eta)\eta d\eta + \Im \tag{3.104}$$

where the two-dimensional integral

$$\Im = 2\int_0^1 i^{\downarrow}(\eta)\eta d\eta \int_0^1 R_{\infty}(\eta,\xi) i^{\uparrow}(-\xi)\xi d\xi \tag{3.105}$$

can be simplified. For this we note that it follows

$$i^{\uparrow}(-\xi) = 2\int_0^1 i^{\downarrow}(\eta) R_{\infty}(\eta,\xi)\eta d\eta \tag{3.106}$$

and, therefore,

$$\Im = \int_0^1 i^{\uparrow 2}(-\xi)\xi d\xi \tag{3.107}$$

or

$$\Im = -\int_{-1}^0 i^{\downarrow 2}(\xi)\xi d\xi. \tag{3.108}$$

Therefore, it follows:

$$M = \mathbb{C}\int_{-1}^1 i^2(\eta)\eta d\eta, \tag{3.109}$$

where

$$\mathbb{C} = \left[\int_0^1 u(\eta) i(\eta) d\eta\right]^{-1}. \tag{3.110}$$

We will use the normalization condition: $\mathbb{C} = 2$. Then one derives:

$$M = 2\int_{-1}^1 i^2(\eta)\eta d\eta. \tag{3.111}$$

We present the equation for M together with other important relationships in Table 3.3. The constant N defined in the property 8 (see Table 3.3) will be used in further derivations devoted to studies of relationships between auxiliary functions defined as

$$P(\tau) = \int_{-1}^1 \eta i(\eta) I(\tau,\eta) d\eta \tag{3.112}$$

and

$$Q(\tau) = \int_{-1}^1 \eta i(-\eta) I(\tau,\eta) d\eta. \tag{3.113}$$

The relationships between functions $P(\tau)$ and $Q(\tau)$ can be used for the derivation of asymptotic equations for reflection and transmission functions valid as $\tau_0 \to \infty$. Let us show this.

Table 3.3. Main equations and constants ($u_0 \equiv u(\omega_0 = 1)$, $R_{0\infty} \equiv R_{0\infty}(\omega_0 = 1)$).

N	Property
1	$(1-k\eta)i(\eta) = \frac{\omega_0}{2}\int_{-1}^{1} p(\eta,\eta')i(\eta')d\eta'$
2	$(1+k\eta)i(-\eta) = \frac{\omega_0}{2}\int_{-1}^{1} p(\eta,\eta')i(-\eta')d\eta'$
3	$\frac{\omega_0}{2}\int_{-1}^{1} i(\eta)d\eta = 1$
4	$i(-\eta) = 2\int_0^1 i(\xi)R_\infty(\xi,\eta)\xi d\xi$
5	$i(\eta) = 2\int_0^1 i(-\xi)R_\infty(\xi,\eta)\xi d\xi + Mu(\eta)$
6	$2\int_0^1 u(\eta)i(\eta)\eta d\eta = 1$
7	$M = 2\int_{-1}^{1} i^2(\eta)\eta d\eta$
8	$N = 2\int_0^1 i(-\eta)u(\eta)\eta d\eta$
9	$\delta = \int_0^1 u_0(\xi)\xi^2 d\xi$
10	$u_0(\xi) = \frac{3}{4}\left(\xi + 2\int_0^1 R_{0\infty}(\xi,\eta)\eta^2 d\eta\right)$
11	$2\int_0^1 u_0(\eta)\eta d\eta = 1$
12	$2\int_0^1 R_{0\infty}(\xi,\eta)\eta d\eta = 1$

First of all, we note that it follows from Eq. (3.90) after multiplication of this equation by $i(\eta)$ and integration from –1 to 1:

$$\frac{dP(\tau)}{d\tau} = -kP(\tau) + P_0(\tau), \tag{3.114}$$

where

$$P_0(\tau) = \int_{-1}^{1} i(\eta)B_0(\tau,\eta)d\eta \tag{3.115}$$

and we used the equality

$$-kP(\tau) = \int_{-1}^{1} B(\tau,\eta)i(\eta)d\eta - \int_{-1}^{1} i(\eta)I(\tau,\eta)d\eta. \tag{3.116}$$

This equality can be obtained from property 2 in Table 3.3. Namely, we have after multiplying Eq. (1) in Table 3.3 by $I(\tau,\eta)$ and integrating this equation from –1 to 1 with respect to η:

$$\int_{-1}^{1} I(\tau,\eta)i(\eta)d\eta - kP(\tau) = \frac{\omega_0}{2}\int_{-1}^{1} d\eta \int_{-1}^{1} I(\tau,\eta)p(\eta,\eta')i(\eta')d\eta' \tag{3.117}$$

or

$$\int_{-1}^{1} I(\tau,\eta)i(\eta)d\eta - kP(\tau) = \int_{-1}^{1} d\eta' B(\tau,\eta')i(\eta')d\eta', \tag{3.118}$$

where we used the property: $p(\eta,\eta') = p(\eta',\eta)$. The last equation coincides with Eq. (3.116).

The next step is to find $P(\tau)$. For this we use the following substitution in Eq. (3.114):

$$P(\tau) = f(\tau)e^{-k\tau}. \tag{3.119}$$

Then it follows:

$$\frac{df(\tau)}{d\tau} = P_0(\tau)e^{k\tau} \tag{3.120}$$

or

$$f\Big|_{\tau_1}^{\tau} = \int_{\tau_1}^{\tau} P_0(t)e^{kt}dt. \tag{3.121}$$

It means that

$$f(\tau) = f(\tau_1) + \int_{\tau_1}^{\tau} P_0(t)e^{kt}dt. \tag{3.122}$$

So we we have:

$$P(\tau) = f(\tau_1)e^{-k\tau} + e^{-k\tau}\int_{\tau_1}^{\tau} P_0(t)e^{kt}dt. \tag{3.123}$$

The value of τ_1 can be found from boundary conditions. In particular, we are interested in the diffuse light. Diffuse light does not enter the medium from above or below ($I(0,\eta) = 0$ for $\eta > 0$ and $I(\tau_0,\eta) = 0$ for $\eta < 0$). Therefore, we have: $\tau_1 = 0$. Then the boundary condition on the upper boundary is satisfied. Finally, it follows

$$P(\tau) = P(0)e^{-k\tau} + \int_{0}^{\tau} P_0(t)e^{k(t-\tau)}dt. \tag{3.124}$$

A similar relationship can be obtained for $Q(\tau,\eta)$. Then we have:

$$\frac{dQ(\tau)}{d\tau} = kQ(\tau) + Q_0(\tau), \tag{3.125}$$

where

$$Q_0(\tau) = \int_{-1}^{1} i(-\eta)B_0(\tau,\eta)d\eta. \tag{3.126}$$

This equation differs for the correspondent equation for P only by the sign before k. So we have:

$$Q(\tau,\eta) = \psi\left(\tau_1^*\right)e^{k\tau} + e^{k\tau}\int_{\tau_1^*}^{\tau} Q_0(t)e^{-kt}dt, \tag{3.127}$$

where it was assumed that

$$Q(\tau) = \psi(\tau)e^{k\tau}. \tag{3.128}$$

The value of τ_1^* can be found from the boundary condition at the lower boundary of a medium. Namely, we have: $\tau_1^* = \tau_0$. Therefore, it follows:

$$Q(\tau) = Q(\tau_0)e^{k(\tau-\tau_0)} + \int_{\tau_0}^{\tau} Q_0(t)e^{-k(t-\tau)}dt. \tag{3.129}$$

This equation is satisfied automatically at $\tau = \tau_0$ due to the accurate account for the boundary conditions.

Summing up, we have the following important relationships:

$$P(\tau) = P(0)e^{-k\tau} + V(\tau), \tag{3.130}$$

$$Q(\tau) = Q(\tau_0)e^{k(\tau-\tau_0)} + W(\tau), \tag{3.131}$$

where

$$V(\tau) = \int_0^{\tau} P_0(t)e^{k(t-\tau)}dt, \tag{3.132}$$

$$W(\tau) = \int_{\tau_0}^{\tau} Q_0(t)e^{-k(t-\tau)}dt. \tag{3.133}$$

These fundamental relationships are valid for any τ and for any light sources represented by B_0 (Sobolev, 1975). They can be used for the derivation of a number of important results of cloud optics.

We will use a particular case at $\tau = 0$ in the first equation and a case $\tau = \tau_0$ in the second equation. Then it follows:

$$P(\tau_0) = P(0)\exp(-k\tau_0) + V(\tau_0), \tag{3.134}$$

$$Q(0) = Q(\tau_0)\exp(-k\tau_0) + W(0), \tag{3.135}$$

where

$$V(\tau_0, \eta) = e^{-k\tau_0}\int_0^{\tau} P_0 dt \int_{-1}^{1} i(\eta)\frac{\omega_0}{4}p(\eta,\xi)e^{-t((1/\xi)-k)}d\xi$$
$$= \frac{1}{2}\left(e^{k\tau_0} - e^{-\tau_0/\xi}\right)\xi i(-\xi), \tag{3.136}$$

and we used property 2 in Table 3.3. Also we have:

$$W(0,\eta) = \int_{\tau_0}^{0} e^{-kt}dt \int_{-1}^{1} i(-\eta)\frac{\omega_0}{4}p(\eta,\xi)e^{-t/\xi}d\eta$$
$$= -\frac{1}{2}\left(1 - e^{-(k+(1/\xi))\tau_0}\right)\xi i(-\xi). \tag{3.137}$$

Therefore, neglecting small numbers $e^{-\tau_0/\xi}$, it follows as $\tau_0 \to \infty$:

$$P(0, \eta) = -P(\tau_0, \eta)e^{k\tau_0} + \frac{1}{2}\xi i(-\xi), \tag{3.138}$$

$$Q(0, \eta) = Q(\tau_0, \eta)e^{-k\tau_0} - \frac{1}{2}\xi i(-\xi). \tag{3.139}$$

These are auxiliary relations we were bound to establish from the very start. They can be also written in the following form:

$$i(\xi) = 2\int_{-1}^{1} T(\eta, \xi, \tau_0)i(\eta)\eta d\eta + 2e^{-k\tau_0}\int_{-1}^{1} R(\eta, \xi, \tau_0)i(-\eta)d\eta, \tag{3.140}$$

$$i(-\xi) = 2\int_{-1}^{1} R(\eta, \xi, \tau_0)i(\eta)\eta d\eta + 2e^{-k\tau_0}\int_{-1}^{1} T(\eta, \xi, \tau_0)i(-\eta)\eta d\eta, \tag{3.141}$$

where T and R are determined as

$$R = \frac{\pi I_R}{\mu_0 F}, \quad T = \frac{\pi I_T}{\mu_0 F}. \tag{3.142}$$

3.6.2 Asymptotic Equations

The general form of functions $R(\eta, \xi, \tau_0)$ and $T(\eta, \xi, \tau_0)$ can be obtained using physical arguments. In particular T should be proportional to $u(\eta)$ (and, actually due to the reciprocity principal also to $u(\xi)$). Therefore, we have:

$$T(\eta, \xi, \tau_0) = \alpha(\tau_0)u(\eta)u(\xi). \tag{3.143}$$

Let us consider now a semi-infinite layer (see Fig. 3.4) and make a cut at a large optical thickness τ_0. Then we can represent $R_\infty(\eta, \xi)$ as a sum of reflection

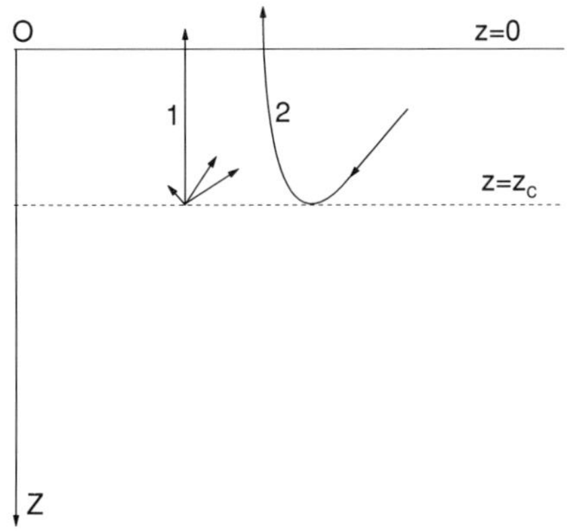

Fig. 3.4. Contributions to upward light flux.

from upper layer $R(\tau_0, \eta, \xi)$ (contribution 1, see Fig. 3.4) and light transmitted by the upper layer and reflected back (contribution 2, see Fig. 3.4). The angular distribution of the transmitted light should be proportional to $u(\eta)$ as it was specified above. So we have:

$$R_\infty(\eta, \xi) = R(\eta, \xi, \tau_0) + \beta(\tau_0)u(\eta)u(\xi). \tag{3.144}$$

Let us find α and β substituting two last equations in asymptotic formulae (3.140), (3.141). Then it follows:

$$i(\xi) = 2e^{k\tau_0}\int_{-1}^{1} \alpha(\tau_0)u(\eta)u(\xi)i(\eta)\eta d\eta + 2\int_{-1}^{1} (R_\infty(\eta, \xi) - \beta(\tau_0)u(\eta)u(\xi))i(-\eta)\eta d\eta, \tag{3.145}$$

$$i(-\xi) = 2\int_{-1}^{1} (R_\infty(\eta, \xi) - \beta(\tau_0)u(\eta)u(\xi))i(\eta)d\eta + 2e^{-k\tau_0}\int_{-1}^{1} \alpha(\tau_0)u(\eta)u(\xi)i(-\eta)\eta d\eta, \tag{3.146}$$

and, therefore,

$$i(\xi) = e^{k\tau_0}\alpha(\tau_0)u(\xi) + i(\xi) - Mu(\xi) - \beta(\tau_0)u(\xi)N, \tag{3.147}$$

$$i(-\xi) = i(-\xi) - \beta(\tau_0)u(\xi) + \alpha N e^{-k\tau_0}u(\xi), \tag{3.148}$$

where we introduced the integral (see Table 3.3)

$$N = 2\int_{-1}^{1} u(\eta)i(-\eta)\eta d\eta. \tag{3.149}$$

It follows:

$$\alpha(\tau_0) - Me^{-k\tau_0} - \beta N e^{-k\tau_0} = 0, \tag{3.150}$$

$$\beta(\tau_0) = \alpha(\tau_0)Ne^{-k\tau_0} \tag{3.151}$$

and, therefore,

$$\alpha(\tau_0) = \frac{Me^{-k\tau_0}}{1 - N^2e^{-2k\tau_0}}. \tag{3.151}$$

Finally, we have (Sobolev, 1975, 1984):

$$R(\eta, \xi, \varphi, \tau_0) = R_\infty(\eta, \xi, \varphi) - T(\xi, \eta, \tau_0)Ne^{-k\tau_0}, \tag{3.152}$$

$$T(\xi, \eta) = \frac{Me^{-k\tau_0}}{1 - N^2e^{-2k\tau_0}}u(\eta)u(\xi), \tag{3.153}$$

where we accounted for the fact that the transmitted light does not depend on the azimuth φ.

These formulae are central equations of the cloud optics. The importance of these equations is due to the fact that the dependence on τ_0 is given explicitly. Our next task is to derive approximate equations for constants k, M, N and functions

$u(\eta)$, $R_\infty(\eta, \xi, \varphi)$ in a number of particular cases. Numerical calculations of these functions have been performed by Nakajima and King (1992). The accuracy of Eqs. (3.152), (3.153) was studied by Konovalov (1975) and also by Melnikova and Vasyliev (2005).

3.6.3 Weak Absorption Limit

Equations given above can be simplified considerably for the case of values of ω_0 close to one. This case is of particular importance for cloud optics due to weak absorption of water in visible and near-infrared. Therefore, we need to find approximate expressions for functions $R_\infty(\eta, \xi)$, $u(\eta)$ and also for parameters k, M, N as $\omega_0 \to 1$. Let us concentrate on this problem now.

3.6.3.1 The constants k, M and the diffuse light field in deep layers

The parameter M depends on the diffuse light intensity $i(\eta)$ in deep layers of a cloud:

$$M = 2\int_{-1}^{1} i^2(\eta)\eta d\eta. \tag{3.154}$$

So we need to study functions $i(\eta)$ as $\omega_0 \to 1$. The radiative transfer equation for the normalized light intensity $i(\eta)$ deep inside of a homogeneous cloudy medium has the following form as it was discussed above:

$$(1 - k\eta)i(\eta) = \frac{\omega_0}{2}\int_{-1}^{1} p(\eta, \eta')i(\eta')d\eta', \tag{3.155}$$

where $p(\eta, \eta')$ is the azimuthally averaged phase function, ω_0 is the single-scattering albedo and k is the diffusion exponent. The normalization condition for $i(\eta)$ has the following form:

$$\frac{\omega_0}{2}\int_{-1}^{1} i(\eta')d\eta' = 1. \tag{3.156}$$

We introduce the following expansions:

$$p(\eta, \eta') = \sum_{n=0}^{\infty} x_n P_n(\eta)P_n(\eta') \tag{3.157}$$

and

$$i(\eta) = \sum_{n=0}^{\infty} \sigma_n P_n(\eta). \tag{3.158}$$

The task is to find σ_n knowing x_n and using Eq. (3.155). So we have from Eq. (3.155):

$$B = \sum_{n=0}^{\infty} \sigma_n P_n(\eta) - k\eta \sum_{n=0}^{\infty} \sigma_n P_n(\eta), \tag{3.159}$$

where

$$B = \frac{\omega_0}{2} \sum_{l=0}^{\infty} \sum_{n=0}^{\infty} \int_{-1}^{1} x_l \sigma_n P_l(\eta) P_l(\eta') P_n(\eta') d\eta' \tag{3.160}$$

or

$$B = \omega_0 \sum_{l=0}^{\infty} \sum_{n=0}^{\infty} \sigma_n x_l \delta_{nl} [2n+1]^{-1} P_n(\eta) \tag{3.161}$$

and after simplifications:

$$B = \omega_0 \sum_{n=0}^{\infty} x_n \sigma_n [2n+1]^{-1} P_n(\eta). \tag{3.162}$$

We used the following orthogonality relationship:

$$\int_{-1}^{1} P_n(\eta) P_n(\eta) d\eta = 2\delta_{nl} [2n+1]^{-1}, \tag{3.163}$$

where δ_{nl} is the Kronecker symbol.

Therefore, it follows:

$$\frac{1}{k} \sum_{n=0}^{\infty} \sigma_n \left\{ 1 - \frac{x_n \omega_0}{2n+1} \right\} P_n(\eta) = \sum_{n=0}^{\infty} \sigma_n \left\{ \frac{n+1}{2n+1} P_{n+1}(\eta) + \frac{n}{2n+1} P_{n-1}(\eta) \right\}, \tag{3.164}$$

where we used the property:

$$P_n(\eta) = \frac{n+1}{2n+1} P_{n+1}(\eta) + \frac{n}{2n+1} P_{n-1}(\eta). \tag{3.165}$$

The expressions for

$$\varsigma(\eta) = \sum_{n=0}^{\infty} \sigma_n \frac{n+1}{2n+1} P_{n+1}(\eta) \tag{3.166}$$

and

$$\upsilon(\eta) = \sum_{n=0}^{\infty} \sigma_n \frac{n}{2n+1} P_{n-1}(\eta) \tag{3.167}$$

can be written as

$$\varsigma(\eta) = \sum_{l=1}^{\infty} \sigma_{l-1} \frac{l}{2l-1} P_l(\eta) \tag{3.168}$$

and

$$\upsilon(\eta) = \sum_{s=0}^{\infty} \sigma_{s+1} \frac{s+1}{2s+3} P_l(\eta), \tag{3.169}$$

where $l = n+1$, $s = n-1$.

Therefore, it follows:

$$\sum_{m=0}^{\infty} \left[\frac{1}{k}\sigma_m - \frac{x_m \omega_0}{(2m+1)k}\sigma_m - \frac{m}{2m-1}\sigma_{m-1} - \frac{m+1}{2m+3}\sigma_{m+1} \right] P_n(\eta) = 0 \tag{3.170}$$

for arbitrary η. This means that

$$\frac{1}{k}\sigma_m - \frac{x_m \omega_0}{(2m+1)k}\sigma_m - \frac{m}{2m-1}\sigma_{m-1} - \frac{m+1}{2m+3}\sigma_{m+1} = 0 \tag{3.171}$$

or

$$\sigma_{m+1} = \frac{(2m+3)(2m - \omega_0 x_m + 1)}{(2m+1)(m+1)k}\sigma_m + \frac{(2m+3)m}{(2m-1)(m+1)}\sigma_{m-1}. \tag{3.172}$$

Let us assume that $m = 0$. Then we have:

$$\sigma_1 = \frac{3\sigma_0(1-\omega_0)}{k}. \tag{3.173}$$

It is easy to derive the analytical expression for the value of σ_0. It follows that

$$\sigma_m = \frac{2m+1}{2} \int_{-1}^{1} i(\eta) P_n(\eta) d\eta \tag{3.174}$$

and, therefore,

$$\sigma_0 = \frac{1}{2} \int_{-1}^{1} i(\eta) d\eta. \tag{3.175}$$

So we can derive: $\sigma_0 = \omega_0^{-1}$ and, therefore,

$$\sigma_1 = \frac{3(1-\omega_0)}{k\omega_0}. \tag{3.176}$$

This allows to obtain the following asymptotic expression from Eq. (3.158):

$$i(\eta) = \omega_0^{-1} \left\{ 1 + 3k^{-1}(1-\omega_0)\eta \right\}, \tag{3.177}$$

where we neglected higher order terms. It follows as $\omega_0 \to 1$:

$$i(\eta) = 1 + 3k^{-1}(1 - \omega_0)\eta. \tag{3.178}$$

Recurrence relations allow us to find σ_m and $i(\eta)$ at any k. We will not consider correspondent results here, however, but rather concentrate on the derivation of the approximate equation for k valid as $\omega_0 \to 1$.

For this we introduce:

$$\Upsilon_m = \frac{\sigma_m}{\sigma_{m-1}}. \tag{3.179}$$

Then it follows:

$$\Upsilon_{m+1} = \frac{(2m+3)(2m - \omega_0 x_m + 1)}{(2m+1)(m+1)k} - \frac{(2m+3)m}{(2m-1)(m+1)\Upsilon_m} \tag{3.180}$$

and

$$\Upsilon_m = \frac{(2m+3)m(2m - \omega_0 x_m + 1)}{(2m-1)(m+1)\left[\frac{(2m+3)(2m+1-\omega_0 x_m)}{(2m+1)(m+1)k} - \Upsilon_{m+1}\right]} \tag{3.181}$$

or

$$\Upsilon_m = \frac{(2m+3)(2m+1)mk}{(2m+3)(2m-1)(2m+1-\omega_0 x_m) - \varepsilon_m}, \tag{3.182}$$

where

$$\varepsilon_m = \frac{(2m+1)k\sigma_{m+1}}{(2m-1)(m+1)}. \tag{3.183}$$

Because we are interested in the asymptotic solution valid as $k \to 0$, we can ignore ε_m and derive at $m = 1$:

$$\Upsilon_1 = \frac{3k}{(3 - \omega_0 x_1)}. \tag{3.184}$$

So it follows:

$$\sigma_1 = \frac{3k}{(3 - \omega_0 x_1)}\sigma_0 \tag{3.185}$$

or

$$\sigma_1 = \frac{3k}{(3 - \omega_0 x_1)\omega_0}. \tag{3.186}$$

Therefore, we finally derive (see also Eq. (3.176)):

$$k = \sqrt{3(1 - \omega_0)(1 - g\omega_0)}, \tag{3.187}$$

where $g = x_1/3$ is the asymmetry parameter. This important equation shows that the intensity in the deep layers of clouds decreases faster for smaller values of g

(less extended in the forward direction phase functions). Our derivations are valid as $\omega_0 \to 1$ only. So we can also write:

$$k = \sqrt{3(1-\omega_0)(1-g)}. \tag{3.188}$$

The approximate expression for the diffusion constant given here is of a great importance for cloud optics studies. In particular, we can derive:

$$i(\eta) = 1 + \sqrt{3}s\eta, \tag{3.189}$$

where

$$s = \sqrt{\frac{1-\omega_0}{1-g}} \tag{3.190}$$

is the similarity parameter. Sometimes the value of $g\omega_0$ instead of g is used in the definition of the similarity parameter. Both formulations coincide as $\omega_0 \to 1$. Surprising result is that the angular pattern $i(\eta)$ does not depend on the choice of the particular light scattering medium if s kept constant. The function $i(\eta)$ is completely determined by the similarity parameter s as $\omega_0 \to 1$. Therefore, cloudy media having different ω_0 and g but the same s have very similar light angular distributions at $\omega_0 \approx 1$.

Let us introduce the ratio $D = i(-1)/i(1)$. This ratio is equal to one for nonabsorbing media. However, it follows from Eq. (3.189) for weakly absorbing media: $D = 1 - \varsigma\sqrt{1-\omega_0}$, $\varsigma = 2\sqrt{3/(1-g)}$. Therefore, measurements of g and D can be used to find the probability of light absorption $\beta = 1 - \omega_0$. Namely, it follows: $\beta = (1-D)^2/\varsigma^2$.

The parameters k and s are of a crucial importance for the theory considered here. We expect that other asymptotic constants and functions must depend on these parameters as well. In particular, we derive using property 7 in Table 3.3:

$$M = \frac{8s}{\sqrt{3}} \tag{3.191}$$

as $k \to 0$.

3.6.3.2 The constant N and the escape function

The expansion of $u(\eta)$ with respect to the diffusion coefficient k can be presented as

$$u(\eta) = \sum_{n=0}^{\infty} k^n u_n(\eta). \tag{3.192}$$

We are interested only in the case of weak absorption. Then it follows:

$$u(\eta) = u_0(\eta) + ku_1(\eta). \tag{3.193}$$

The task is to calculate the function $u_1(\eta)$. This will be performed in two steps. First of all we note that the weak absorption of light does not alter single-scattering law considerably. The angular distribution of emerging light $u(\eta)$ is determined largely by the scattering processes. So it is safe to assume that $u(\eta) \approx u_0(\eta)$ as $k \to 0$ or $u_1(\eta) = bu_0(\eta)$, where the constant b should be determined. Clearly, due to physical reasons we should have: $u(\eta) < u_0(\eta)$ and $b < 0$. Therefore, absorption plays the role of a veil in this case. It reduces the contrast but it does not change details of the scattering pattern. We start from the expression:

$$2\int_0^1 u(\eta)i(\eta)\eta d\eta = 1. \tag{3.194}$$

Let us substitute the following expansions in this formula:

$$u(\eta) = u_0(\eta)(1 + bk) \tag{3.195}$$

and

$$i(\eta) = 1 + ak\eta, \tag{3.196}$$

where $a = (1 - g)^{-1}$ as underlined above.

Then it follows (see Eqs. (3.188), (3.189)):

$$2\int_0^1 u_0(\eta)\eta d\eta + 2bk\int_0^1 u_0(\eta)\eta d\eta + 2ak\int_0^1 u_0(\eta)\eta^2 d\eta = 1 \tag{3.197}$$

or

$$b = -2a\int_0^1 u_0(\eta)\eta^2 d\eta, \tag{3.198}$$

where we accounted for the fact that (see property 6 at $\omega_0 = 1$ ($i = 1$) in Table 3.3)

$$2\int_0^1 u_0(\eta)\eta d\eta = 1. \tag{3.199}$$

Finally, it follows:

$$b = -\frac{2\delta}{1 - g}, \tag{3.200}$$

where we accounted for the fact that $a = (1 - g)^{-1}$ and

$$\delta = \int_0^1 u_0(\eta)\eta^2 d\eta \tag{3.201}$$

is the second moment of the escape function. Therefore, one finally derives:

$$u(\eta) = \left(1 - \frac{2\delta k}{1-g}\right) u_0(\eta). \tag{3.202}$$

This equation allows to find the constant N as $k \to 0$. Namely, it follows from property 8 in Table 3.3:

$$N = 2\int_{-1}^{1} d\eta u_0(\eta) \left\{1 - \frac{2\delta k}{1-g}\right\}\left\{1 + \frac{k\eta}{1-g}\right\} \tag{3.203}$$

or

$$N = 1 - \frac{3\delta k}{2(1-g)}, \tag{3.204}$$

where we neglected terms of the second order with respect to k. We can also write:

$$N = 1 - \frac{3\sqrt{3}s}{2}. \tag{3.205}$$

Note that functions $u_0(\mu)$ enter asymptotic formulae in the combination: $F(\eta, \xi) = Mu(\eta)u(\xi)$. This means that one can use the following approximation valid as $k \to 0$:

$$F(\eta, \xi) = \frac{8s}{\sqrt{3}} u_0(\eta) u_0(\xi). \tag{3.206}$$

3.6.3.3 The reflection function of a semi-infinite layer $R_\infty(\xi, \eta)$

The last point in our derivations of asymptotics as $\omega_0 \to 1$ is the derivation of the weak absorption approximation for the reflection function of a semi-infinite medium $R_\infty(\xi, \eta)$. This will be done in two steps.

Step 1.
The expression for a plane albedo of a semi-infinite medium is written by a definition as

$$r_d(\xi) = 2\int_0^1 R_\infty(\xi, \eta)\eta d\eta. \tag{3.207}$$

We will use the following expansion of $R_\infty(\xi, \eta)$ with respect to k:

$$R_\infty(\xi, \eta) = R_{0\infty}(\xi, \eta) - kR_{1\infty}(\xi, \eta), \tag{3.208}$$

where $R_{1\infty}(\xi, \eta)$ is the function we need to derive. The minus signifies the fact that $R_\infty(\xi, \eta) \le R_{0\infty}(\xi, \eta)$ by definition. One can see that

$$r_d(\xi) = 1 - kJ(\xi), \tag{3.209}$$

where

$$J(\xi) = 2\int_0^1 R_1(\xi, \eta)\eta d\eta \tag{3.210}$$

and we used the property:

$$2\int_0^1 R_{0\infty}(\xi, \eta)\eta d\eta = 1. \tag{3.211}$$

Step 2.
We derive now the asymptotic equation for $r_d(\xi)$ as $k \to 0$ using other set of equations. This will allow us to give a relationship between $J(\xi)$ and $u_0(\xi)$. We start from the following equation derived above (see Table 3.3):

$$i(-\xi) = 2\int_0^1 i(\eta)R_\infty(\xi, \eta)\eta d\eta. \tag{3.212}$$

Substituting expansions with respect to k and ignore high-order terms, we obtain:

$$1 - \frac{k\xi}{1-g} = 2\int_0^1 \left(1 + \frac{k\eta}{1-g}\right)(R_{0\infty}(\xi, \eta) - kR_{1\infty}(\xi, \eta))\eta d\eta. \tag{3.213}$$

This means that

$$1 - \frac{k\xi}{1-g} = 1 - kJ + \frac{2k}{1-g}\int_0^1 R_{0\infty}(\xi, \eta)\eta^2 d\eta \tag{3.214}$$

or

$$1 - \frac{\xi}{1-g} = 1 - J + \frac{2}{1-g}\int_0^1 R_{0\infty}(\xi, \eta)\eta^2 d\eta, \tag{3.215}$$

where [see Eq. (3.209)]

$$J = \frac{1 - r_d(\xi)}{k}. \tag{3.216}$$

This means that

$$(1-g)(1 - r_d(\xi))k^{-1} = 2\int_0^1 R_{0\infty}(\xi, \eta)\eta^2 d\eta + \xi \tag{3.217}$$

or

$$r_d(\xi) = 1 - \frac{k}{1-g}\left\{\xi + 2\int_0^1 R_{0\infty}(\xi, \eta)\eta^2 d\eta\right\}. \tag{3.218}$$

On the other hand, we have (see Table 3.4):

$$i(\xi) = Mu(\xi) + 2\int_0^1 i(-\eta)R_\infty(\xi,\eta)\eta d\eta \tag{3.219}$$

or as $k \to 0$:

$$1+\frac{k\xi}{1-g} = \frac{8ku_0(\xi)}{3(1-g)} + 2\int_0^1 \left(1-\frac{k\eta}{1-g}\right)(R_{0\infty}(\xi,\eta)-kR_{1\infty}(\xi,\eta))\eta d\eta. \tag{3.220}$$

This means that

$$1+\frac{k\xi}{1-g} = \frac{8ku_0(\xi)}{3(1-g)} + 1 - \frac{2k}{1-g}\int_0^1 R_{0\infty}(\xi,\eta)\eta^2 d\eta - kJ(\xi) \tag{3.221}$$

or

$$\xi = \frac{8}{3}u_0(\xi) - 2\int_0^1 R_{0\infty}(\xi,\eta)\eta^2 d\eta - (1-g)J(\xi). \tag{3.222}$$

We remind that

$$J(\xi) = 2\int_0^1 R_{1\infty}(\xi,\eta)\eta d\eta = (1-r_d(\xi))k^{-1}. \tag{3.223}$$

Therefore, it follows:

$$(1-r_d(\xi))k^{-1} = (1-g)^{-1}\left(\xi + 2\int_0^1 R_{0\infty}(\xi,\eta)\eta^2 d\eta\right) \tag{3.224}$$

and

$$J(\xi) = (1-g)^{-1}\left(\xi + 2\int_0^1 R_{0\infty}(\xi,\eta)\eta^2 d\eta\right) \tag{3.225}$$

or

$$\xi = \frac{8u_0(\xi)}{3} - \xi - 4\int_0^1 R_{0\infty}(\xi,\eta)\eta^2 d\eta \tag{3.226}$$

and

$$\frac{8u_0(\xi)}{3} = 2\xi + 4\int_0^1 R_{0\infty}(\xi,\eta)\eta^2 d\eta. \tag{3.227}$$

This allows us to derive the following important relationship:

$$u_0(\xi) = \frac{3}{4}\left[\xi + 2\int_0^1 R_{0\infty}(\xi,\eta)\eta^2 d\eta\right]. \tag{3.228}$$

Comparing this formula with Eq. (3.224), we obtain:

$$r_d(\xi) = 1 - \frac{4ku_0(\xi)}{3(1-g)}. \tag{3.229}$$

But also we derived:

$$r_d(\xi) = 1 - kJ(\xi). \tag{3.230}$$

This means that

$$J(\xi) = 2\int_0^1 R_{1\infty}(\xi,\eta)\eta d\eta = \frac{4u_0(\xi)}{3(1-g)}. \tag{3.231}$$

The function $R_{1\infty}(\xi,\eta)$ must be symmetric with respect to the pair (ξ,η). Therefore, it follows:

$$R_{1\infty}(\xi,\eta) = cu_0(\xi)u_0(\eta). \tag{3.232}$$

Substituting this formula in Eq. (3.231) given above, we derive for the constant c:

$$c = \frac{4}{3(1-g)}, \tag{3.233}$$

where we used the property

$$2\int_0^1 u_0(\eta)\eta d\eta = 1. \tag{3.234}$$

Finally, we have:

$$R_\infty(\xi,\eta) = R_{0\infty}(\xi,\eta) - \frac{4k}{3(1-g)}u_0(\xi)u_0(\eta) \tag{3.235}$$

or

$$R_\infty(\xi,\eta) = R_{0\infty}(\xi,\eta) - 4su_0(\xi)u_0(\eta). \tag{3.236}$$

This also means that (see Table 3.1)

$$r = 1 - \frac{4}{\sqrt{3}}s \tag{3.237}$$

and

$$r_d(\xi) = 1 - \frac{4}{\sqrt{3}}su_0(\xi). \tag{3.238}$$

Main asymptotic equations valid as $k \to 0$ are given in Table 3.4. It follows that R and T in the case of weak absorption can be easily calculated if the functions $u_0(\xi)$ and $R_{0\infty}(\xi,\eta)$ are known. Moreover, Eq. (3.228) can be used to find $u_0(\xi)$ from $R_{0\infty}(\xi,\eta)$.

The choice of the normalization condition for the function $u(\xi)$ (and also for $i(\xi)$ and derived parameters like M) is arbitrary. We followed notations of Sobolev (1975). They differ from correspondent equations used by van de Hulst (1980). For

Table 3.4. Asymptotic equations valid as $k \to 0$ $\left(\delta = \int_0^1 u_0(\xi)\xi^2 d\xi,\ s = \sqrt{\frac{1-\omega_0}{1-g}}\right)$.

$R_\infty(\xi, \eta, \varphi)$	$R_{0\infty}(\xi, \eta, \varphi) - \dfrac{4k}{3(1-g)} u_0(\xi)u_0(\eta)$
$u(\xi)$	$\left(1 - \dfrac{\delta k}{2}\right) u_0(\xi)$
M	$\dfrac{8k}{3(1-g)}$
N	$1 - \dfrac{\delta k}{1-g}$
$Mu(\xi)u(\eta)$	$\dfrac{8k}{3(1-g)} u_0(\xi)u_0(\eta)$
k	$\sqrt{3(1-\omega_0)(1-g)}$
r	$1 - 4s/\sqrt{3}$
$r_d(\xi)$	$1 - 4su_0(\xi)/\sqrt{3}$

instance, van de Hulst's diffusion pattern $P(\xi)$ must be divided by ω_0 to yield $i(\xi)$. His escape function $K(\xi)$ must be multiplied by ω_0 to yield $u(\xi)$, and his Mequals to that used by Sobolev multiplied by ω_0^2. These differences do not lead to any extra factors in main equations. However, they must be remembered. Clearly, all differences between normalizations disappear for the case of nonabsorbing clouds. This case will be considered in the next Section.

3.6.4 Nonabsorbing Optically Thick Clouds

3.6.4.1 Main equations

Let us assume that there is no absorption in the medium (e.g., water clouds in the visible). Then it follows using Table 3.4 and Eqs. (3.152) and (3.153):

$$R(\xi, \eta, \varphi) = R_{0\infty}(\xi, \eta, \varphi) - tu_0(\xi)u_0(\xi) \tag{3.239}$$

and

$$T(\xi, \eta) = tu_0(\xi)u_0(\xi), \tag{3.240}$$

where we accounted for the fact that the transmittance of a thick cloud layer does not depend on the azimuth φ and

$$t = \frac{1}{\alpha + 0.75(1-g)\tau}, \tag{3.241}$$

where $\alpha = 3\delta$. The plane albedo $r_d(\xi)$, the spherical albedo r, the diffuse transmittance $t_d(\xi)$ and the global transmittance $t$are given as (see Table 3.1)

$$r_d(\xi) = 2\int_0^1 R(\xi, \eta)\eta d\eta, \tag{3.242}$$

$$r = 2\int_0^1 r_d(\xi)\xi d\xi, \tag{3.243}$$

$$t_d(\xi) = 2\int_0^1 T(\xi, \eta)\eta d\eta, \tag{3.244}$$

$$t = 2\int_0^1 t_d(\xi)\xi d\xi. \tag{3.245}$$

This allows to derive the following analytical relationships:

$$r_d(\xi) = 1 - tu_0(\xi), r = 1 - t, t_d(\xi) = tu_0(\xi) \tag{3.246}$$

and also confirm that $t$given by Eq. (3.241) coincides with the global transmittance. It follows that the calculation of reflection and transmission functions of optically thick nonabsorbing cloud layers is reduced to the calculation of the reflection function of a semi-infinite nonabsorbing cloud $R_{0\infty}(\xi, \eta, \varphi)$. The functions $u_0(\xi)$ can be calculated from Eq.(3.228) and the parameter α is defined as

$$\alpha = 3\int_0^1 u_0(\eta)\eta^2 d\eta. \tag{3.247}$$

We will show later that $\alpha \approx 1.07$ independent of the phase function (Kokhanovsky et al., 2004a). The function $R_{0\infty}(\xi, \eta, \varphi)$ can be derived from the numerical solution of the correspondent integral equation (Ambartsumian, 1943).

In the next section we introduce useful approximations for $R_{0\infty}(\xi, \eta, \varphi)$ and also for $u_0(\xi)$. The important property of these functions is the fact that they do not depend on the pair (ω_0, τ) and that they are completely determined by the phase function. Moreover, the dependence on the phase function is rather weak because functions $u_0(\xi)$, $R_{0\infty}(\xi, \eta, \varphi)$ are related to the problems involving light diffusion in semi-infinite nonabsorbing media. So multiple light scattering is quite important in this case. It leads to the averaging of the scattering features characteristic for a single-scattering event. This also means that a good starting point for the derivation of approximate solutions for $u_0(\xi)$, $R_{0\infty}(\xi, \eta, \varphi)$ valid at arbitrary g is the case of $g = 0$ (isotropic scattering).

3.6.4.2 *Auxiliary functions*

We start the consideration of auxiliary functions from the well studied case of isotropic scattering. Then the nonlinear integral equation for the reflection function of a nonabsorbing semi-infinite medium can be presented in the following form (Ambartsumian, 1943):

$$R_{0\infty}(\xi,\eta) = \frac{1 + 2\xi \int_0^1 R_{0\infty}(\eta,\eta')d\eta' + 2\eta \int_0^1 R_{0\infty}(\eta,\eta')d\eta' + G(\xi,\eta)}{4(\xi+\eta)}, \tag{3.248}$$

where

$$G(\xi,\eta) = 4\xi\eta \int_0^1 \int_0^1 R_{0\infty}(\xi,\eta')R_{0\infty}(\eta,\eta'')d\eta' d\eta''. \tag{3.249}$$

The inspection of this equation shows that it can be reduced to the following more simple form:

$$R_{0\infty}(\xi,\eta) = \frac{H(\xi)H(\eta)}{4(\xi+\eta)} \tag{3.250}$$

with

$$H(\xi) = 1 + 2\xi \int_0^1 R_{0\infty}(\xi,\eta')d\eta'. \tag{3.251}$$

The last two equations allow to formulate the integral equation for the function $H(\xi)$:

$$H(\xi) = 1 + 0.5\xi H(\xi) \int_0^1 \frac{H(\eta)}{\xi+\eta} d\eta. \tag{3.252}$$

It follows immediately: $H(0) = 1.0$. Numerical calculations show that the function $H(\xi)$ can be approximated by the linear function [$H(\xi) = 1 + 2\xi$ (see Fig. 3.5)]. The substitution of this linear approximation into the expression for $R_{0\infty}(\xi,\eta)$ gives:

$$R_{0\infty}(\xi,\eta) = \frac{1 + 2(\xi+\eta) + 4\xi\eta}{4(\xi+\eta)}. \tag{3.253}$$

This is a rather accurate approximation of the reality in the isotropic scattering case. Further, we note that the value of $R_{0\infty}(\xi,\eta)$ can be separated in two parts:

$$R_{0\infty}(\xi,\eta) = R^s_{0\infty}(\xi,\eta) + R^m_{0\infty}(\xi,\eta), \tag{3.254}$$

where the first term is due to single scattering [$R^s_{0\infty}(\xi,\eta) = 0.25(\xi+\eta)^{-1}$, see Eq. (3.62) at $p = 1,\ \omega_0 = 1,\ \tau_0 \to \infty$] and the second one ($R^m_{0\infty}(\xi,\eta) = [0.5 + \xi\eta(\xi+\eta)^{-1}]$) is due to multiple light scattering.

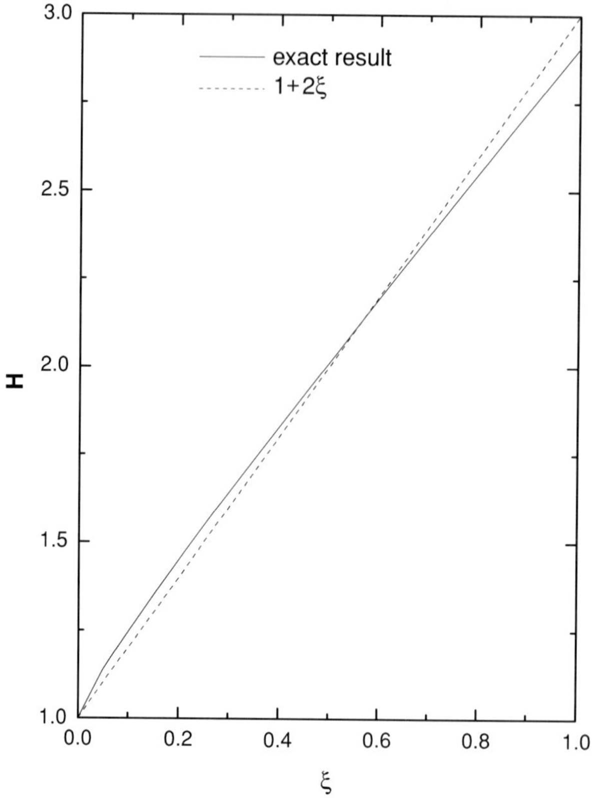

Fig. 3.5. Dependence of the H-function on the cosine of incidence angle using exact calculations and approximation $H(\xi) = 1 + 2\xi$.

We make the same separation for the nonisotropic ($g \neq 0$) scattering case. Then we have:

$$R^{s}_{0\infty}(\xi, \eta) = 0.25p(\theta)(\xi + \eta)^{-1} \tag{3.255}$$

and we assume that it holds for multiple nonisotropic light scattering:

$$R^{m}_{0\infty}(\xi, \eta) = \frac{A + B(\xi + \eta) + C\xi\eta}{4(\xi + \eta)}, \tag{3.256}$$

where A, B and C are constants to be determined. There are different ways to get these constants. In particular integral relationships involving the function $R_{0\infty}(\xi, \eta)$ can be used.

They can be also found using the following fitting technique. The function $R_{0\infty}(\xi, \eta, \varphi)$ is calculated using the exact radiative transfer equation and then functions $\Xi(\xi, \eta, \varphi) = 4(\xi + \eta)\tilde{R}_{0\infty}(\xi, \eta, \varphi)$, where $\tilde{R}_{0\infty}(\xi, \eta, \varphi) =$

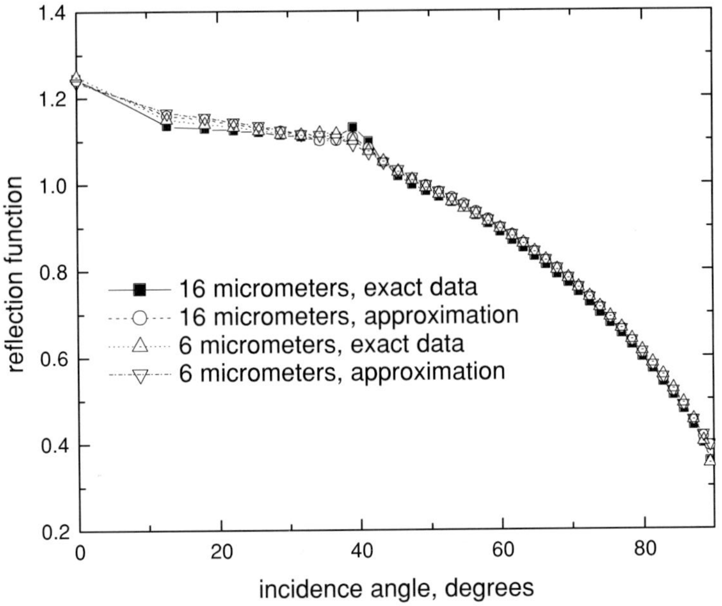

Fig. 3.6. Dependence of reflection function at the nadir observation of a semi-infinite cloud on the solar zenith angle at $a_{ef} = 4,\ 16$ μm and $\lambda = 0.55$ μm. The gamma PSD with $\mu = 6$ was used in calculations. Results both exact and approximate calculations are shown.

$R_{0\infty}(\xi,\eta,\varphi) - R^s_{0\infty}(\xi,\eta,\varphi)$, are fitted by linear functions of the argument assuming, e.g., $\eta = 1$.This technique gives: $A = 3.944$, $B = -2.5$, $C = 10.664$ for water clouds, and $A = 1.247$, $B = 1.186$, $C = 5.157$ for ice clouds as discussed by Kokhanovsky (2004b, 2005). The results of numerical calculations of $R_{0\infty}(\xi,\eta,\varphi)$ are given in Fig. 3.6. It follows that the reflection function of clouds having different values of a_{ef} almost coincide at $\omega_0 = 1$ [although phase functions do depend on the size of particles (see Fig. 3.7)]. Note that the shape of particles has some influence on functions $R_{0\infty}(\xi,\eta,\varphi)$ as shown in Fig. 3.8.

The next point is to derive the correspondent equation for the function $u_0(\xi)$. This can be done in the following way.

It was shown above that the following relationship holds:

$$u_0(\xi) = \frac{3}{4}\left[\xi + 2\int_0^1 R_{0\infty}(\xi,\eta)\eta^2 d\eta\right]. \tag{3.257}$$

Let us substitute $R_{0\infty}(\xi,\eta)$ for the isotropic case in this equation. Then it follows:

$$u_0(\xi) = \frac{3}{4}\left[\xi + \frac{1}{2}\int_0^1 \frac{H(\xi)H(\eta)}{\xi+\eta}\eta^2 d\eta\right]. \tag{3.258}$$

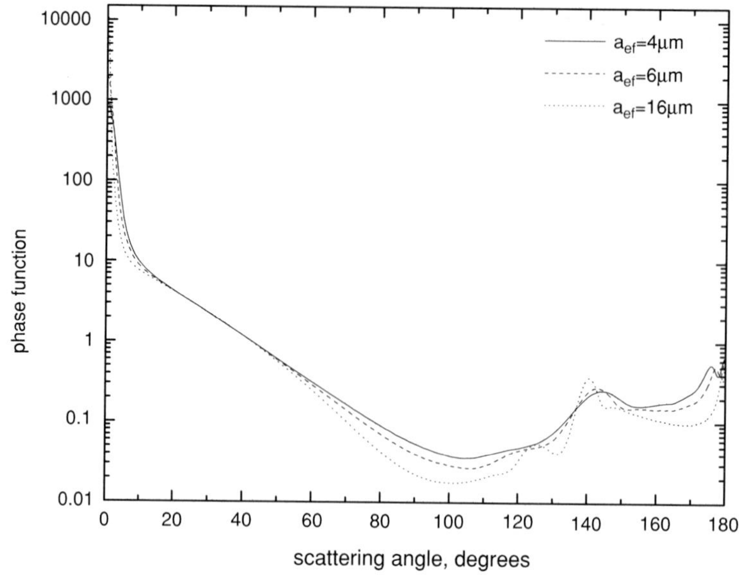

Fig. 3.7. Dependence of the cloud phase function on the effective size of water droplets at $\lambda = 0.55$ μm. The gamma PSD with $\mu = 6$ was used in calculations.

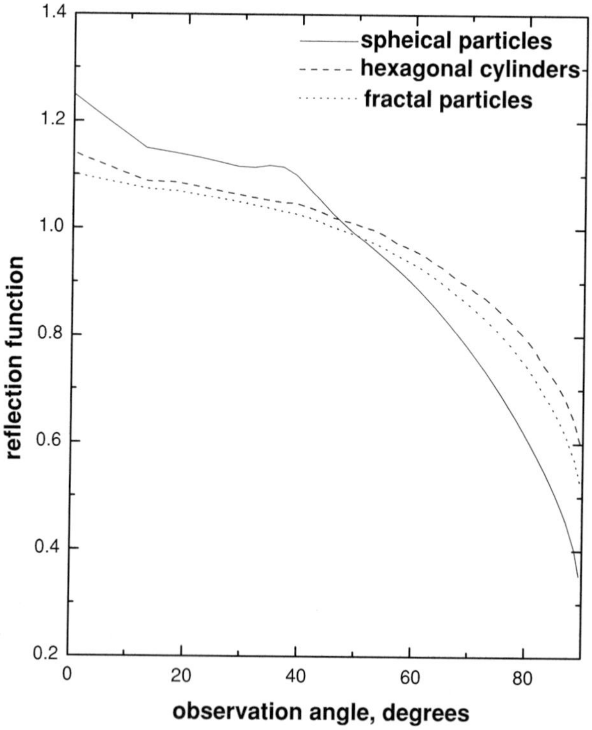

Fig. 3.8. Dependence of reflection function at the nadir observation of a semi-infinite cloud on the solar zenith angle for spherical particles (gamma PSD, $a_{ef} = 6$ μm, $\mu = 6$), hexagonal randomly oriented ice cylinders with aspect ratio 1.0 and ice fractal particles calculated using exact radiative transfer code (Mishchenko et al., 1999) at $\lambda = 0.55$ μm.

We can represent $H(\eta)\eta(\xi+\eta)^{-1}$ as $H(\eta)(1-\xi(\xi+\eta)^{-1})$. So we have:

$$u_0(\xi) = \frac{3}{4}\left[\xi + \frac{1}{2}H(\xi)\int_0^1 H(\eta)\eta d\eta - \frac{H(\xi)\xi}{2}\int_0^1 \frac{H(\eta)}{\xi+\eta}\eta d\eta\right]. \quad (3.259)$$

This can be written as

$$u_0(\xi) = \frac{3}{4}\left[\xi + \frac{1}{2}\mathbb{C}H(\xi) - \Psi\xi\right], \quad (3.260)$$

where

$$\mathbb{C} = \int_0^1 H(\eta)\eta d\eta \quad (3.261)$$

and (see Eq. (3.250))

$$\Psi = 2\int_0^1 R_{0\infty}(\xi,\eta)\eta d\eta. \quad (3.262)$$

But we have, due to property 11 (see also property 4) in Table 3.3 (the conservation energy law): $\Psi = 1$ and

$$u_0(\xi) = \frac{3\mathbb{C}}{8}H(\xi). \quad (3.263)$$

This means that the function $u_0(\xi)$ is proportional to $H(\xi)$. The constant $\mathbb{C}$ can easily be derived for the isotropic scattering. For this we multiply the last equation by 2ξ and integrate with respect to ξ. Then it follows:

$$\mathbb{C} = \frac{2}{\sqrt{3}}, \quad (3.264)$$

where we used normalization conditions. Therefore, we establish an important relationship:

$$u_0(\xi) = \frac{\sqrt{3}}{4}H(\xi). \quad (3.265)$$

Surprisingly, two completely separate radiative transfer problems [for the determination of $H(\xi)$ and $u_0(\xi)$] have shown themselves to be interrelated in the case under study. This important theoretical result, valid for isotropic scattering allows to derive approximate equations for $u_0(\xi)$ just using the correspondent equations for $H(\xi)$. A number of parameterizations and approximations can be derived in such a way.

We will use the fact that $H(\xi)$ is well approximated by the function $1+2\xi$ (see Fig. 3.5). Then it follows:

$$u_0(\xi) = \mathbb{Q}(1+2\xi), \quad (3.266)$$

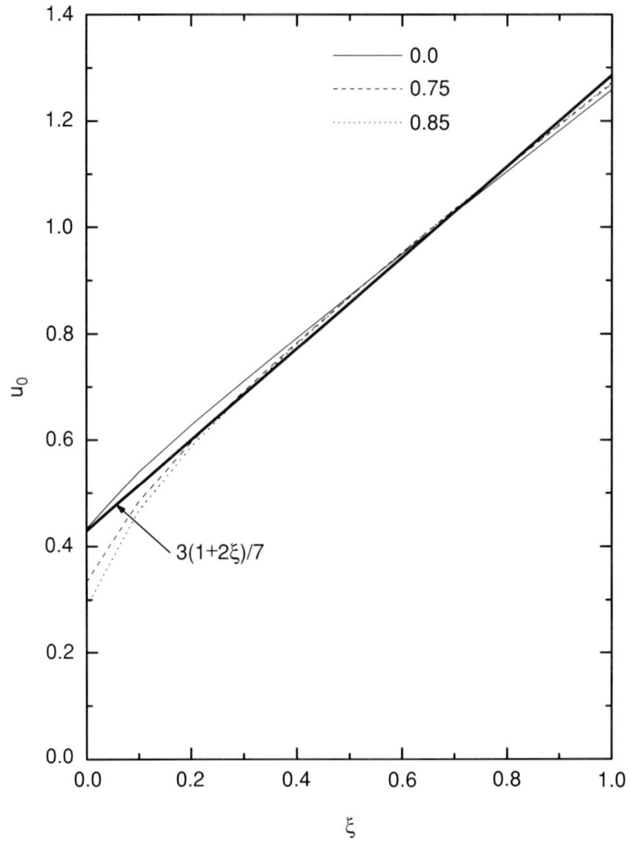

Fig. 3.9. Dependence of the escape function on the cosine of the zenith observation angle at $g =$ 0, 0.75, and 0.85. Results of calculations obtained using the approximation $u_0(\xi) = 3(1 + 2\xi)/7$ are also shown.

where $\mathbb{Q} = \sqrt{3}/4 \approx 3/7$. We use the approximate equality (the error is under 1%) here to satisfy the normalization condition (see property 11 in Table 3.3). So finally, we have:

$$u_0(\xi) = \frac{3}{7}(1 + 2\xi). \tag{3.267}$$

Although this result is strictly valid only for isotropic scattering, we find that the error of this approximation is below 2% at $\xi \geq 0.2$ (see Fig. 3.9). This also allows to derive the value of $\delta = 15/42$ (see property 9 in Table 3.3). Therefore, one can use for α [see Eq. (3.247)]:$\alpha \equiv 3\delta \approx 1.07$. This completes our derivations for the nonabsorbing case.

3.6.5 Exponential Approximation

3.6.5.1 Statistical physics approach

Asymptotic solutions for weak absorption derived above allow for the consideration of the influence of light absorption on cloud radiative characteristics for small values of the probability of photon absorption $\beta = 1 - \omega_0$, if correspondent characteristics are known for the nonabsorbing case. The results are limited to a very narrow range of β (typically, $\beta \leq 0.001$). There are two possibilities to avoid this problem. One is related to the derivation of higher order corrections to the results given in Table 3.4 (generally, following the same path as described above).

Yet another approach is based on the exponential approximation often used in the diffusion theory. To demonstrate this technique, we represent the spherical albedo as a series with respect to ω_0 :

$$r(\omega_0) = \sum_{n=1}^{\infty} a_n \omega_0^n, \tag{3.268}$$

where

$$r(\omega_0) = 2\int_0^1 r_d(\xi)\xi d\xi. \tag{3.269}$$

with

$$r(1) = \sum_{n=1}^{\infty} a_n. \tag{3.270}$$

However, it also follows by the definition: $r(1) = 1$, which is due to the energy conservation law. Thus, one obtains that

$$\sum_{n=1}^{\infty} a_n = 1 \tag{3.271}$$

and numbers a_n can be interpreted in terms of the probability theory. Such an approach is often used in statistical physics. In particular, the value of a_1 gives us the probability that a photon will be singly scattered before escaping a cloud. The probabilities of scattering events $a_1, a_2, a_3, \ldots$ do not depend on each other. The theorem of adding independent probabilities brings us to Eq. (3.268) as well. Let us substitute the following exact expansion in Eq. (3.268):

$$\omega_0^n \equiv (1-\beta)^n = \sum_{j=0}^{n} (-1)^j \binom{n}{j} \beta^j, \tag{3.272}$$

where

$$\binom{n}{j} \equiv \frac{n!}{j!(n-j)!}. \tag{3.273}$$

Then it follows from Eq. (3.268):

$$r(\beta) = \sum_{n=1}^{\infty} a_n \sum_{j=0}^{n} (-1)^j \binom{n}{j} \beta^j \tag{3.274}$$

or in the explicit form:

$$r(\beta) = \sum_{n=1}^{\infty} a_n \left[1 - \beta n + \frac{\beta^2 n(n-1)}{2} - \frac{\beta^3 n(n-1)(n-2)}{6} + \ldots\right], \tag{3.275}$$

where we accounted for equalities:

$$\binom{n}{0} = 1, \binom{n}{1} = n, \binom{n}{2} = \frac{n(n-1)}{2}, \binom{n}{3} = \frac{n(n-1)(n-2)}{6}. \tag{3.276}$$

Equation (3.275) can be rewritten in the following form:

$$r = 1 - \beta\bar{n} + \frac{\beta^2 \overline{n(n-1)}}{2} - \frac{\beta^3 \overline{n(n-1)(n-2)}}{6} + \ldots, \tag{3.277}$$

where we used the normalization condition (3.271) and defined the following averages:

$$\bar{n} = \sum_{n=1}^{\infty} n a_n, \overline{n(n-1)} = \sum_{n=1}^{\infty} n(n-1) a_n, \overline{n(n-1)(n-2)} = \sum_{n=1}^{\infty} n(n-1)(n-2) a_n. \tag{3.278}$$

and so on. Here $\bar{n}$ is the average number of scattering events in the medium.

Equation (3.277) is an exact formula. We did not make any approximations so far. Now we should make some assumptions to have a possibility to sum series in Eq. (3.277). First of all, we assume that the value of $\bar{n}$ is large and, consequently, $\overline{n(n-1)} \approx \overline{n^2}$, $\overline{n(n-1)(n-2)} = \overline{n^3}$ and so on. Clearly, such an approximation is valid as $\beta \to 0$ only. This gives us instead of Eq. (3.277):

$$r = 1 - \beta\overline{n} + \frac{\beta^2}{2}\overline{n^2} - \frac{\beta^3}{6}\overline{n^3} + \ldots \tag{3.279}$$

or

$$r = \overline{\exp(-\beta n)}, \tag{3.280}$$

where we used the expansion

$$\exp(-\beta n) = \sum_{k=0}^{\infty} \frac{(-1)^k (\beta n)^k}{k!}. \tag{3.281}$$

Thus, the value of r is given by

$$r = \sum_{n=1}^{\infty} \exp(-\beta n) a_n. \tag{3.282}$$

Applying the sum formula

$$\sum_{n=1}^{\infty} f(n) = \int_0^{\infty} f(x) dx, \tag{3.283}$$

we have:

$$r = \int_0^{\infty} \exp(-\beta x) a(x) dx. \tag{3.284}$$

or using the mean value theorem:

$$r = \exp(-\beta \bar{x}). \tag{3.285}$$

We also used the integral form of the normalization condition (3.271):

$$\int_0^{\infty} a(x) dx = 1. \tag{3.286}$$

We did not specify any specific laws of a scattering event in the derivation of Eq. (3.277). Thus, Eq. (3.285) can be applied in a much broader context than just scattering of photons by particles in clouds. The comparison of Eqs. (3.277) and (3.285) shows us that $\bar{x} \to \bar{n}$ as $\beta \to 0$. However, generally speaking, $\bar{x} \neq \bar{n}$. This is due to the differences $\sigma_2 = \overline{n^2} - \bar{n}^2$, $\sigma_3 = \overline{n^3} - \bar{n}^3$. and so on, which are not necessary to be exact zeros.

The problem we face now is the determination of the parameter $\bar{x}$. For this we will use the well-known asymptotic result of the radiative transfer theory derived above (see Table 3.4):

$$r = 1 - \sqrt{\frac{16\beta}{3(1-g)}}, \tag{3.287}$$

which is valid as $\beta \to 0$.

Eq. (3.285) takes the following form as $\beta \to 0$:

$$r = 1 - \beta \bar{x}. \tag{3.288}$$

So, comparing Eqs. (3.287) and (3.288) we have

$$\overline{x} = \frac{4}{k}, \tag{3.289}$$

where $k = \sqrt{3(1-g)\beta}$ is the diffusion exponent of the radiative transfer theory. Therefore, it follows:

$$r = \exp\left(-4\sqrt{\frac{1-\omega_0}{3(1-g)}}\right). \tag{3.290}$$

We notice that the combination of local optical characteristics, given by

$$y = 4\sqrt{\frac{1-\omega_0}{3(1-g)}} \tag{3.291}$$

completely determines the spherical albedo. The value of $y = 4s/\sqrt{3}$ can be also measured experimentally ($y = \ln(1/r)$).

By analogy, relationships similar to Eq. (3.290) must be valid also for other asymptotic parameters. This allows to derive the following relationships using Table 3.4 (Zege et al., 1991):

$$R_{\infty}(\xi,\eta,\varphi) = R_{0\infty}(\xi,\eta,\varphi)\exp(-yu_0(\xi)u_0(\eta)R_{0\infty}^{-1}(\xi,\eta,\varphi)), \tag{3.292.a}$$

$$r_{d\infty}(\xi) = \exp(-yu_0(\xi)), \tag{3.292.b}$$

$$M = 1 - \exp(8k/(3(1-g)), \quad N = \exp(-\delta k/(1-g)), \tag{3.292.c}$$

$$Mu(\xi)u(\eta) = (1 - \exp(-8k/3(1-g)))u_0(\xi)u_0(\eta). \tag{3.292.d}$$

We obtain using these relationships:

$$R(\xi,\eta,\varphi) = R_{0\infty}\exp(-yu(\xi,\eta,\varphi)) - t\,\exp(-x-y)u_0(\xi)u_0(\eta), \tag{3.293}$$

$$T(\xi,\eta) = tu_0(\xi)u_0(\eta), \tag{3.294}$$

where we introduced a new parameter $x = k\tau$. The global transmittance t is given by:

$$t = \frac{\sinh y}{\sinh(\alpha y + x)}. \tag{3.295}$$

We can also derive the analytical results for the plane and spherical albedos and the diffuse transmittance. Correspondent equations are summarized in Table 3.5.

Equation (3.293) transforms into Eq. (3.239) [and also Eq. (3.294) transforms into Eq. (3.240)] as $\beta = 0$. However, Eq. (3.292a) unlike Eq. (3.239) allows to consider absorbing media as well. It is important that no new angular functions arise in Eq. (3.292a) as compared to Eq. (3.239). This is in contrast with Eq. (3.152), where

Table 3.5. Radiative transfer characteristics in the framework of the exponential approximation ($r_{d\infty}(\xi) = \exp(-yu_0(\xi))$, $u_0(\xi) = \frac{3}{7}(1 + 2\xi)$, $r_\infty = e^{-y}$, $x = k\tau$, $y = k\tau$, $k = \sqrt{3(1-\omega_0)(1-g)}$, $\alpha = 1.07$).

Radiative characteristic	Symbol	Equation
Plane albedo	$r_d(\xi)$	$r_{d\infty}(\xi) - (r_\infty - r)u_0(\xi)$
Spherical albedo	r	$r_\infty - t\exp(-x-y)$
Diffuse transmittance	$t_d(\xi)$	$tu_0(\xi)$
Global transmittance	t	$\dfrac{\sinh(y)}{\sinh(x+\alpha y)}$

parameters and functions have an implicit and complex dependence on the probability of photon absorption, β. Equation (3.293) can be used for the rapid estimations of light reflection from cloudy media and also for the speeding up cloud retrieval algorithms (Kokhanovsky et al., 2003).

The range of applicability of the exponential approximation (3.293) with respect to cloud optical thickness can be extended using correction terms derived from the numerical solution of the radiative transfer equation. In particular, we find that the accuracy of Eq. (3.293) for cloudy media can be increased using the following substitutions: $u_0 \to u_0(1 - 0.05y)$ (in the argument of the exponent in Eq. 3.293), $t \to t - \Delta$, where

$$\Delta = \frac{a + b\mu\xi + c\mu^2\mu_0^2}{\tau^3}\exp(x) \tag{3.296}$$

and $a = 4.86$, $b = -13.08$, $c = 12.76$. Therefore, the final equation for the cloud reflection function can be written as

$$R(\eta, \xi, \varphi, \tau) = R_{0\infty}\exp(-y(1-0.05y)u_0(\eta, \xi, \varphi)) - (t - \Delta)\exp(-x-y) \times u_0(\eta)u_0(\xi). \tag{3.297}$$

Equation (3.297) is called the Modified Exponential Approximation (MEA). We show the accuracy of the MEA given by Eq. (3.297) in Figs. 3.10 and 3.11 for the nadir observation conditions, the solar zenith angle 60° and wavelengths 865 and 2130 nm. These wavelengths are often used in cloud retrieval techniques. Note that the single-scattering albedo is equal to 1.0 and 0.9872 at these wavelengths, respectively. The asymmetry parameter is 0.8435 for the smaller wavelength. It is 0.8054 for the wavelength 2130 nm. Exact data shown in Fig. 3.10 are obtained using the vector radiative transfer code based on the discrete ordinate approach and thoroughly tested against tabular results presented by Siewert (2000). It follows that the accuracy of the approximation is better than 6% for the cloud optical thickness $\tau \geq 4$ in the case considered.

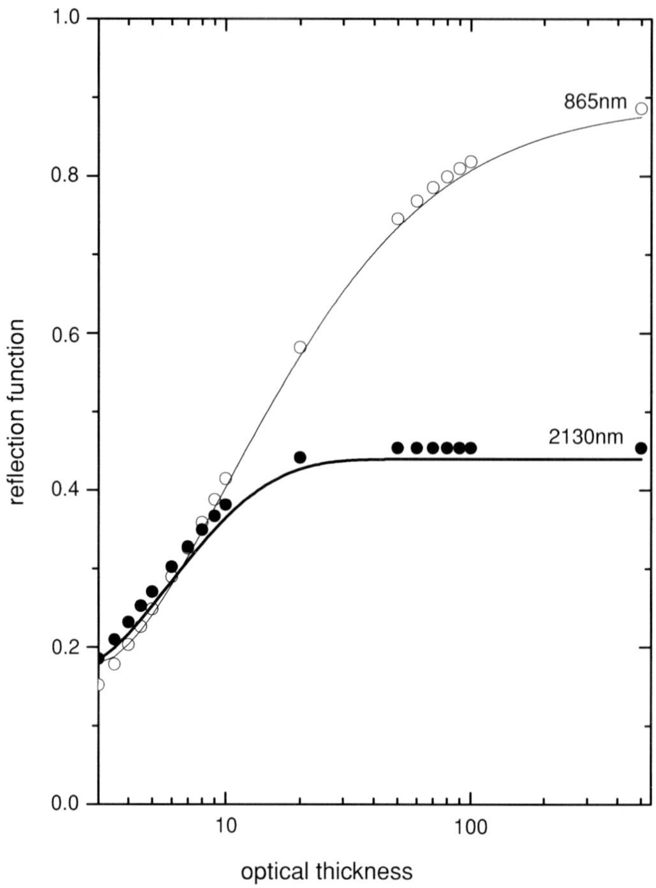

Fig. 3.10. Dependence of reflection function on the optical thickness at wavelengths 865 and 2130 nm calculated using the Cloud C1 PSD model. Solid lines give the approximate results according to exponential approximation and points are obtained solving the exact radiative transfer equation using SCIATRAN (Rozanov et al., 2005).

Calculations performed for other angles and cloud optical thickness show (see Figs. 3.12–3.18) that the accuracy only weakly depends on the geometry, providing that grazing observation and illumination conditions are excluded (Kokhanovsky and Rozanov, 2003). It means that the top-of-atmosphere reflectance over cloudy scenes can be accurately modelled in the framework of the MEA (even as compared to the vector radiative transfer model). It follows from Fig. 3.11 that the accuracy of the MEA could be increased if the exact result for the reflection function of a semi-infinite layer is used in calculations. Note that we used in Eq. (3.297) the following simple formula valid for the nadir observation conditions only (Kokhanovsky,

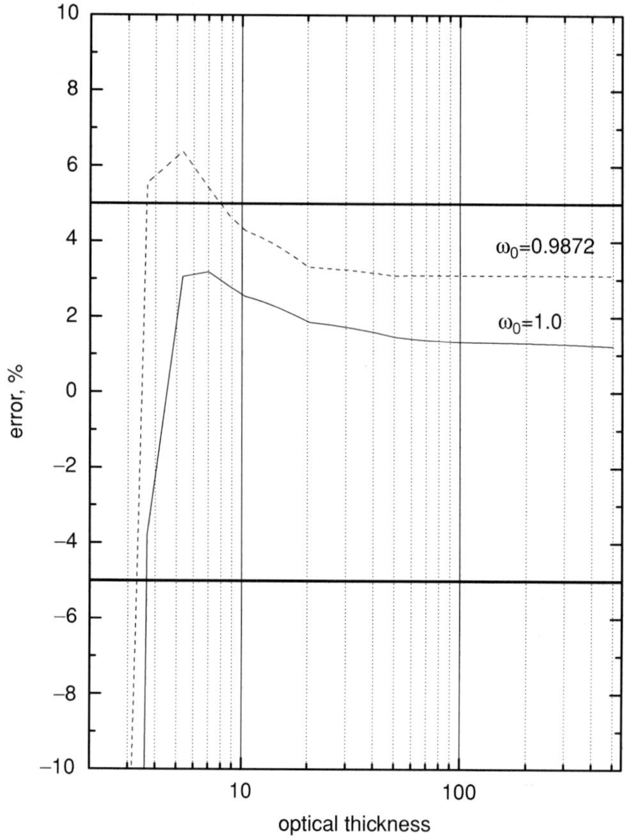

Fig. 3.11. Errors of exponential approximation calculated using data shown in Fig. 3.10. Upper curve corresponds to the wavelength 2130 nm, where single-scattering albedo is equal to 0.9872. Lower curve corresponds to a nonabsorbing cloud at $\lambda = 865$ nm.

2002):

$$R_{0\infty}(\eta, \xi, \varphi) = \frac{0.37 + 1.94\xi}{1 + \xi}. \tag{3.298}$$

The accuracy of Eq. (3.298) can be further increased adding the function $F = 0.25p(1 - \arccos(\mu_0))$ to the nominator.

The analysis of Figs. 3.12–3.18 shows that the exponential approximation is applicable for $\tau \geq 5$ and most solar angles for the nadir observation. Also these figures show main dependencies related to light reflection from cloud media, like the increase in reflection with τ (see Fig. 3.12) and a high sensitivity of reflectances in the infrared to the size of particles (see Fig. 3.17). This sensitivity is almost negligible in the visible (see Fig. 3.12).

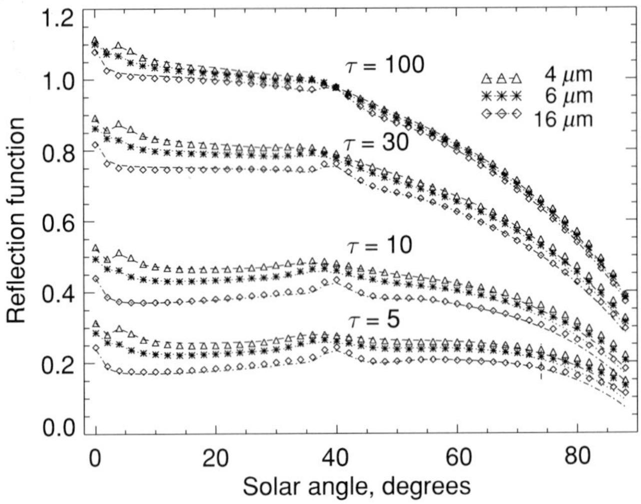

Fig. 3.12. Dependence of cloud reflection on the solar zenith angle at $\tau = 5, 10, 30, 100$ and $a_{ef} = 4, 6, 16$ μm (the gamma PSD with $\mu = 6$) for nadir observations. Symbols show exact calculations and lines are obtained using the approximate result at $\lambda = 0.55$ μm.

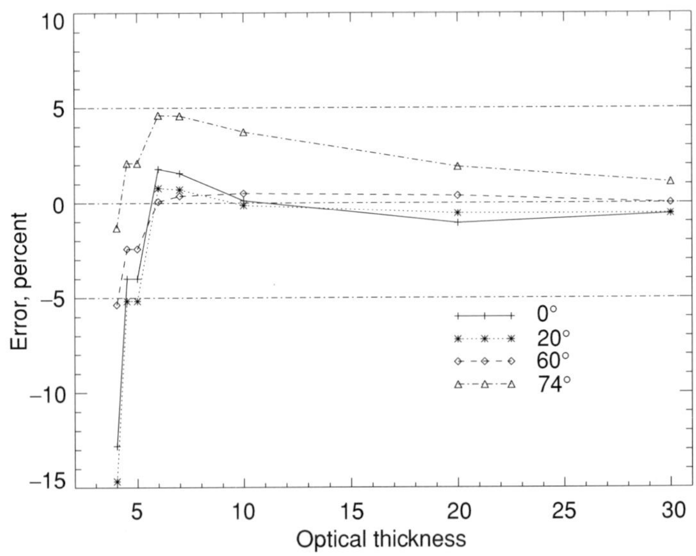

Fig. 3.13. Relative errors of the approximation as functions of the optical thickness for several solar zenith angles calculated using data shown in Fig. 3.12 at $a_{ef} = 6$ μm.

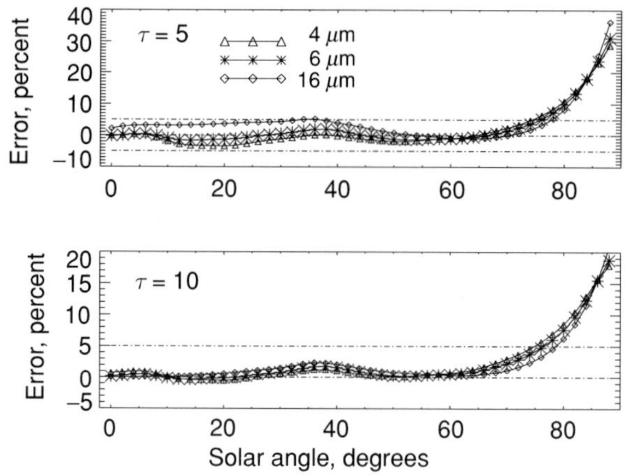

Fig. 3.14. Relative errors of the approximation as functions of the solar zenith angle at the optical thickness 5 and 10 and $a_{ef} = 4,\ 6,\ 16$ μm calculated using data shown in Fig. 3.12.

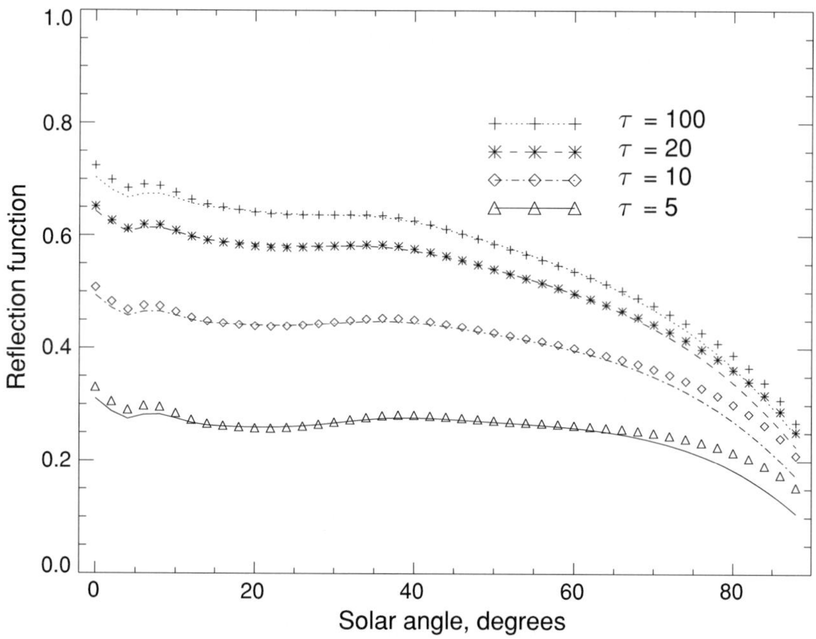

Fig. 3.15. Dependence of cloud reflection on the solar zenith angle at $\tau = 5, 10, 20, 100$ and $a_{ef} = 6$ (the gamma PSD with $\mu = 6$) for nadir observations at $\lambda = 1.55$ μm. Symbols show exact calculations and lines are obtained using the approximate result.

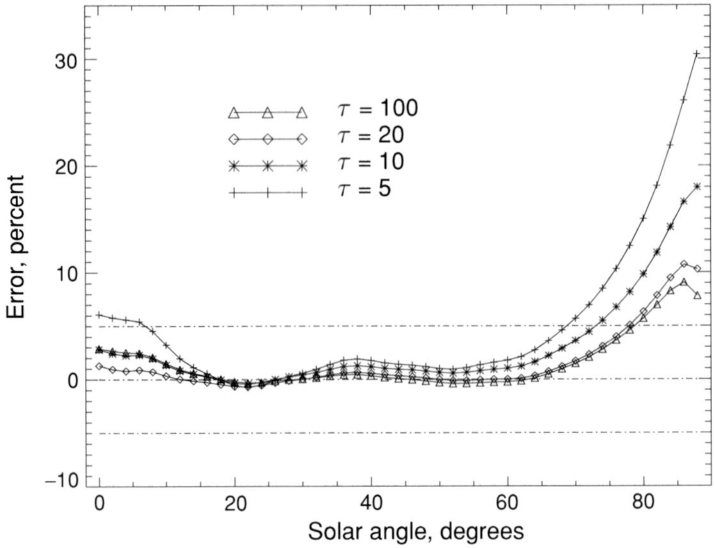

Fig. 3.16. Relative errors of the approximation as functions of the solar zenith angle at the optical thickness 5, 10, 20, and 100 and $a_{ef} = 6$ µm calculated using data shown in Fig. 3.15.

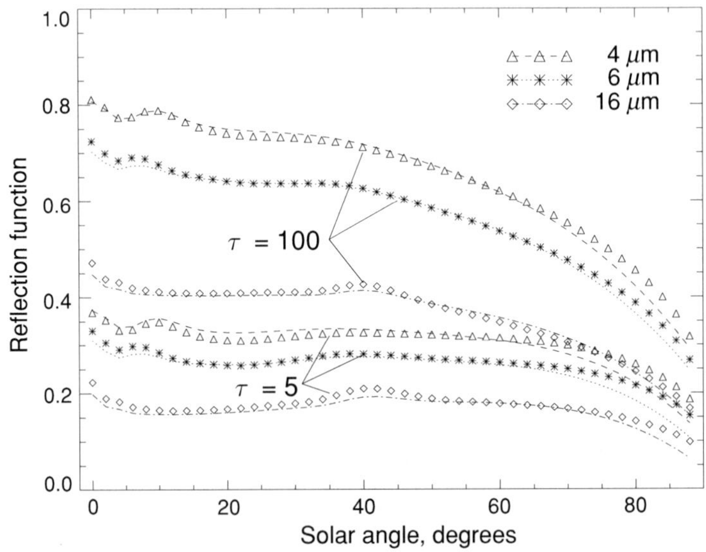

Fig. 3.17. Dependence of cloud reflection on the solar zenith angle at $\tau = 5$, 100 and $a_{ef} =$ 4, 6, 16 µm (the gamma PSD with $\mu = 6$) for nadir observations at $\lambda = 1.55$ µm. Symbols show exact calculations and lines are obtained using the approximate result.

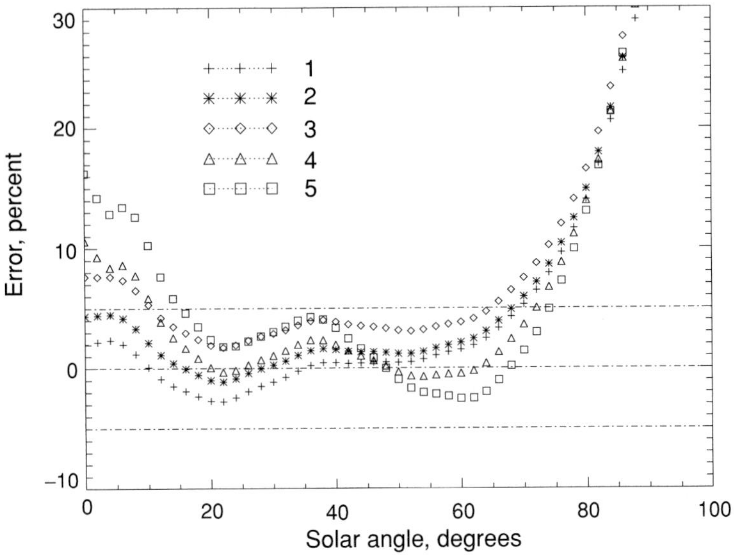

Fig. 3.18. Errors of the approximation at various values of the single-scattering albedo as functions of the solar zenith angle at the nadir observation and $\tau = 5$. The curves 1–5 were obtained using phase functions and single-scattering albedos derived from the Mie theory for the C1 PSD at the wavelength 1.55 μm assuming that the refractive index is equal to $1.3109 - i\chi$ with $\chi = 0.00001(1)$, $0.00005(2)$, $0.0001(3)$, $0.0005(4)$, and $0.001(5)$. The values of $\beta = 1 - \omega_0$ were equal to 0.0005(1), 0.0034(2), 0.0047(3), 0.0226(4), and 0.0435(5).

3.6.5.2 *The radiative transfer in the gaseous absorption band*

The exponential approximation presented above can be easily extended to account for the gaseous absorption. Then one should use the following substitutions in equations given above: $\tau \to \tau + \tau_g$, $\beta \to (\sigma_{abs} + \sigma_{abs,g})/(\sigma_{ext} + \sigma_{abs,g})$, where the subscript '$g$' relates the correspondent value to the gaseous absorption process. The phase function does not need to be modified because we ignore molecular scattering. This could easily be accounted for if necessary. However, we account for the additional light absorption in the atmosphere above a cloud. Therefore, it follows for the cloud reflection function $\bar{R}$ in the gaseous absorption band: $\bar{R} = T_1 R T_2$, where we omitted arguments for the sake of simplicity. The value of R is given by Eq. (3.297) and $T_j = \exp(-m_j \tau_{abs})$, $j = 1, 2$, where $m_1 = 1/\xi$, $m_2 = 1/\eta$, and

$$\tau_{abs} = \sum_{i=1}^{N} \int_{z_1}^{z_2} C_{abs,i}(z) \varsigma_i(z) dz, \qquad (3.299)$$

where $C_{abs,i}$ is the ith gas absorption cross section, N is the total number of gases present and $\varsigma_i(z)$ is the concentration of the ith gas at a given height. The integration extends from the upper cloud boundary position z_1 to the height of the optical instrument z_2. The accuracy of the MEA for the gaseous absorption band can be increased if the single-scattering contribution in the signal from the atmospheric layer above the cloud R_s (Kokhanovsky and Rozanov, 2004) is also taken into account. Then it follows:

$$\bar{R} = T_1 R T_2 + R_s \tag{3.300}$$

We checked the accuracy of Eq. (3.300) by performing exact calculations using the radiative transfer code SCIATRAN (Rozanov et al., 2005) for the oxygen absorption A-band located at wavelengths 758–768 nm. The atmospheric model used in calculations coincides with that described by Kokhanovsky and Rozanov (2004). The values of $\bar{R}$ are averaged with respect to the Gaussian instrument response function with the half-width of 0.225 nm. The absorption by the oxygen was accounted for by using the HITRAN 2000 (Rothman et al., 2003) database in conjunction with the correlated k-distribution approximation (Kokhanovsky and Rozanov, 2004). To increase the accuracy of the model, we accounted for light scattering and absorption below the cloud layer using the approximate technique developed by Kokhanovsky and Rozanov (2004). Results are given in Fig. 3.19

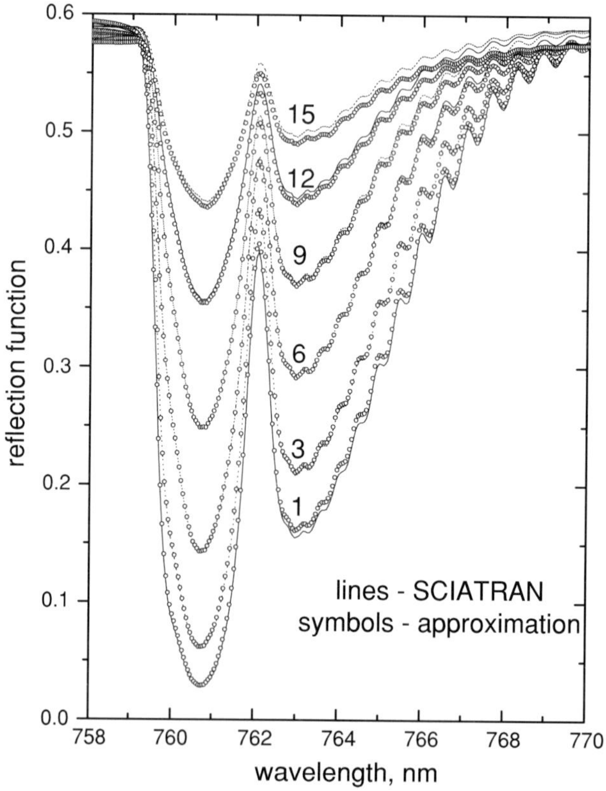

Fig. 3.19. Oxygen A-band cloud reflectance spectrum calculated using the approximate theory and SCIATRAN solver at $\tau = 20$, the cloud geometrical thickness equal to 1 km and the cloud top altitudes 1, 3, 6, 9, 12, and 15 km for the nadir observation and the solar zenith angle 60°.

for different cloud top heights. It follows from the analysis of the data presented that the accuracy of approximate calculations is better than 5% (see Fig. 3.20) in most cases. The errors increase for low clouds having larger values of τ due to the simplicity of our model, which accounts for the cloud – upper atmospheric layer interaction in a first coarse approximation only (Kokhanovsky and Rozanov, 2004). This interaction becomes more important for lower thick clouds (see Fig. 3.20).

The parameterization developed here is useful in cloud top altitudes retrievals from airborne and satellite-based optical sensors. We see that the depth of the absorption line depends on cloud altitude. Although other gaseous absorption bands (see Fig. 3.21) can be used for this purpose, the vertical contribution of the gaseous concentration must be known in advance. This is a difficult task, e.g., for water vapour, which has a lot of absorption bands in the optical range. However, unlike oxygen, this gas does not belong to the category of well-mixed gases with stable vertical profiles and concentrations.

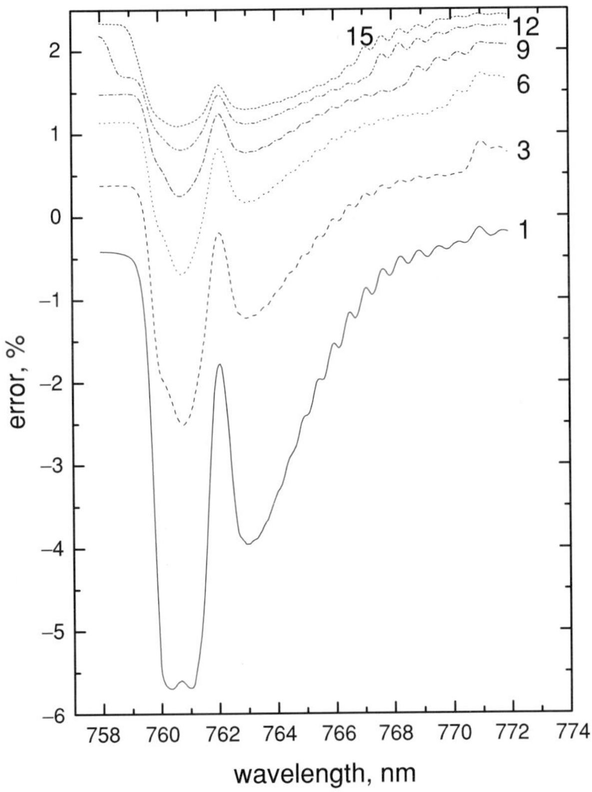

Fig. 3.20. Relative errors of the approximation obtained using Fig. 3.18.

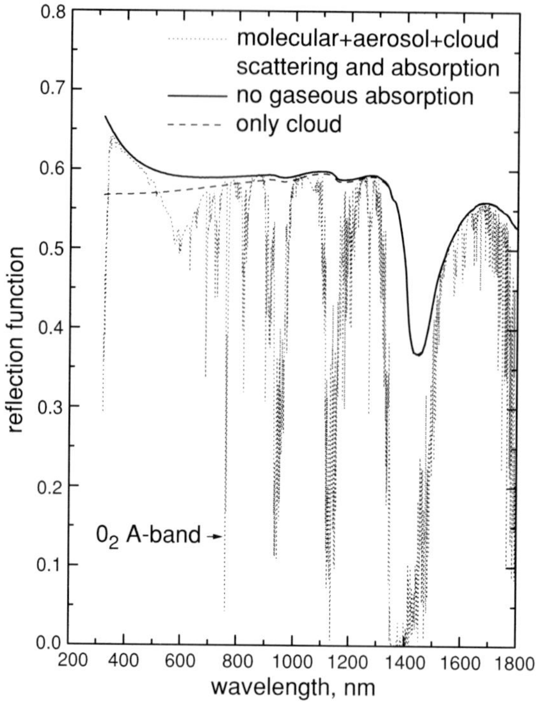

Fig. 3.21. The cloud reflection function calculated with SCIATRAN (Rozanov et al., 2005) accounting for molecular and aerosol scattering and absorption using forward model described by Kokhanovsky and Rozanov (2004) at the nadir observation, the solar incidence angle 60°, and $\tau = 20$. Results of calculations for the artificial case with no gaseous absorption are also shown. Dashed line corresponds to the artificial case, when only cloud presents in atmosphere. Rayleigh scattering enhances the reflection function at short wavelengths. The decrease in the reflection function (dashed line) in UV is due to the gaseous absorption.

Note that approximations for optically thick layers discussed here are of great importance for cloud optics, because such clouds are very common (see Fig. 3.22).

3.6.6 Polarization of Light by Optically Thick Clouds

Equations for optically thick clouds presented in the previous section can be generalized to account for polarization. Correspondent equations for azimuthally averaged reflection $\hat{R}$ and transmission $\hat{T}$ matrices were obtained by Domke (1978a,b). They have the following forms for isotropic symmetric light scattering media:

$$\hat{R}(\xi, \eta) = \hat{R}_\infty(\xi, \eta) - N\hat{T}(\xi, \eta)\exp(-k\tau), \tag{3.301}$$

$$\hat{T}(\xi, \eta) = \frac{M\exp(-k\tau)}{1 - N^2\exp(-2k\tau)}\vec{K}(\eta)\vec{K}^T(\xi), \tag{3.302}$$

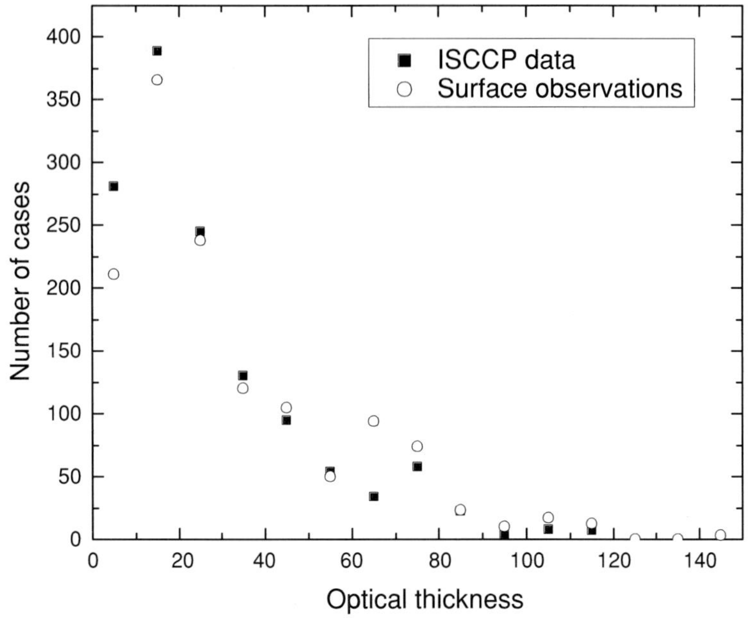

Fig. 3.22. Frequency of a given cloud optical thickness as observed using ground and satellite observations. (Trishchenko and Liu, 2001)

where only two-dimensional matrices and vectors are involved. Other components of generally four-dimensional matrices and vectors vanish due the azimuthal averaging. Note that this is also the case for a normal illumination of the scattering layer. Then the azimuth does not enter theory at all. Here $\hat{R}_{\infty}(\xi, \eta)$ is the azimuthally averaged reflection matrix of a semi-infinite medium with the same optical characteristics as a finite slab under study. One can also introduce the vector $\vec{\Theta}(\mu)$, which describes the intensity and degree of light polarization in deep layers of a semi-infinite scattering medium (in a so-called asymptotic regime, when the intensity and polarization angular distributions are symmetrical with respect to the normal to a scattering layer and exponentially decrease with the depth $\sim \exp(-k\tau)$). Functions $\hat{R}_{\infty}(\xi, \eta)$ and $\vec{\Theta}(\eta)$ determine all parameters in the equations given above. The explicit equations for their calculation are discussed by Domke (1978a,b).

We see, therefore, that intensity and polarization characteristics of reflected and transmitted light for optically thick turbid media are determined by the reflection matrix of a semi-infinite layer and the angular distribution of the light intensity and polarization in deep layers of the same medium. This reduction of a problem for a finite optical thick slab to the case of a semi-infinite medium is of a general importance for the radiative transfer theory. Note that the matrix $\hat{R}_{\infty}$ and vector $\vec{\Theta}$ are obtained from solutions of the well-known integral equations, which can be found elsewhere (van de Hulst, 1980; de Rooij, 1985).

Equations (3.301) and (3.302) are valid only for the azimuthally averaged matrices. In practice, however, measurements are performed for a fixed azimuth. The transmission matrix is azimuthally independent in the case of optically thick layers. The azimuthal dependence in the reflected light disappears in some specific cases (e.g., for the case of normal illumination of an isotropic light scattering plane-parallel slabs).

Equations (3.301) and (3.302) are simplified for nonabsorbing media. Then it follows:

$$\hat{R}(\xi, \eta) = \hat{R}_{0\infty}(\xi, \eta) - \hat{T}(\xi, \eta), \tag{3.303}$$

$$\hat{T}(\xi, \eta) = \frac{4}{3(\tau + 2q_0)(1 - g)} \vec{K}_0(\xi) \vec{K}_0^T(\eta), \tag{3.304}$$

where

$$q_0 = \frac{2}{1 - g} \int_0^1 d\eta \eta^2 \vec{K}_0^T(\eta) \vec{j}. \tag{3.305}$$

Here

$$\vec{j} = \begin{pmatrix} 1 \\ 0 \end{pmatrix} \tag{3.306}$$

is the unity vector,

$$g = \frac{1}{4} \int_0^\pi p(\theta) \sin 2\theta d\theta \tag{3.307}$$

is the asymmetry parameter,

$$\vec{K}_0(\eta) = \frac{3}{4} \left[\eta + 2 \int_0^1 d\xi \xi^2 \hat{R}_{0\infty}(\xi, \eta) \right] \vec{j} \tag{3.308}$$

and $\hat{R}_{0\infty}(\mu, \mu_0)$ is the azimuthally averaged reflection matrix of a semi-infinite nonabsorbing medium. This matrix is completely determined by the phase matrix $\hat{P}$, introduced above. It does not depend on the single-scattering albedo and optical thickness by definition. Clearly, the first component of the vector $\vec{K}_0$ coincides with the escape function u_0 discussed above.

Eqs. (3.101), (3.102) are simple in form. However, they can be used only if auxiliary functions and parameters are known. Their calculations, however, can be quite a complex procedure.

However, it appears that for weakly absorbing media, when single-scattering albedo $\omega_0 = \sigma_{sca}/\sigma_{ext}$ is close to one, simplifications are possible. Then it follows (Kokhanovsky, 2003a,b):

$$\hat{R}(\xi, \eta) = \hat{R}_{0\infty}(\xi, \eta) \exp(-y\hat{D}(\xi, \eta)) - \hat{T}(\xi, \eta) \exp(-x - y), \tag{3.309}$$

$$\hat{T}(\xi, \eta) = t\vec{K}_0(\xi) \vec{K}_0^T(\eta). \tag{3.310}$$

where

$$x = k\tau, y = 4\sqrt{\frac{1-\omega_0}{3(1-g)}}, k = \sqrt{3(1-\omega_0)(1-g)},$$
$$\hat{D}(\xi,\eta) = \hat{R}_{0\infty}^{-1}(\xi,\eta)\vec{K}_0(\xi)\vec{K}_0(\eta),$$

$t = \dfrac{\sinh y}{\sinh(x+\alpha y)}$ is the global transmittance of a scattering layer, $\alpha = \frac{1}{2}\int_0^1 u_0(\eta)\eta^2 d\eta \approx 1.07$ and $\hat{R}_{0\infty}(\mu,\mu_0)$ is the reflection matrix of a semi-infinite nonabsorbing layer with the same phase matrix as an absorbing layer of a finite thickness under study. The two-dimensional vector $\vec{K}_0(\mu)$ describes the polarization and intensity of light in the Milne problem for nonabsorbing semi-infinite media (Wauben, 1992). The components $K_{01}(\mu)$ and $K_{02}(\mu)$ of this vector were calculated by Chandrasekhar (1950) for Rayeigh particles ($g = 0$) and by Wauben (1992) for spherical particles with the refractive index $n = 1.44$ and the gamma particle size distribution (1.5) with $\mu = 11.3, a_0 = 0.83$ μm. The wavelength λ was equal to 0.55 μm. Note, that the model of spheres with $\mu = 11.3, r_0 = 0.83$ μm and $n = 1.44$ is generally used to characterize particles in clouds on Venus (Hansen and Hovenier, 1974). It follows for the effective size a_{ef}, the effective variance Δ_{ef}, and the asymmetry parameter g, respectively, in this case: $a_{ef} = 1.05$ μm, $\Delta_{ef} = 0.07$ μm, $g = 0.718$. It was found that the ratio $p_l = -K_{02}/K_{01}$, which gives the degree of polarization for transmitted light is very low. It changes from zero to 1.2% while the escape angle changes from 0 till 90°. Note that for Rayleigh scattering we have a change from 0 till 11.7% for the same conditions. This means that light transmitted by thick clouds is almost unpolarized. It is possible to understand this on general grounds. Indeed, the polarization of unpolarized solar light occurs due to single-scattering events. Multiple light scattering leads to an increase of entropy and the reduction of initial polarization arising in single-scattering events.

Note that the ellipticity is equal to zero in this case and that $P_l \leq 0$, which means that light is polarized in the plane perpendicular to the meridional plane.

Formulae (3.309) and (3.310) can be simplified for nonabsorbing media (y= 0):

$$\hat{R}(\xi,\eta) = \hat{R}_{0\infty}(\xi,\eta) - \hat{T}(\xi,\eta), \tag{3.311}$$
$$\hat{T}(\xi,\eta) = t\vec{K}_0(\xi)\vec{K}_0^T(\eta), \tag{3.312}$$

where

$$t = \frac{1}{\alpha + \frac{3}{4}\tau(1-g)} \tag{3.313}$$

is the global transmittance.

Let us apply Eq. (3.311) to a particular problem, namely, to the derivation of a relation between the spherical albedo $r = 1 - t$ and the degree of polarization of reflected light $p_l(\eta)$ at the illumination along the normal to the scattering layer

($\xi = 1$) by a wide, unidirectional unpolarized light beam. The value of $p_l(\mu)$ is given simply by $-R_{21}(1, \mu)/R_{11}(1, \mu)$ in this case. Thus, it follows from Eq. (3.311):

$$p_l(\eta) = \frac{p_{l\infty}(\eta)}{1 - (1 - r)\mathrm{N}(\eta)}, \tag{3.314}$$

where

$$\mathrm{N}(\eta) = \frac{u_0(1)u_0(\eta)}{R_{0\infty}(1, \mu)} \tag{3.315}$$

$$p_{l\infty}(\eta) = -\frac{R_{\infty 21}(1, \eta)}{R_{\infty}(1, \eta)} \tag{3.316}$$

and we accounted for the equality: $K_{02}(1) = 0$.

Our calculations show that the value of $\mathrm{N}(\eta)$ is close to 1 for most of observation angles, which implies the inverse proportionality between the brightness of a turbid medium and the degree of polarization of reflected light ($rp_l \approx p_{l\infty}$). The accuracy of Eq. (3.314) is shown in Fig. 3.23.

This inverse proportionality between the spherical albedo r and the degree of polarization p_l was discovered experimentally by Umow (1905). Equation (3.314) can be considered as a manifestation of this important law, which has important applications in reflectance spectroscopy (Hapke,1993).

Equation (3.314) is easily generalized to account for the absorption of light in a medium using the exponential approximation described above. Namely, it follows:

$$p_l(\eta) = \frac{p^*_{l\infty}(\eta)}{1 - \mathrm{N}^*(\eta)t\exp(-x - y)}, \tag{3.317}$$

where

$$t = \frac{\sinh y}{\sinh(x + \alpha y)} \tag{3.318}$$

and

$$\mathrm{N}^*(\eta) = \frac{u_0(1)u_0(\eta)}{R^*_{\infty}(1, \eta)}. \tag{3.319}$$

Values of $p^*_{l\infty}(\eta)$ and $R^*_{\infty}(1, \eta)$ represent the degree of polarization and reflection function of a semi-infinite weakly absorbing medium at the nadir illumination. Note, that Eq. (3.317) can be written in the following form:

$$p_1(\eta) = c(\eta, \tau)p^*_{l\infty}(\eta), \tag{3.320}$$

where

$$c(\eta, \tau) = \frac{1}{1 - \mathrm{N}^*(\eta)t\exp(-x - y)} \tag{3.321}$$

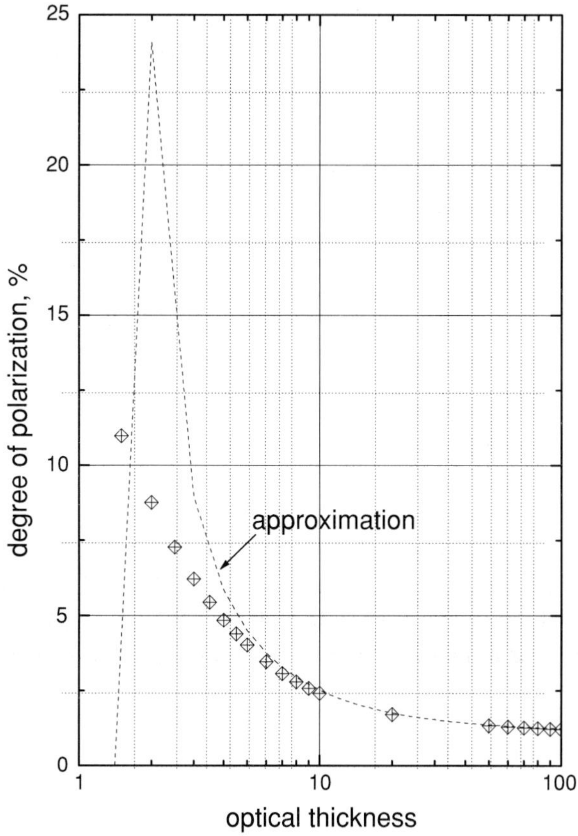

Fig. 3.23. Degree of linear polarization of light reflected from a cloudy atmosphere as the function of the cloud optical thickness at $a_{ef} = 6\mu\text{m}$ (the gamma PSD with $\mu = 6$) for nadir observations at $\lambda = 0.55\mu\text{m}$. The solar zenith angle is equal to $60°$. Both exact (symbols) and approximate (lines) results are shown.

can be interpreted as the polarization enhancement factor, which is solely due to a finite cloud depth. It follows for semi-infinite layers that the transmittance t is equal to zero and $c = 1$ as it should be. Also it follows from Eq. (3.320) that zeroes of polarization curves for semi-infinite and optically thick finite layers almost coincide, which is supported by numerical calculations with the radiative transfer code (see Fig. 3.24). This is due to the fact that the function $N^*(\eta)$ only weakly depends on the angle.

Multiple light scattering fails to produce the polarization of incident unpolarized light. It only diminishes the polarization of singly scattered light. Thus, the angles where polarization is equal to zero for semi-infinite layers are almost equal to those for the case of single light scattering.

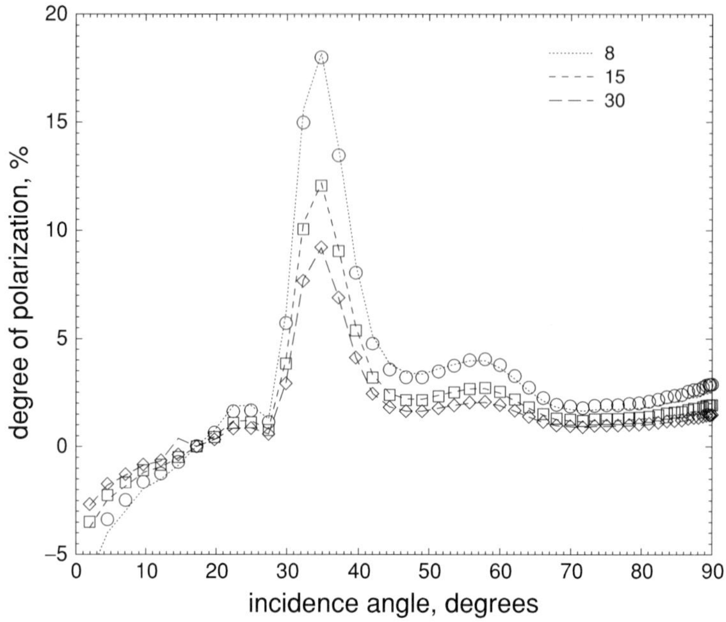

Fig. 3.24. Degree of linear polarization of light reflected from a cloudy atmosphere as the function of the solar zenith angle at the cloud optical thickness 8, 15, and 30, $a_{ef} = 6$ μm (the gamma PSD with $\mu = 6$) for nadir observations at $\lambda = 0.55$ μm. The solar zenith angle is equal to 60°. Both exact (symbols) and approximate (lines) results are shown.

The exact calculations for the case of a cloudy atmosphere takes much more computer time as compared to molecular scattering. This is not related to the larger optical thickness of clouds, which can reach 100 and more, but solely due to the peaked phase functions of water clouds. This means that one must account for a lot of Legendre polynomials in the correspondent expansions. This also leads to a large number N of qaussian quadrature points required to solve this problem ($N \approx L/2$). This obstacle can be avoided using the so-called delta-M approximation (Nakajima and Tanaka, 1988; Min and Duan, 2004).

We show the cloud reflection function calculated using exact vector code in Fig. 3.25a. As one might expect, the reflection function increases with τ. It reaches an asymptotic value for a semi-infinite cloud at $\tau \approx 11$ for wavelength 2130 nm. For a nonabsorbing wavelength the asymptotic value is reached at much higher values of τ ($\tau \sim 500$ at wavelength 865 nm, see Fig. 3.25a). The single-scattering approximation (SSA) works quite well for $\tau \approx 0.03$ and below. So it can be used for the estimation of scattering characteristics of subvisual Cirrus, but not for most water and ice clouds present in the troposphere. The performance of the SSA for the polarization difference $D = -\pi Q_r/F_0\mu_0$ is much better. It is valid at least

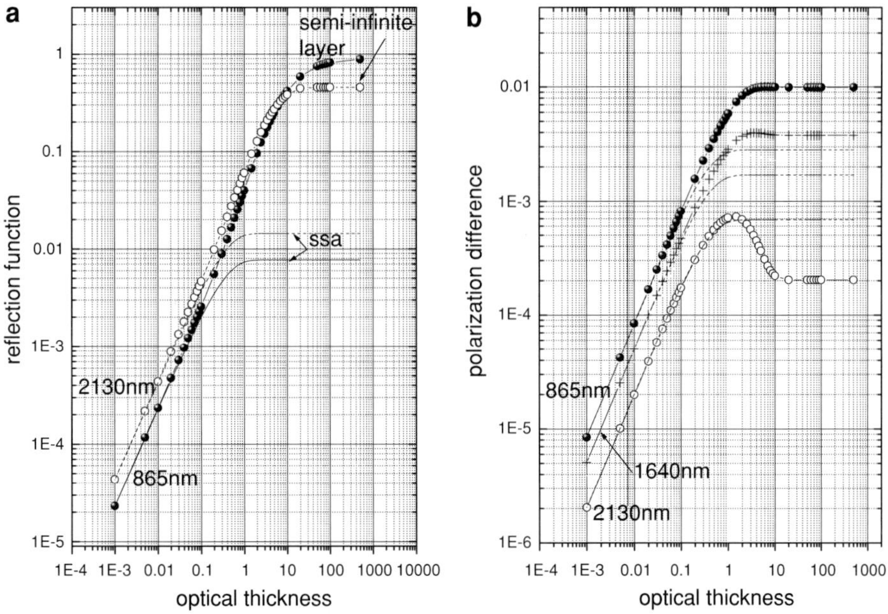

Fig. 3.25. Dependence of the cloud reflection function on the cloud optical thickness at wavelengths 865 and 2130 nm. The results for the semi-infinite layer and the single-scattering approximation are also shown. The solar zenith angle is 60°. The observation zenith angle is 0°. The phase function was calculated using Mie theory for the Deirmendjian's Cloud C1 model (Deirmendjian, 1969). (b) Dependence of the polarization difference D on the cloud optical thickness at wavelengths 865, 1640, and 2130 nm. The results for the semi-infinite layer and the single-scattering approximation are also shown. The solar zenith angle is 60°. The observation zenith angle is 0°. The phase function was calculated using Mie theory for the Deirmendjian's Cloud C1 model.

up to $\tau = 0.1$ (and even up to $\tau = 1.0$ for the wavelength $\lambda = 2130$ nm, see Fig. 3.25b).

There is a peculiarity in the behavior of the function $D(\tau)$ at $\lambda = 2130$ nm shown in Fig. 3.25b. In particular, there is a maximum around $\tau = 1$. Such maxima are not particularly pronounced for other solar angles shown in Fig. 3.26. The existence of the maximum cannot be explained on physical grounds because the difference D does not have a direct physical meaning. Physically based quantities I_l, I_r, R, and also the degree of polarization

$$p = \frac{D}{R} \tag{3.322}$$

all behave in a monotonous way [see, e.g., Fig. 3.27)]. Yet another peculiarity of the function $D(\tau)$ is that it reaches its asymptotical value for a semi-infinite cloud D_∞ at relatively small values of $\tau = \tau_\infty$ ($\tau_\infty = 1 - 4$, depending on the

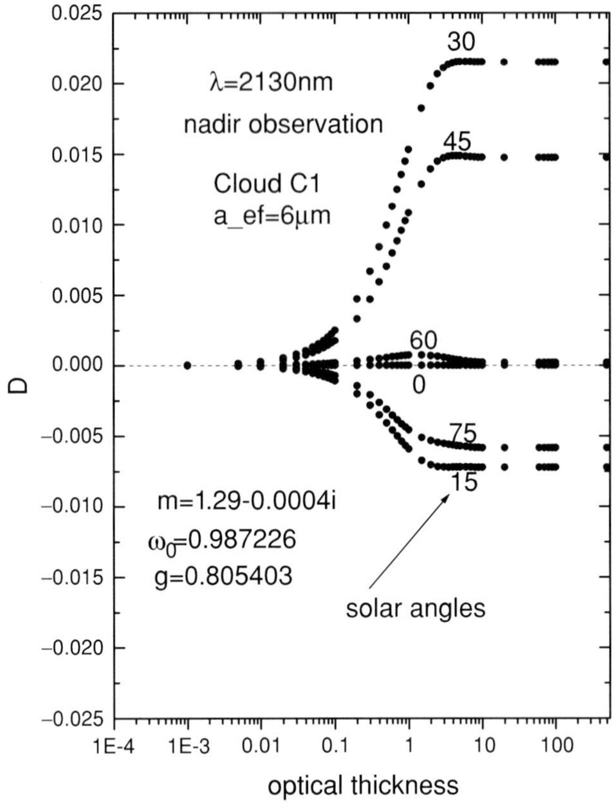

Fig. 3.26. Dependence of D on τ for various solar zenith angles at the wavelength 2130 nm. The refractive index used in calculations of the phase function using Mie theory for the Deirmendjian's Cloud C1 model is also shown. Values of ω_0 and g give correspondent values of the single-scattering albedo and the asymmetry parameter.

wavelength, see Fig. 3.26), which is not the case for R and P (see Figs. 3.25a and 3.27). The optical thickness of clouds is usually larger than τ_∞. It means that the value of D is *a priori* known for a given wavelength and the effective radius of droplets. Such a peculiarity can be used to estimate the sub-pixel cloud fraction K from remote sensing measurements. Indeed, the values of R and D can be presented as

$$R = KR_c + (1 - K)R_a, \quad D = KD_c + (1 - K)D_a, \tag{3.323}$$

for a partially cloudy scene, where the symbol c shows that the correspondent characteristic is related to the cloudy portion of the pixel and a denotes the characteristic of a cloudless atmosphere. It is known that values of R_a and D_a are determined mostly by the molecular scattering in the UV region of the electromagnetic spectrum

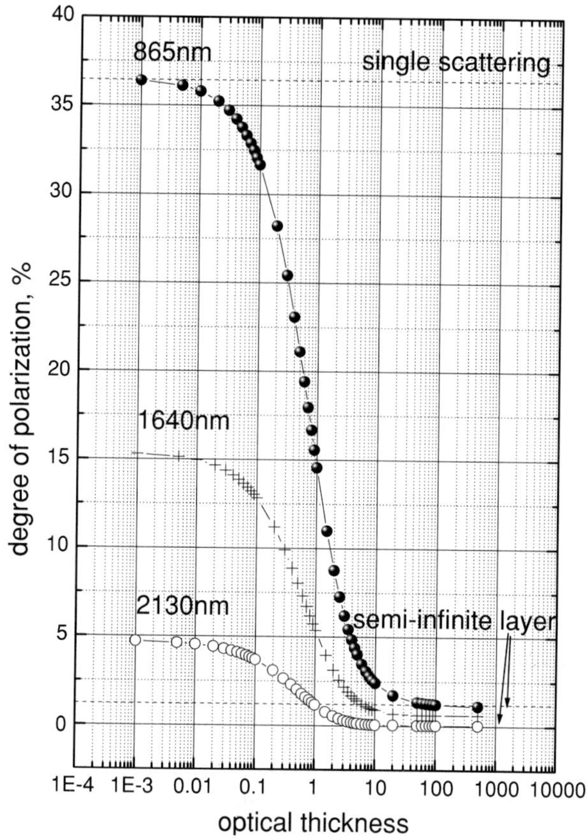

Fig. 3.27. The same as in Fig. 3.25b but for the degree of polarization in percent.

($R_a \to R_m$, $D_a \to D_m$, where m denotes the pure molecular scattering case (no clouds and aerosols)). This contribution is known a priori due to the relative stability of the Rayleigh optical thickness for a given wavelength. Then it follows in the UV:

$$K = \frac{D - D_m}{D_\infty - D_m}, \; R_c = K^{-1}R - (K^{-1} - 1)R_m, \tag{3.324}$$

which allows also to find the cloud optical thickness in a partially cloudy scene. Another interesting possibility arises at geometries, where $D_c = 0$ (and, therefore, $P = 0$). Then it follows from Eq. (3.324): $K = 1 - DD_m^{-1}$. So if the measured value of D is equal to D_m, then we have: $K = 0$. K is equal to one for a completely cloudy atmosphere ($D = 0$ for clouds at the chosen favorable geometry). It follows from Fig. 3.24 that $D_c \approx 0$ at the solar angle 22° and the nadir observation. This corresponds to a scattering angle of 158°. So if one constructs the device capable to

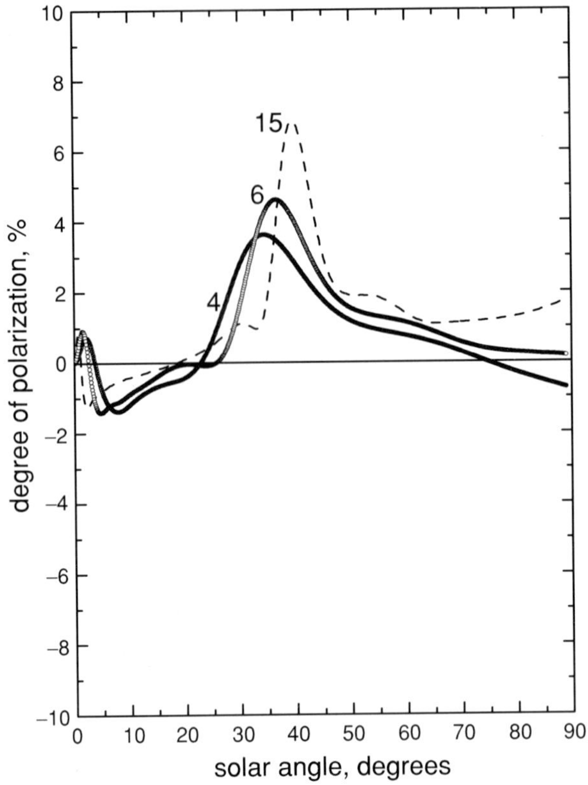

Fig. 3.28. Dependence of the degree of polarization P on the solar zenith angle at the optical thickness 500, nadir observation, and several effective radii of droplets. The phase function was calculated using Mie theory for the Deirmendjian's Cloud C1 model but with different effective radii $a_{ef} = 4$, 6, and 15 μm at the wavelength 865 nm. The maximum corresponds to the size dependent rainbow scattering.

measure D at this scattering angle, one will be able to determine the cloud fraction K quite accurately. The origin of a minimum at $\theta = 158^\circ$ is due to peculiarities of single scattering by water droplets. Note that multiple scattering hardly moves the positions, where the degree of polarization of singly scattered light vanishes (Kokhanovsky, 2003a).

The degree of polarization of light reflected from clouds for several values of a_{ef} is shown in Fig. 3.28. We see that the degree of polarization is more pronounced at the rainbow geometry.

Studies of polarization characteristics of solar light transmitted and reflected by cloudy media have a long history. However, the real burst of research in this area was given by a launch of the POLDER (Polarization and Directionality of Earth

Reflectances) instrument on board of Japanese ADEOS-I and ADEOS-II satellites. The POLDER (Goloub et al., 2000) was able to transmit to the Earth a huge amount of information about polarization characteristics of light reflected from cloudy media, aerosols and underlying surfaces at several wavelengths. Specifically, the first three components of the Stokes vector $\vec{S}_r(I, Q, U, V)$ have been measured for wavelengths λ equal to 443, 670 and 865 nm. There is no doubt that even more advanced polarimeters with wide spectral coverage will appear on board of different satellites in future, which makes further theoretical studies of polarization characteristics of cloudy media extremely important. This is due to potential possibilities for global retrievals of cloud microstructure, the shape of particles and the optical thickness of clouds based on polarization measurements.

Of course, similar information can be obtained from reflected intensity measurements. However, it could well appear that the degree of polarization can be used as a source of additional information about cloud particle size distributions close to the top of a cloud. Indeed, the high proportion of photons scattered from a thin upper layer of a cloud in creating light polarization is quite understandable. Multiply scattered light fluxes from deep layers are hardly polarized at all. Radiative characteristics, on the other hand, represent the cloud as a whole. Thus, the effective radius derived from radiative measurements is an average of large ensembles of possibly very different particle size distributions, presented in different parts of cloudy media.

The polarization characteristics of cloudy media can be studied applying numerical codes, based on the vector radiative transfer equation solution. However, one can also use the fact that cloud fields are optically thick in most cases. This allows to apply asymptotic analytical relations, derived for optically thick disperse media with arbitrary phase functions and absorption. These solutions help us to explain physical mechanisms and main features behind the polarization change due to the increase of the size of droplets or the thickness of a cloud. Analytical solutions also provide an important tool for the simplification of the inverse problem. They can be used, e.g., in studies of the information content of polarimetric measurements (Deschamps et al., 1994).

3.7 Clouds Over Reflective Surfaces

Equations for R and T derived above can be easily generalized to account for the underlying surface reflection.

Light intensity observed in a direction specified by the pair (ϑ, φ) can be considered as composed of two parts: due to the cloud itself (I_1) and due to surface contribution (I_2). The contribution I_2 can be also separated into two terms (I_{21}, I_{22}), namely

$$I_{21} = I_s t(\eta) \tag{3.325a}$$

for the contributions of the surface in the diffused light ($\eta = \cos\vartheta$) and

$$I_{22} = I_s e^{-\tau/\eta} \tag{3.325b}$$

for the contribution of the surface in the direct light.

Summing up, we have:

$$I(\eta, \varphi) = I_1(\eta, \varphi) + I_s t(\eta) + I_s e^{-\tau/\eta}, \tag{3.326}$$

where we assumed that the surface is Lambertian. This means that the upward intensity I_s for the light emerging from the ground surface does not depend on angle. Let us relate I_s to the albedo A of underlying Lambertian surface. For this we note that the upward flux density is

$$F_u = \int_{2\pi} I_s \cos\vartheta \, d\Omega = \int_0^{2\pi} d\varphi \int_0^{\pi/2} d\vartheta \, I_s \cos\vartheta \sin\vartheta = \pi I_s. \tag{3.327}$$

We have for the ideally reflecting Lambertian surface ($A = 1$): $F_u = F_d$ or $F_d = \pi I_s$, where F_d is the downward flux density. F_d is composed of three components: the direct transmission component $F_{dir} = \xi F_0 e^{-\tau/\xi}$, the diffuse transmission component $F_{dif} = \xi F_0 t(\xi)$ and the component coming from the surface but reflected by a scattering layer back to the underlying surface: $F_{ref} = r F_u$, where r is the spherical albedo of a scattering layer under illumination from below. Obviously, for the underlying surface with arbitrary ground albedo A, we have:

$$F_u = A F_d \tag{3.328}$$

and, therefore,

$$\pi I_s = A\left[\xi F_0 \left(t(\xi) + e^{-r/\xi}\right) + \pi r I_s\right]. \tag{3.329}$$

The intensity I_s can be easily found from this equation. Namely, it follows:

$$I_s = \frac{A t^*(\xi) \xi F_0}{\pi(1 - Ar)} \tag{3.330}$$

where

$$t^*(\xi) = t(\xi) + e^{-\tau/\xi} \tag{3.331}$$

is the total transmittance. Therefore, we have (Liou, 2002):

$$I(\eta, \varphi) = I_1(\eta, \varphi) + \frac{A t^*(\eta) t^*(\xi) \xi F_0}{\pi(1 - Ar)} \tag{3.332}$$

or

$$R(\eta, \xi\varphi) = R_b(\eta, \xi\varphi) + \frac{At^*(\eta)t^*(\xi)}{1 - Ar}, \tag{3.333}$$

where $R_b(\eta, \xi, \varphi) \equiv R(\eta, \xi, \varphi)$ at $A = 0$. All functions presented in this equation have been studied in the previous section. A similar simple account for the Lambertian underlying surface can be performed also for the transmitted component. Namely, we have then:

$$I_{tr}(\eta, \varphi) = I_{1tr}(\eta, \varphi) + I_s r_d(\eta), \tag{3.334}$$

where the first component is due to light transmission by a cloud itself and the second component accounts for the reflection of the diffuse light (I_s) coming from the surface, $R(\eta)$ is the plane albedo illumination from below. Finally, we have:

$$I_{tr}(\eta, \varphi) = I_{1tr}(\eta, \varphi) + \frac{At^*(\xi)r_d(\eta)\xi F_0}{\pi(1 - Ar)} \tag{3.335}$$

or for the transmission function:

$$T(\xi, \eta, \varphi) = T_b(\xi, \eta, \varphi) + \frac{Ar_d(\xi)t^*(\eta)}{1 - Ar}, \tag{3.336}$$

where $T_b(\xi, \eta, \varphi) \equiv T(\xi, \eta, \varphi)$ at $A = 0$.

3.8 Vertically Inhomogeneous Clouds

Vertically inhomogeneous disperse media are of frequent occurrence both in nature (e.g., multi-level cloud systems, snow deposited at different times at a given place, terrestrial atmosphere and ocean, biological tissues, etc.) and technological applications (multi-layered painted surfaces, paper, etc.). This explains a great interest in studies of radiative transfer in vertically inhomogeneous media. Recent advances in this area have been summarized by Yanovitskij (1997). A great number of exact (see, e. g., Minin, 1988; Rozanov et al., 2005) and approximate (Germogenova and Konovalov, 1974; Minin, 1988; Melnikova and Vasilyev, 2005) techniques has been developed. At present there is no problem to account for an arbitrary vertical inhomogeneity of a horizontally homogeneous plane-parallel light scattering medium. However, it appears that one needs to perform quite complex numerical calculations with the use of up-to-date computer technology. On the other hand, the practical work requires simple approximate solutions, which can be used to perform rapid estimations of the influence of vertical inhomogeneity on light reflection and transmission by cloudy media. Such results are presented in this section. To derive approximate solutions, we make an assumption that the

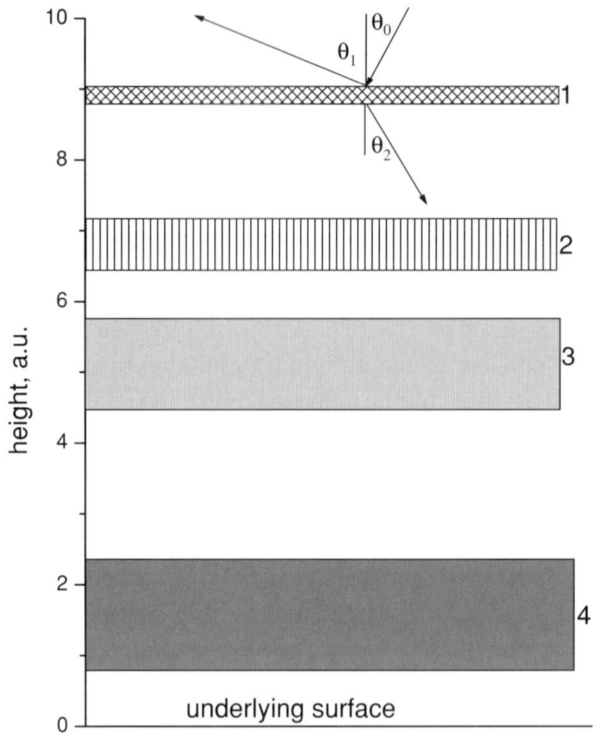

Fig. 3.29. Geometry of the problem.

probability of light absorption by droplets in a cloud is small and the exponential approximation can be applied.

Let us assume that a cloudy medium is composed of several cloud layers having different light scattering characteristics (e.g., the phase function $p(\theta)$, the absorption σ_{abs} and extinction σ_{ext} coefficients). The geometry of the problem is shown in Fig. 3.29. Light enters a disperse medium at the angle θ_0. The reflected light is observed at the angle θ_1 and the diffusely transmitted light is observed in the direction specified by the angle θ_2. We need to model the intensity of diffusely reflected and transmitted beams taking into account effects of vertical inhomogeneity of the medium under consideration.

Note that we put no limitations on the number of layers in a cloudy medium. The region which separates clouds is assumed to be free of light scattering and absorption. This means that the situation we consider is equivalent to the touching light scattering layers (e.g., several snow layers). This assumption, however, is not crucial for the theory developed here [e.g., absorption and scattering of light by gases and aerosols between clouds can be accounted for, if needed (see, e.g., Kokhanovsky and Rozanov, 2004)]. We assume that all layers are optically thick

($l_i \gg \ell_i$, where l_i is the geometrical thickness of i-layer and ℓ_i is the photon free path length in the layer, which is equal to the inverse value of the extinction coefficient in a given layer) and weakly absorbing (the probability of photon absorption $\beta = \sigma_{abs}/\sigma_{ext} \to 0$). We need these assumptions to apply the exponential approximation as discussed above. There is no limitation on the type of the cloud phase function, however. Although our assumptions severely restrict the applicability of the model to many natural clouds, they do provide an accurate approach to the solution of a number of important problems.

The starting point is the expression for the reflection function of a single homogeneous, optically thick, weakly absorbing layer. This can be written in the following form for a scattering layer above a Lambertian surface with albedo A:

$$R_A(\xi, \eta, \tau) = R(\xi, \eta, \tau) + \frac{At_d(\xi)t_d(\eta)}{1 - Ar}, \tag{3.337}$$

where

$$R(\xi, \eta, \tau) = R_\infty(\xi, \eta) - t \exp(-x - y)u_0(\xi)u_0(\eta) \tag{3.338}$$

is the reflection function of a scattering layer for the black ($A = 0$) underlying surface, $R_A(\xi, \eta, \tau)$ is the same function but for the scattering layer–underlying surface system, $x = k\tau$, $k = \sqrt{3\beta(1 - g)}$ is the diffusion exponent, g is the asymmetry parameter, $\tau = \sigma_{ext} l$ is the optical thickness, l is the geometrical thickness of a scattering layer, $y = 4k/3(1 - g)$, r is the spherical albedo of a scattering layer for an illumination from below at $A = 0$, $t_d(\xi)$ is the diffuse transmittance for the illumination in the direction $\theta_0 = \cos^{-1}(\xi)$, $T(\xi, \eta, \tau)$is the transmission function at $A = 0$. The function $t_d(\eta)$ is the diffuse transmittance for the illumination in the direction $\theta_2 = \cos^{-1}(\eta)$. The dependence of the reflection function in Eq. (3.338) on the azimuth is omitted for simplicity. Also we neglect the direct light transmittance, which takes rather small values for optically thick layers. Convenient approximate equations for functions $t_d(\xi)$, $R_\infty(\xi, \eta)$ and r are presented above.

Let us proceed further now. It is known that the reflection of light from an optically thick weakly absorbing strongly light scattering layer is rather close to that of a Lambertian reflector. This is due to well-developed multiple light scattering in the medium in this case. Therefore, to find the reflectance from a multi-layered system as given in Fig. 3.29, we propose to use Eq. (3.337) with A substituted by the spherical albedo r_2^* of the multi-layered system below the upper layer (see Fig. 3.29). Other functions in Eq. (3.337) then refer to the first layer from the illumination side. Such an approach was proposed and successfully used by Melnikova and Minin (1977) for studies of light fluxes in a cloudy atmosphere.

The spherical albedo r_2^* can be easily found integrating Eq. (3.337). Indeed, it follows from Eq. (3.337):

$$r_2^* = r_2 + \frac{t_2^2 r_3^*}{1 - r_2 r_3^*}. \tag{3.339}$$

Here r_3^* is the spherical albedo of the system starting from the 3rd layer down (see Fig. 1), t_2 is the global transmittance of the second layer, r_2 is the spherical albedo of the second scattering layer for an illumination from below. The spherical albedo r_3^* can be found using equation similar to Eq. (3.339):

$$r_3^* = r_3 + \frac{t_3^2 r_4^*}{1 - r_3 r_4^*} \tag{3.340}$$

with meaning of all parameters similar to those in Eq. (3.339). Clearly, we need to repeat this procedure till the underlying surface is reached. Then we have:

$$r_n^* = r_n + \frac{t_n^2 A}{1 - r_n A} \tag{3.341}$$

and the procedure is complete.

Let us check the accuracy of the straightforward procedure outlined above using the exact solution of the radiative transfer equation with SCIATRAN (Rozanov et al., 2005) for a two-layered disperse system over a black surface. Then A in Eq. (3.339) should be substituted by the spherical albedo of a lower layer r_2. All other parameters in Eq. (3.339) refer to an upper layer. Note that SCIATRAN is a well-documented and thoroughly tested radiative transfer code based on the discrete ordinates approach. Its accuracy is better than 1%.

The results of comparisons are shown in Figs. 3.30a–c. In particular, we give the dependence of the reflection function of a disperse medium on the incidence angle for the nadir observation. Due to the reciprocity principle, our calculations are also valid for the nadir illumination and varying observation angles.

The middle curves in Figs. 3.30a–c correspond to a two-layered cloud system with an optical thickness of the bottom layer $\tau_b = 30$ and an optical thickness of the upper layer $\tau_u = 10$. We assume that the upper layer does not absorb incident radiation. The single-scattering albedo of the bottom layer is equal to 0.9945 (Fig. 3.30a), 0.9982 (Fig. 3.30b), 0.9681 (Fig. 3.30c).

The upper curves in these figures correspond to a single layer having an optical thickness $\tau = \tau_b + \tau_u$ and the single-scattering albedo equal to 1. Lower curves in Figs. 3.30a–c correspond to a single layer having an optical thickness $\tau = \tau_b + \tau_u$ and the single-scattering albedo equal to 0.9945 (Fig. 3.30a), 0.9982 (Fig. 3.30b), 0.9681 (Fig. 3.30c). Exact results are shown by symbols. Lines correspond to calculations according to the approximation developed here.

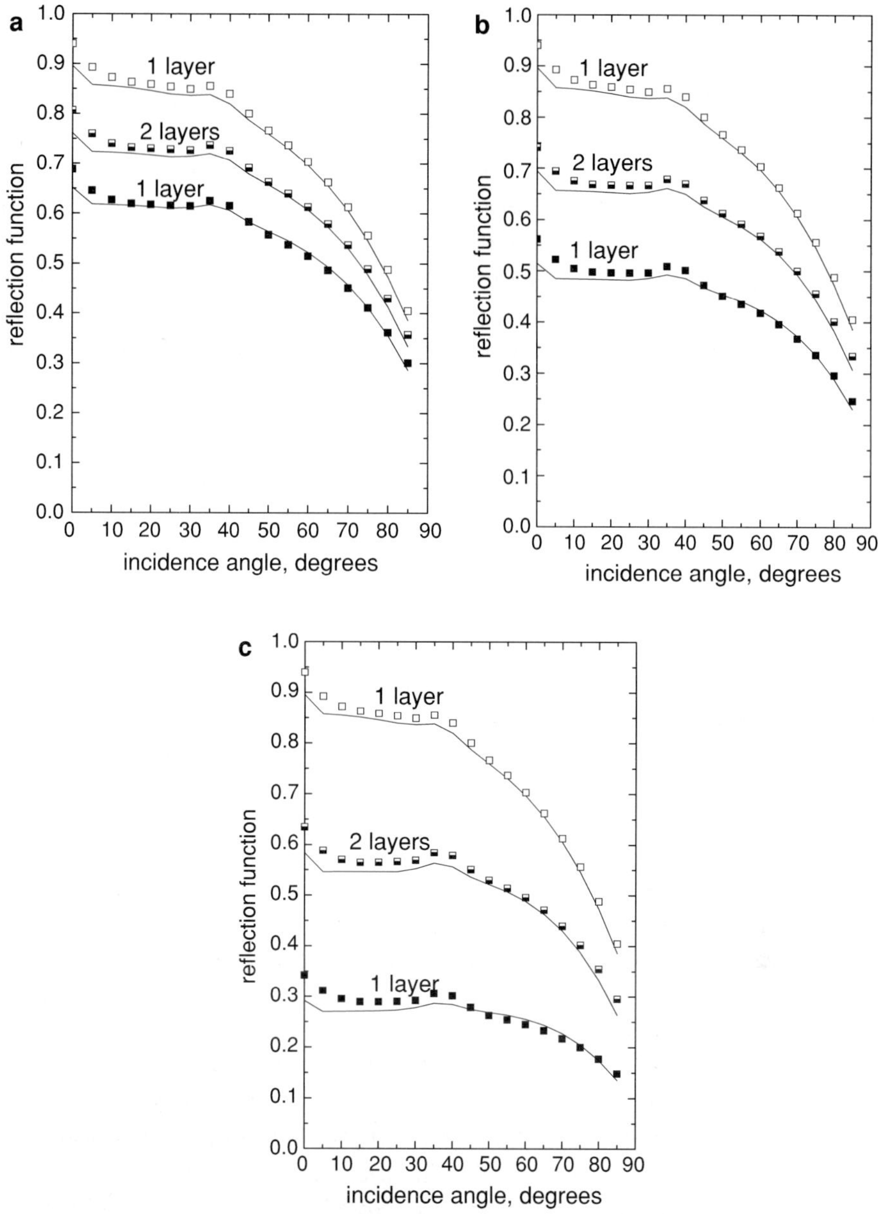

Fig. 3.30. (a) The dependence of the reflection function of a cloud medium on the incidence angle for the nadir observation. Symbols give exact results and lines are due to the approximation (see details in text). The curve in the middle was obtained for an upper nonabsorbing layer and the absorbing layer with $\omega_0 = 0.9945$ at the bottom. Upper curves correspond to a nonabsorbing single layer. Lower curves correspond to an absorbing single layer. The total optical thickness is kept constant for all calculations ($\tau = 40$). The phase function was calculated using Mie theory for the gamma droplet distribution with $a_{ef} = 10$ μm, $\mu = 1/9$, $\lambda = 0.65$ μm, and the refractive index $1.331 - i\chi$, $\chi = 0$ (upper curve) and $\chi = 0.00005$ (lower curve). (b) The same as in Fig. 3.30a except $\chi = 0.0001$ for a lower curve and $\omega_0 = 0.9892$. (c) The same as in Fig. 3.30a except $\chi = 0.0003$ for a lower curve and $\omega_0 = 0.9691$.

The phase function was found using the narrow gamma droplet size distribution with the effective radius 10 μm and the effective variance equal to 1/9 at wavelength $\lambda = 0.65\,\mu$m for water droplets with the refractive index m equal to 1.331-0i (Figs. 3.29a–c, upper curves), 1.331–0.00005i (Fig. 3.30a, lower curves), 1.331–0.0001i (Fig. 3.30b, lower curves), 1.331–0.0003i (Fig. 3.30c, lower curves).

Note that the variation of the imaginary part of the refractive index allows us to model various levels of cloud pollution (e.g., due to black carbon). The phase function differs not significantly for all layers. For instance, the asymmetry parameter is equal to 0.85 at $m = 1.331\text{–}0i$, 0.8513 at $m = 1.331\text{–i}0.00005i$, 0.8525 at $m = 1.331\text{–}0.0001i$ and 0.8571 at $m = 1.331\text{–}0.0003i$. This is according to the general fact that the phase function of weakly absorbing particles is not particularly affected by the level of light absorption. However, note that there is a slight tendency to the general increase of the asymmetry parameter with the imaginary part of the refractive index.

It follows from Figs. 3.30a–c that the accuracy of our simple approximation is quite high for the case considered. In fact it is comparable with the accuracy of correspondent equations for a single layer or even higher than that. This paradox is explained by the fact that the accuracy of the approximation is highly influenced by the value of ω_0. The average value of ω_0 is, however, lower for a two-layered system (with a nonabsorbing upper layer) as compared to a single absorbing layer having the same optical thickness.

We see that the two-layered system with total optical thickness 40 has values of R intermediate between those for an upper layer (at $\tau = 40$) and lower layer (also at $\tau = 40$). This can be expected on general grounds as well. Errors are generally below 5% but they increase for oblique incidence angles. The accuracy decreases with β. The value of $\beta \approx 0.03$ can be considered as an upper boundary for the application of this theory. Although it can be applied to slightly larger values of β if the accuracy is not a primary concern (e.g., for rapid estimations of vertical inhomogeneity effects).

It is interesting to see the performance of equations for larger and smaller values of τ_b, τ_u. This is shown in Figs. 3.31a and 3.31b for an absorbing lower layer and nonabsorbing upper layer. We see that the accuracy of our equations is robust against change of the turbid layer thickness. Note that the variation of the optical thickness of a lower layer (see Fig. 3.31a) does not change the reflection function very much. This is due to the fact that the spherical albedo of a lower absorbing cloud does not depend strongly on τ_b. On the other hand, the variation of the upper layer optical thickness τ_u (see Fig. 3.31b) changes the result considerably. Clouds become much brighter with a larger thickness of the upper layer. Obviously, for a very thick upper layer the sensitivity of the reflection function to the presence of a turbid layer at the bottom is lost. This is similar to the effect of the disappearance of objects in a heavy fog. Therefore, we conclude that high nonabsorbing clouds

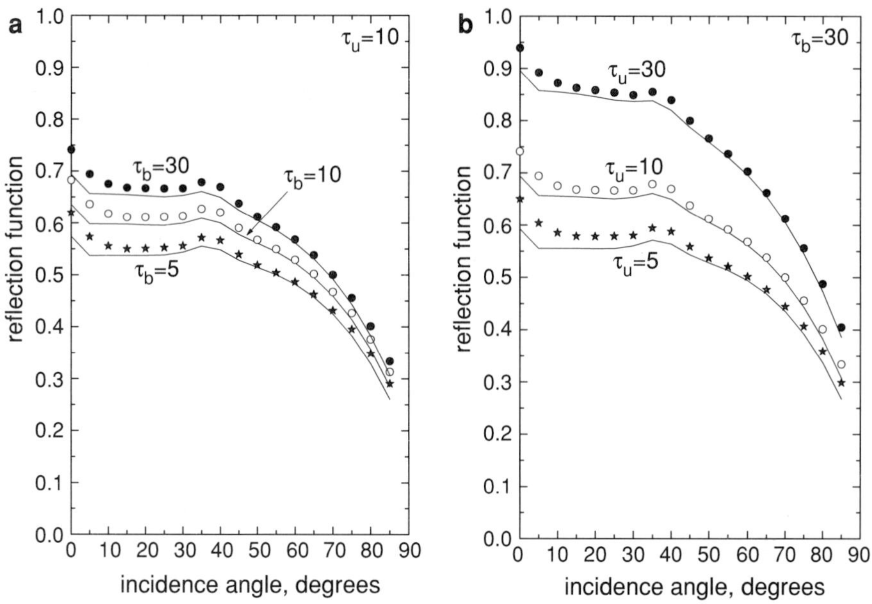

Fig. 3.31. (a) The dependence of the reflection function of a two-layered cloud medium on the incidence angle for the nadir observation at $\tau_u = 10$ and $\tau_b = 5, 10, 30$. The upper layer does not absorb radiation. The bottom layer is characterized by the single-scattering albedo 0.9892. Symbols give exact results and lines are due to the approximation (see details in the text). (b) The dependence of the reflection function of a two-layered cloud medium on the incidence angle for the nadir observation at $\tau_u = 5, 10, 30$ and $\tau_b = 30$. The upper layer does not absorb radiation. The bottom layer is characterized by the single-scattering albedo 0.9892. Symbols give exact results and lines are due to the approximation (see details in the text).

can shield lower (and possibly) polluted clouds. This can lead to important climatic effects not accounted for in Global Circulation Models at the moment.

To make this point more clear, we present the reflection function of a single absorbing cloud layer with the optical thickness 30 in Fig. 3.32. Then we add a nonabsorbing cloud at the higher level in the atmosphere. It follows that the reflection of the system considerably increases both for warm water and cold ice upper-level clouds. The phase function of an ice cloud was taken from a study by Mishchenko et al. (1999) (the fractal particle model) and the phase function of the water cloud was calculated as indicated above (at $m = 1.331 - 0.0001i$ for a lower cloud and $m = 1.331$ for an upper cloud in the two-layered system). The increase in the reflection is much more pronounced for crystalline clouds. It means that ice clouds not only warm the system by trapping terrestrial radiation. They also may shield lower polluted cloud systems (e. g., in urban areas) and increase general reflection of the surface–atmosphere system. This indicates the complexity of the issue of cloud influence on the climate.

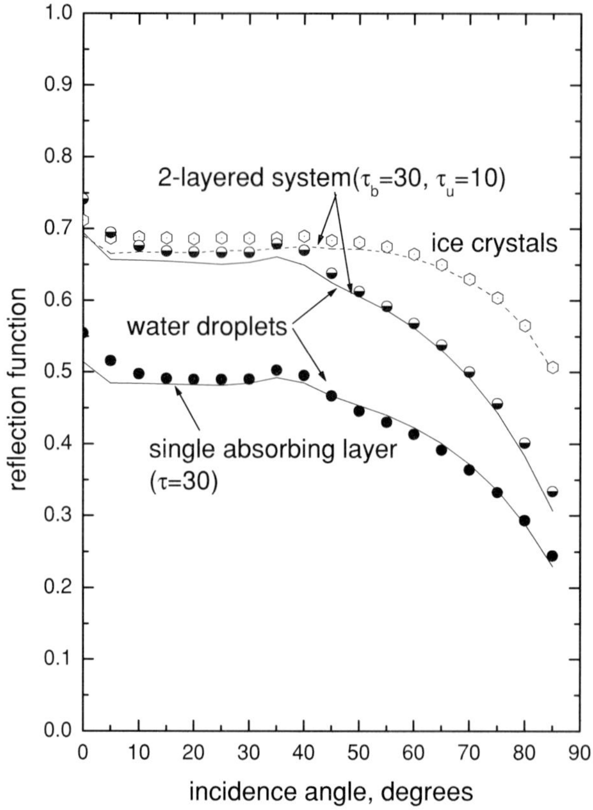

Fig. 3.32. The dependence of the reflection function on the incidence angle for the nadir observation for a two-layered cloud medium at $\tau_u = 10$, $\tau_b = 30$ (upper curves) and for an absorbing single cloud layer with $\tau = 30$ (lower curves). The upper layer does not absorb radiation and composed either of water droplets or ice crystals. The bottom layer and a single absorbing layer is characterized by the single-scattering albedo 0.9892. Symbols give exact results and lines are due to the approximation (see details in text).

The question arises if the model presented above can be applied to studies of light transmission. The answer on this question is positive. Indeed, we have for the transmission function of a single homogeneous disperse layer over a Lambertian surface with surface albedo A :

$$T_A(\xi, \eta, \tau) = T(\xi, \eta, \tau) + \frac{At_d(\xi) r_d(\eta)}{1 - Ar}. \qquad (3.342)$$

Here $r_d(\eta)$ is the plane albedo of the layer for the illumination from below. Again assuming that the layers below the upper one can be substituted by a Lambertian reflector, we have for the transmission function $\bar{T}_1(\xi, \eta)$ of the first layer in the

n-layer system:

$$\bar{T}_1(\xi, \eta) = T_1(\xi, \eta) + \frac{t_{d1}(\xi) r_{d1}(\eta) r_2^*}{1 - r_1 r_2^*}, \tag{3.343}$$

where we omitted the dependence on the optical thickness and r_2^* is found using the iterative procedure starting from the ground surface as underlined in the previous section. Functions $T_1(\xi, \eta)$, $t_{d1}(\xi)$, r_1 and $r_{d1}(\eta)$ have the same meaning as in Table 3.1 but for the first layer.

Let us consider now the transmittance under the second layer. The second layer is illuminated by the diffuse light transmitted by the first layer. It follows that the diffuse transmittance $\bar{t}_{d1}(\xi)$is given by:

$$\bar{t}_{d1}(\xi) = t_{d1}(\xi) + \frac{t_{d1}(\xi) r_1 r_2^*}{1 - r_1 r_2^*} \tag{3.344}$$

or

$$\bar{t}_{d1}(\xi) = \frac{t_1(\xi)}{1 - r_1 r_2^*}. \tag{3.345}$$

Also we have for the global transmittance:

$$\bar{t}_1 = \frac{t_1}{1 - r_1 r_2^*}. \tag{3.346}$$

Therefore, the transmission function $\bar{T}_2$ after the second layer in the n-layered system is given as

$$\bar{T}_2(\xi, \eta) = \bar{t}_{d1}(\xi) \bar{t}_{d2}(\xi)(\eta), \tag{3.347}$$

where

$$\bar{t}_{d2}(\eta) = \frac{\bar{t}_{d2}(\eta)}{1 - r_2 r_3^*}. \tag{3.348}$$

Following this procedure, we can obtain the transmission function under the third layer:

$$\bar{T}_3(\xi, \eta) = \bar{t}_{d1}(\xi) \bar{t}_2 \bar{t}_{d3}(\eta), \tag{3.349}$$

where $\bar{t}_2 = t_2/(1 - r_2 r_3^*)$ and we accounted for the fact that the second layer is illuminated from above by diffuse light and that it also serves as a diffuse light source for the third layer. Repeating this procedure for each layer, we can arrive finally to the transmission function of a whole system:

$$\bar{T}_n(\xi, \eta) = \bar{t}_{d1}(\xi) \bar{t}_2 \bar{t}_3 \ldots \bar{t}_{n-1} \bar{t}_{dn}(\eta), \tag{3.350}$$

where $\bar{t}_j = t_j/(1 - r_j r^*_{j+1})$ and $r_{n+1} \equiv A$. Interestingly, Eq. (3.350) can be written in the form similar to that for a homogeneous layer:

$$\bar{T}_n(\xi, \eta) = t_{ef} K_0(\xi) K_0(\eta), \tag{3.351}$$

where the effective global transmittance is given by:

$$t_{ef} = \frac{\prod_{j=1}^{n} t_j}{\prod_{j=1}^{n} \left(1 - r_j r^*_{j+1}\right)} \tag{3.352}$$

with all parameters defined in the theory for a single layer. Note that we have used here the equality: $t_{dj} = t_j u_0(\xi)$, where $u_0(\xi)$ is the escape function.

Let us check the applicability of our assumptions making comparisons with exact radiative transfer calculations using SCIATRAN (Rozanov et al., 2005) for a special case of a two-layered medium over a black surface. Then Eq. (3.350) is reduced to the following form:

$$\bar{T}_2(\xi, \eta) = \frac{t_1 t_2 u_0(\xi) u_0(\eta)}{1 - r_1 r_2}, \tag{3.353}$$

where we accounted for the fact that $t_{ef} = t_1 t_2 (1 - r_1 r_2)^{-1}$ in this case. Note that if neither of both layers absorb radiation, the sensitivity of transmitted light to the vertical inhomogeneity is low and in a good approximation one can use the reflection function for a single layer having the optical thickness equal to the sum of optical thicknesses of both layers and the average value of the asymmetry parameter (Sobolev, 1972).

The results of calculations using simple approximation (3.353) are shown in Figs. 3.33–3.37 as functions of the observation angle together with correspondent errors and outcome of exact calculations for the incident angle equal to 60° and the azimuth equal to 0°. In particular, we give the dependence of the transmission function on the observation angle for a single layer having the optical thickness equal to 40 at $\omega_0 = 0.9945$ and $\omega_0 = 1.0$ in Fig. 3.33. The results of computations for a two-layered medium having the total optical thickness 40 but $\omega_0 = 0.9945$ in the bottom layer ($\tau_b = 30$) and $\omega_0 = 1.0$ in upper layer ($\tau_u = 10$) are also given in the same figure (the middle line). Note that phase functions in all calculations given here are very close to each other. So the change of transmission is mostly due to the absorption effect. As expected the largest transmission is observed for a nonabsorbing single layer. It is reduced considerably if absorption is introduced in the bottom part of a layer. Of course, the minimum of transmission occurs for a single absorbing layer (see, e.g., a lower line in Fig. 3.33). It follows that exact and approximate results are quite close to each other for observation angles smaller than 70°. Then the error of approximation is smaller than 5%. The error increases for slabs having larger absorption, however. This is illustrated in Fig. 3.34, where we show a dependence similar to that in Fig. 3.33 but now for the increased

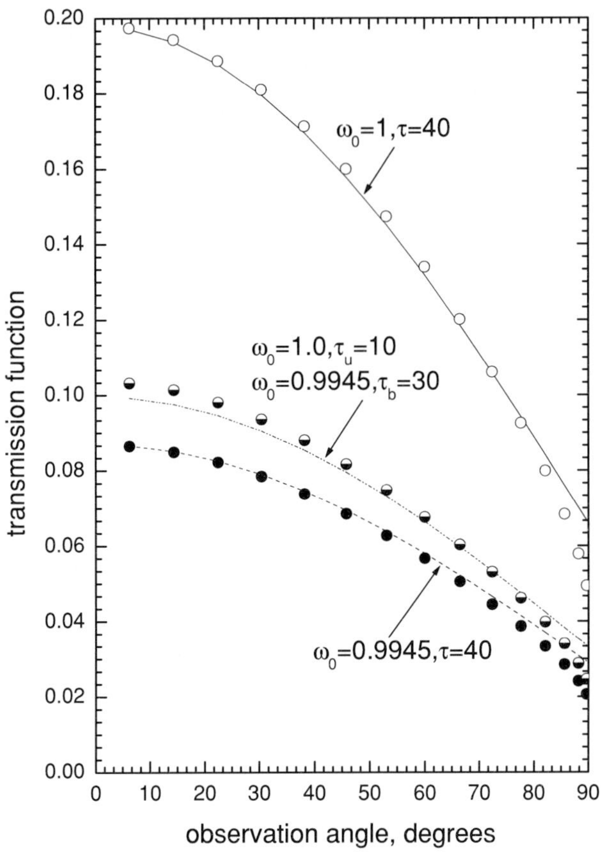

Fig. 3.33. The dependence of the transmission function on the zenith observation angle at the zenith incidence angle equal to 60° and the azimuth equal to 0° for a single homogeneous layer and a two – layered turbid medium with the total optical thickness equal to 40. The single-scattering albedo is equal to 0.9945. Lines correspond to Eq. (3.353) and symbols are obtained from exact calculations. Further explanations are given in the text.

absorption ($\omega_0 = 0.9892$). It is interesting that the error of Eq. (3.353) for a two-layered medium is smaller than that for a single layer with the optical thickness 40 and $\omega_0 = 0.9892$. This points to the fact that the accuracy of our technique is mostly influenced by the total light absorption and transmission and not by a number of layers. Note that the accuracy of reflected light calculation is generally higher than that for the diffusely transmitted light (for a given level of absorption). $\tau_u = 5, 10, 30$

Calculations for a two-layered turbid slab with the optical thickness of a lower absorbing layer equal to 30 for various thicknesses of a nonabsorbing upper layer are shown in Fig. 3.35. The middle lines in this figure coincide with the middle line in Fig. 3.33. We see that the accuracy is better than 5% in this case. The decrease of

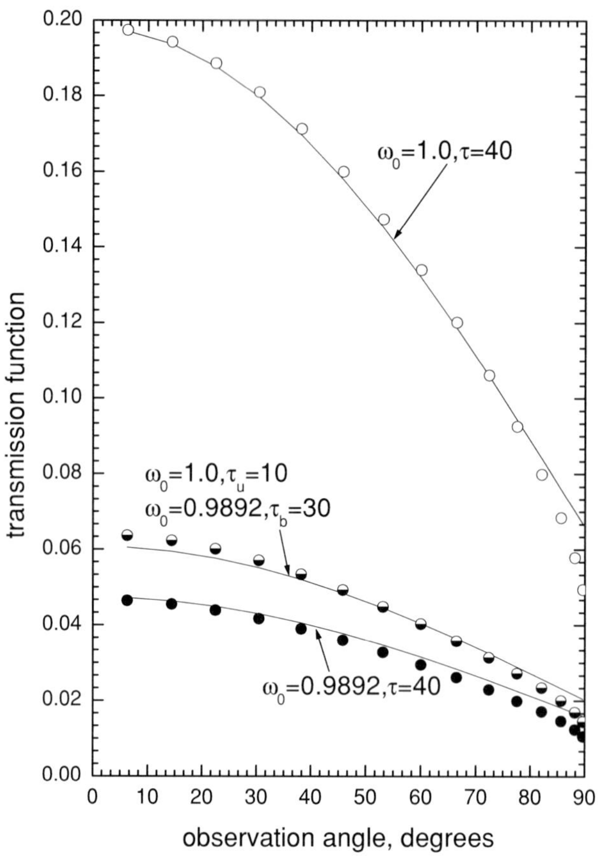

Fig. 3.34. The same as in Fig. 3.33 but at $\omega_0 = 0.9892$.

the optical thickness of an upper layer leads to the increase of the light transmission as one might expect.

Fig. 3.36 is similar to Fig. 3.35, but now the optical thickness of an absorbing layer at the bottom is varied from 5 till 30. The optical thickness of an upper layer is fixed and equal to 10. Clearly, the error approximation increases for thinner layers, which is in accordance with general assumptions of our approximation, which is valid only for weakly absorbing optically thick layers (Kokhanovsky and Rozanov, 2003). However Eq. (3.353) has a comparatively high accuracy even at such comparatively small values of τ as 5 (see Fig. 3.36).

In conclusion, we show the transmission function of a single absorbing layer having optical thickness 40 at $\omega_0 = 0.9945$ in Fig. 3.37 (middle line) in comparison with transmission functions of a two-layered system having $\omega_0 = 0.9945$ in the upper layer ($\tau_u = 30$) and $\omega_0 = 0.9892$ in the layer at the bottom ($\tau_b = 10$) (lower line

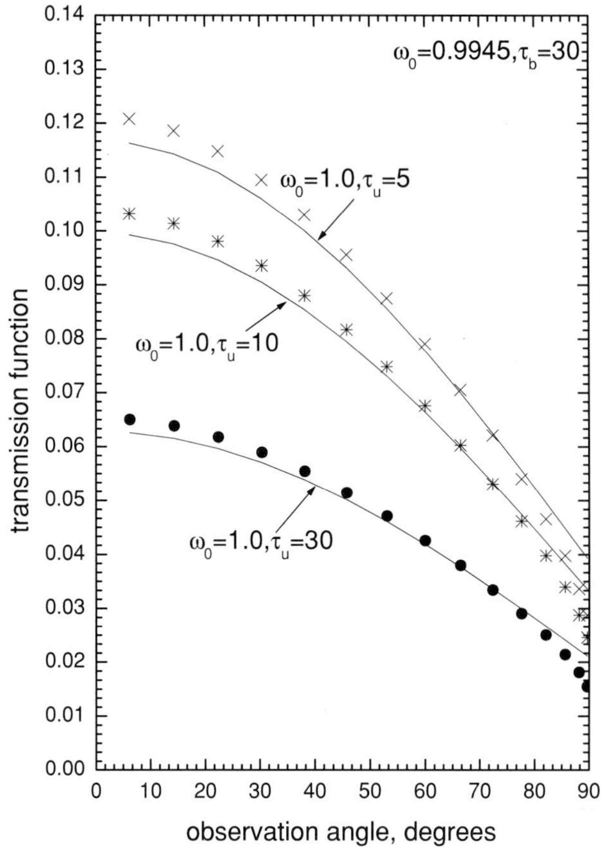

Fig. 3.35. The same as in Fig. 3.33 but for other values of optical thickness of an upper layer.

in Fig. 3.37). Clearly, in the later case the transmission should be lower. Figure 3.37 quantifies this decrease. The upper line in Fig. 3.37 corresponds to a two-layered system with total optical thickness equal to 40 and local optical characteristics of a lower layer equal to that of a single layer shown by the middle line in Fig. 3.37 but having a nonabsorbing scattering layer at the top of the system ($\tau_u = 10$). Then, due to the general decrease of absorption in the system, transmission should increase. This is confirmed by Fig. 3.37. It follows from Fig. 3.37 that the error of Eq. (3.353) is smaller than 5%, which is acceptable for a broad range of applications (e.g., rapid estimations of stratification effects on diffusely transmitted and reflected light fields).

Detailed studies of radiative transfer in vertically inhomogeneous clouds including those with smooth vertical profiles of scattering and absorption characteristics have been performed by Yanovitskij (1997).

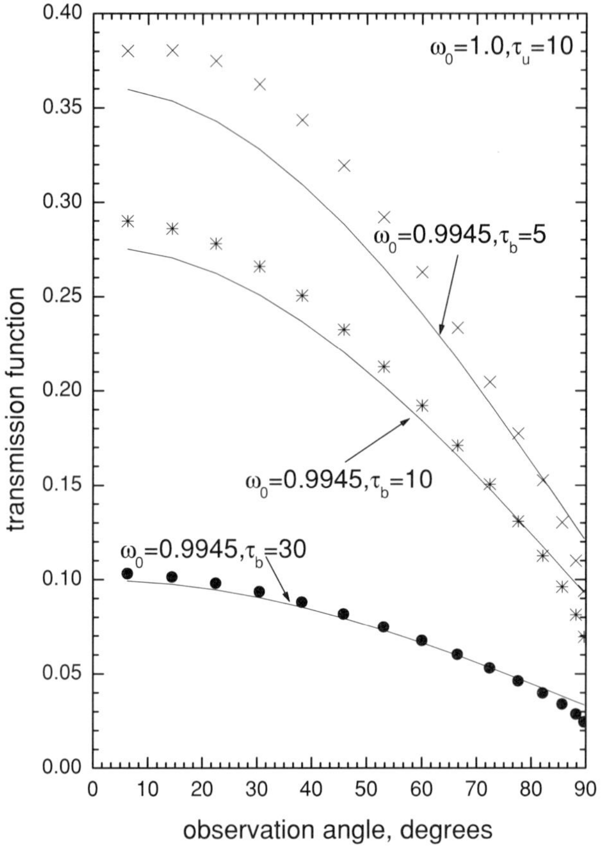

Fig. 3.36. The same as in Fig. 3.33 but for other values of optical thickness of a bottom layer.

3.9 Horizontally Inhomogeneous Clouds

3.9.1 Independent Pixel Approximation

The influence of the horizontal inhomogeneity of clouds on their radiative characteristics is a major subject of modern cloud optics studies (Cahalan et al., 1994, 2001; Barker et al., 1996; Loeb and Davies, 1996; Marshak et al., 1998; Platnick, 2001; Scheirer and Macke, 2001; Davis and Marshak, 2002). In particular, it was found that the horizontal inhomogeneity of clouds effects their abilities to absorb, reflect and transmit solar light (Feigelson, 1981; Scheirer and Macke, 2001). Thus, cloud remote sensing techniques, based on the spectral reflectance method (Kondratyev and Binenko, 1984; Arking and Childs, 1985; Nakajima and King, 1990; Nakajima et al., 1991; King et al., 1992; Han et al., 1994; Rossow et al.,

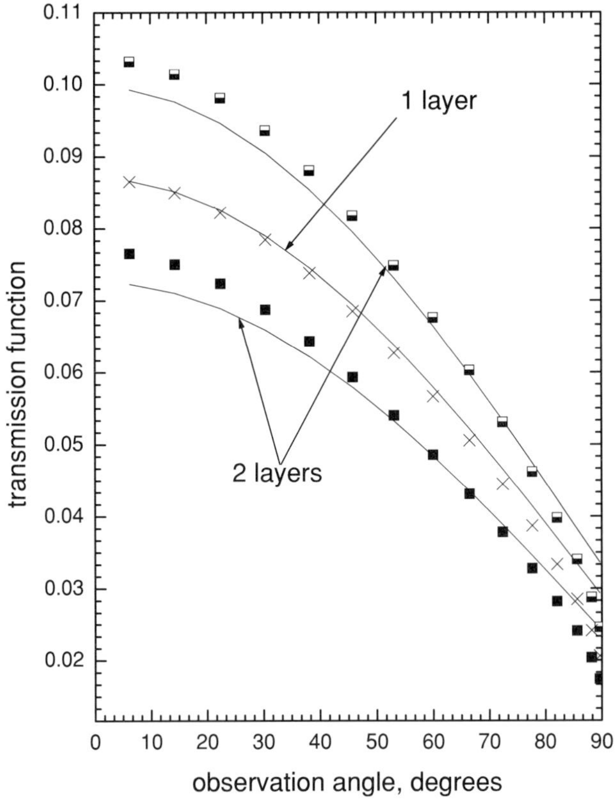

Fig. 3.37. The dependence of the transmission function on the zenith observation angle at the zenith incidence angle equal to 60° and the azimuth equal to 0° for a single homogeneous layer (middle line, $\omega_0 = 0.9945$, $\tau = 40$) and a two-layered turbid medium [$\omega_0 = 0.9945$, $\tau = 30$ for a lower layer and $\omega_0 = 1.0$, $\tau = 10$ for an upper layer (upper line) and $\omega_0 = 0.9892$, $\tau = 10$ for a lower layer and $\omega_0 = 0.9945$, $\tau = 30$ for an upper layer (lower line)]. Lines correspond to results obtained with Eq. (3.353) and symbols are obtained from exact calculations. Further explanations are given in text.

1989; Rossow and Schiffer, 1999), must account for the sub-pixel cloud horizontal inhomogeneity. This is not generally the case so far.

It is known that pixels with inhomogeneous clouds are darker than pixels with homogeneous cloud layers, having the same average optical thickness (Cahalan et al., 1994). This leads to the underestimation of cloud optical thickness by modern satellite retrieval techniques. There are semi-empirical approaches to overcome this problem. They are based on the artificial increase of the reflection function measured to account for the horizontal inhomogeneity of a cloud field under study. The correct magnitude of these adjustments, however, cannot be assumed *a priori*.

So they lack a physical basis. This issue is discussed in detail by Pincus and Klein (2000). Also cloud inhomogeneity could lead to unphysical dependencies of cloud optical thickness retrieved on illumination and viewing geometry (Loeb and Davies, 1996).

Another way to solve the problem is to use 3-D Monte Carlo calculations (see, e.g., Scheirer and Macke, 2001). However, they are time-consuming for realistic clouds and can be used mostly for theoretical studies and not as a core of operational cloud satellite retrieval algorithms. Monte Carlo calculations have shown, however, that in some cases a high accuracy can be achieved if a 3-D cloud field is substituted by N noninteracting vertical columns or cells. A cloud field in each cell is modelled as a horizontally homogeneous plane-parallel layer of an infinite horizontal extent. The optical thickness (and, possibly, microstructure) of each cell varies, depending on its position in a cloud field. Such an approach is called the Independent Column Approximation (ICA) or the Independent Pixel Approximation (IPA). The range of applicability of the ICA was studied by Davis et al. (1997) and Scheirer and Macke (2001).

Effectively, the ICA reduces the 3-D radiative transfer problem to N standard radiative transfer problems for homogeneous media. The number N can be large. Thus, the problem remains computationally very expensive.

It can be simplified, however, if one applies approximate solutions of the radiative transfer problem for each cloud cell. This is done usually in the framework of the two-stream approximation (Barker, 1996; Barker et al., 1996; Barker and Fu, 2000). In particular, one can assume that the cloud optical thickness for a given cloud field obeys the probability distribution law (e.g., as those given in Fig. 3.38). However, the accuracy of two-stream approximations is rather low as compared to exact radiative transfer calculations (King and Harshvardhan, 1986; Thomas and Stamnes, 1999). In particular, for some cases errors introduced by the approximation can be larger than differences of radiative fluxes for horizontally homogeneous and inhomogeneous cloud fields themselves. Also this approximation does not allow consideration of the bi-directional reflection function of clouds, which is routinely measured by various radiometers and spectrometers on satellite platforms. With this in mind, Kokhanovsky (2003b) proposed to use the exponential approximation of the radiative transfer theory to solve each of N standard radiative problems, discussed above. Then it follows for the whole cloud field under observation:

$$\langle R(\xi, \eta, \phi) \rangle = \int_0^\infty R(\xi, \eta, \phi, \tau) f(\tau) d\tau, \tag{3.354}$$

where $f(\tau)$ is the cloud optical thickness distribution and $R(\xi, \eta, \phi, \tau)$ is the reflection function of a cloud with given τ. This equation and also similar formulae for other cloud radiative characteristics allow us to study the influence of cloud

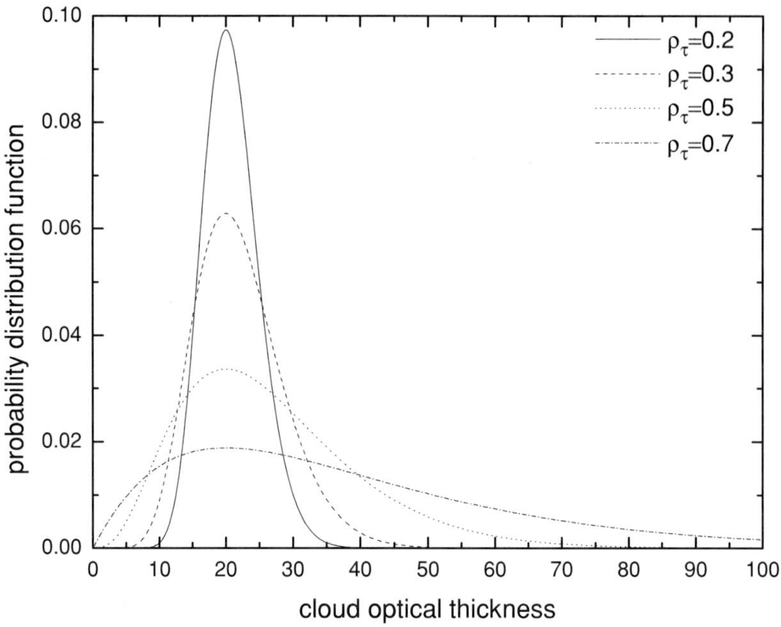

Fig. 3.38. Spatial optical thickness distribution modeled using the gamma distribution for different values of ρ_τ defined as $\Delta/\langle\tau\rangle$, where Δ is the standard deviation of the optical thickness and $\langle\tau\rangle$ is the average optical thickness.

inhomogeneity parameters on the measured reflection function for a given cloudy scene in a simple way. In particular, it follows for a nonabsorbing case:

$$\langle R(\xi, \eta, \phi)\rangle = R_\infty(\xi, \eta, \phi) - \langle t\rangle u_0(\xi)u_0(\eta), \tag{3.355}$$

where $R_\infty(\xi, \eta, \phi)$ does not depend on τ by definition (see Fig. 3.39) and

$$\langle t\rangle = \int_0^\infty t(\tau)f(\tau)d\tau. \tag{3.356}$$

Let us assume that $\tau \to \infty$. Then we have:

$$t(\tau) \approx \frac{4}{3\tau(1-g)} \tag{3.357}$$

and

$$\langle t\rangle = \frac{4\left(1+\frac{1}{\mu}\right)}{3\langle t\rangle(1-g)}, \tag{3.358}$$

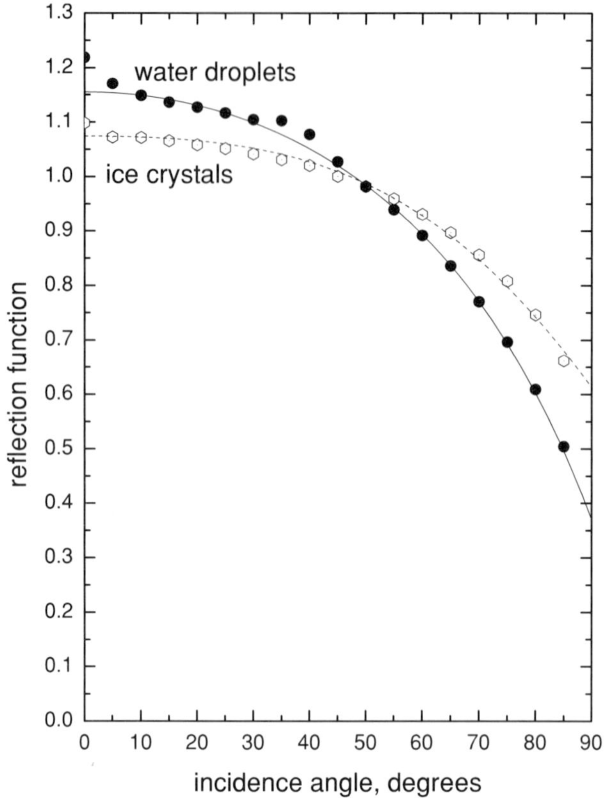

Fig. 3.39. Dependence of reflection functions of a semi-infinite nonabsorbing water and ice cloud on the incidence angle at the nadir observation. Symbols give exact results and lines are due to the approximation (Kokhanovsky, 2004b, 2005).

where we used the following distribution:

$$f(\tau) = \Lambda \tau^{\mu} \exp\left(-\mu \frac{\tau}{\tau_0}\right) \tag{3.359}$$

with

$$\Lambda = \frac{\mu^{\mu+1}}{\tau_0^{\mu+1}\Gamma(\mu+1)} \tag{3.360}$$

and $\langle\tau\rangle = \tau_0(1 + (1/\mu))$ is the average optical thickness defined as

$$\langle t\rangle = \int_0^{\infty} \tau f(\tau) d\tau. \tag{3.361}$$

So we have:

$$\langle t(\tau)\rangle / t(\langle t\rangle) \approx \left(1 + \frac{1}{\mu}\right) \tag{3.362}$$

as $\langle\tau\rangle \to \infty$. It means that $\langle t(\tau)\rangle / t(\langle\tau\rangle) \geq 1$, where we accounted for the fact that $\mu > 1$. Also because of this we have: $\langle R(\xi, \eta, \phi, \tau)\rangle < R(\xi, \eta, \phi, \langle\tau\rangle)$. It means that reflectances of inhomogeneous cloud fields, calculated assuming that the optical thickness is equal to the average optical thickness, give larger values as compared to measured reflectances. So we conclude that inhomogeneity leads to darkening of correspondent pixels as compared to the case of homogeneous clouds with the same optical thickness $\langle\tau\rangle$. On the other hand, the transmission of light by inhomogeneous clouds increases as compared to the case of a horizontally homogeneous cloud with the cloud optical thickness equal to $\langle\tau\rangle$.

3.9.2 Multidimensional Radiative Transfer in Clouds

3.9.2.1 General remarks

Radiative transfer in a cloudy atmosphere is usually studied in the framework of the plane-parallel approximation or the IPA as shown above. Then the diffuse light field changes only along the vertical direction for wide solar beam illumination conditions. There is no any change in the radiation field in the horizontal direction. Although this approximation, which is often called 1-D case, is very important for the case of extended cloudiness (e.g., extended fields of Stratocumulus clouds), it cannot be applied for the majority of cloudy scenes. Indeed cloudiness has a horizontal structure (e.g., holes between clouds).

These effects can be accounted for in the framework of the 3-D radiative transfer equation, where the spatial variation of local optical properties is fully accounted for. Various approaches to deal with 3-D clouds are known (Liou, 2002). The most popular techniques are the Monte Carlo method (Marchuk et al., 1980; Shreier and Macke, 2001), the diffusion approximation (Liou, 2002), the spherical harmonics discrete ordinate method (Evans, 1998), and the iteration technique (Nikolaeva et al., 2005).

In the iteration technique, the phase function in the scattering integral is represented by the spherical harmonics and the integral is replaced by a quadrature sum. Spatial grids are introduced and obtained partial differential equations are approximated by the system of linear algebraic equations. To solve it, the successive-orders-of-scattering (SOS) approach is applied.

Each partial differential equation is integrated along its characteristic throughout the whole calculation region in the framework of the well-known Evans's algorithm (Evans, 1998). These methods have some advantages and some deficiencies.

In particular, some of them may be non-conservative, e.g., do not conserve the number of photons in the transport problem. This defect can lead to significant errors in the solution obtained.

Other spherical harmonics discrete ordinate methods use the local approximation for partial differential equations. Such methods were widely used in various neutron and photon transport problems in the last 50 years. They are conservative and economic since they use very simple equations and do not apply complicated logic (all spatial meshes are calculated successively). In particular, such methods have been incorporated in the RADUGA 3-D solver described by Nikolaeva et al. (2005).

3.9.2.2 *The three-dimensional radiative transfer equation*

The 3-D radiative transfer equation can be written in the following form:

$$\xi\frac{\partial I(x,y,z,\theta,\phi)}{\partial x}+\eta\frac{\partial I(x,y,z,\theta,\phi)}{\partial y}+\beta\frac{\partial I(x,y,z,\theta,\phi)}{\partial z}+\sigma_{ext}(x,y,z)$$
$$\times I(x,y,z,\theta,\phi)=\frac{1}{4\pi}\sigma_{sca}(x,y,z)\int_0^{\pi}\sin\theta' d\theta'\int_0^{2\pi}I(x,y,z,\theta',\phi')p(x,y,z,$$
$$\chi(\theta,\phi,\theta',\phi'))d\phi'+\frac{1}{4\pi}\sigma_{sca}(x,y,z)F_0p(x,y,z,\chi(\theta,\phi,\Theta,\Phi))\exp(-t), \tag{3.363}$$

where σ_{sca} and σ_{ext} are scattering and extinction coefficients, $p(x, y, z, \chi)$ is the phase function. The function $I(x, y, z, \theta, \phi)$ is the diffuse light intensity at the point (x, y, z) propagated in the direction (θ, ϕ), see Fig. 3.40. Also we have: $\xi = \sin\theta\cos\phi$, $\eta = \sin\theta\sin\phi$, $\beta = \cos\theta$. Scalar product χ is defined by the following relation

$$\chi(\theta,\phi,\theta',\phi')=\cos\theta\cos\theta'+\sin\theta\sin\theta'\cos(\phi-\phi'). \tag{3.364}$$

The value ot t in Eq. (3.363) is the optical path between two points defined by radius vector $\vec{r}_0$ and $\vec{r}$: $t = \int_0^d \sigma_{ext}(\vec{r}_0 + \zeta\vec{r})d\zeta$, where $d = |\vec{r} - \vec{r}_0|$.

The vector $\vec{r}_0$ defines the crossing point of the light beam with the boundary of the scattering medium under study.

We assume that there is no diffuse light entering the medium. Therefore, boundary conditions have the following form:

$$I(\vec{r},\vec{\Omega})=0 \text{ at } \vec{\Omega}\vec{n}(\vec{r})<0 \text{ for all } \vec{r}(x,y,z)\in G_{\text{finite}},$$
$$I(\vec{r},\vec{\Omega})=I(\vec{r}^*,\vec{\Omega}) \text{ at } \vec{\Omega}\vec{n}(\vec{r})<0 \text{ for all } \vec{r}(x,y,z)\in G_{\text{infinite}},$$

Here G_{finite} is the finite part of the medium boundary (on z), G_{infinite} is the infinite one (on x and y), $\vec{r}^*$ is an inner point for the periodic boundary conditions.

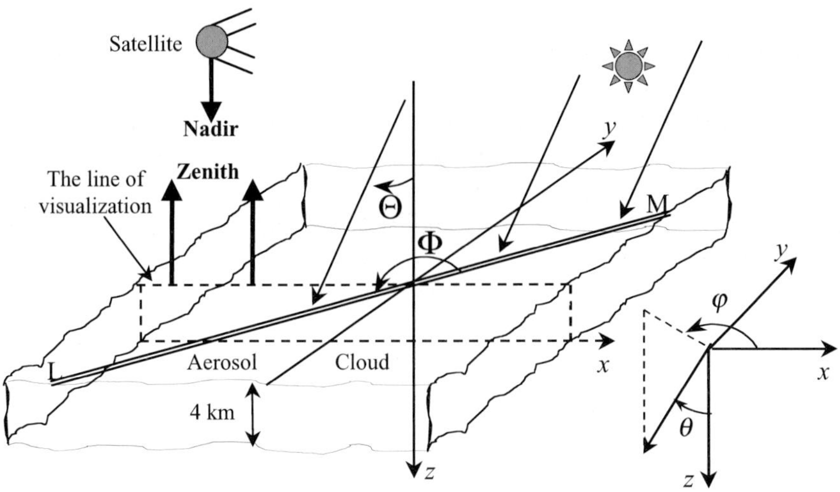

Fig. 3.40. Geometry of the problem.

This full 3-D transport equation is reduced to 2-D equation in some particular cases. For instance, let us assume that the line LM (see Fig. 3.40) coincides with the visualization line. This means that the solution becomes invariant in respect to the coordinate y. Then we can drop the dependence on y in Eq. (3.363) and arrive at the following simplified 2-D transport equation:

$$\xi \frac{\partial I(x, y, z, \theta, \phi)}{\partial x} + \beta \frac{\partial I(x, y, z, \theta, \phi)}{\partial z} + \sigma_{ext}(x, z) I(x, z, \theta, \phi) = \hat{F} I, \tag{3.365}$$

where

$$\hat{F} I \equiv \frac{1}{4\pi} \sigma_{sca}(x, z) \int_0^{\pi} \sin\theta' d\theta' \int_0^{2\pi} I(x, z, \theta', \phi') p(x, z, \chi(\theta, \phi, \theta', \phi')) d\phi'$$
$$+ \frac{1}{4\pi} \sigma_{sca}(x, z) F_0 p(x, z, \chi(\theta, \phi, \Theta, \Phi)) \exp(-t). \tag{3.366}$$

Equation (3.365) can be solved using the method of successive orders of scattering. Namely, at first we neglect the integral term in Eq. (3.366) and calculate the diffuse intensity $I(x, z, \theta, \phi)$ from the solution of the partial differential equation. Then the obtained diffuse intensity is substituted in the scattering integral in Eq. (3.366) and the next approximation for $I(x, z, \theta, \phi)$ is found from the solution of the partial differential equation (3.365). The algorithm is stopped when the convergence is reached.

Therefore, the problem at hand is reduced to the solution of the following transport equation:

$$\xi \frac{\partial I(x, y, \theta, \phi)}{\partial x} + \beta \frac{\partial I(x, y, \theta, \phi)}{\partial z} + \sigma_{ext}(x) I(x, z, \theta, \phi) = \Im(x, z, \theta, \phi), \tag{3.367}$$

where $I(x, z, \theta, \phi)$ is the diffuse light intensity at the point $\vec{r}(x, z)$ propagated in the direction (θ, ϕ), $\Im(x, z, \theta, \phi) = \hat{F}\tilde{I}(x, z, \theta, \phi)$ and $\tilde{I}(x, z, \theta, \phi)$ is a known function obtained from the previous iteration as described above.

We introduce an angular quadrature and replace functions $I(x, z, \theta, \phi)$, $\tilde{I}(x, z, \theta, \phi)$ and $\Im(x, z, \theta, \phi)$ by their values in quadrature nodes. Integral $\Im(x, z, \theta, \phi)$ is calculated using the following standard steps:

- the expansion of function $\tilde{I}(x, z, \theta, \phi)$ in terms of spherical harmonics,
- the representation of integral $\Im(x, z, \theta, \phi)$ by spherical harmonics,
- the calculation of $\Im(x, z, \theta, \phi)$ values in quadrature nodes.

A standard grid method to approximate the partial differential equation of the first order is used. In particular, grids with respect to spatial variables x and z are introduced:

$$x_{1/2} < \ldots < x_{k+1/2} < \ldots < x_{k+1/2}, z_{1/2} < \ldots < z_{\ell+1/2} < \ldots < z_{L+1/2}.$$

A single two-dimensional cell (k, l) has the following dimensions: $[x_{k-1/2}, x_{k+1/2}] \times [z_{\ell-1/2}, z_{\ell+1/2}]$. Correspondingly, its size is $[\Delta x_k] \times [\Delta z_l]$, where $\Delta x_k = x_{k+1/2} - x_{k-1/2}$ and $\Delta z_k = z_{K+1/2} - z_{k-1/2}$. Also the integral operator

$$\hat{\Lambda}_{k,\ell} = \frac{1}{\Delta x_k \Delta y_\ell} \int_{x_{k-1/2}}^{x_{k+1/2}} dx \int_{x_{\ell-1/2}}^{z_{\ell+1/2}} dz$$

is applied to both parts of Eq. (3.365). Then it follows:

$$\xi(I_{k+1/2,\ell} - I_{k-1/2,\ell})/\Delta x_k + \beta(I_{k,\ell+1/2} - I_{k,\ell-1/2})/\Delta z_\ell + \sigma_{ext}^{k,\ell} I_{k,\ell} = \Im_{k,\ell}, \tag{3.368}$$

where

$$I_{k,\ell} = \frac{1}{\Delta x_k \Delta z_\ell} \int_{x_{k-1/2}}^{x_{k+1/2}} dx \int_{x_{\ell-1/2}}^{z_{\ell+1/2}} dz I(x, z), \Im_{k,\ell} = \frac{1}{\Delta x_k \Delta z_\ell} \int_{x_{k-1/2}}^{x_{k+1/2}} dx \times \int_{x_{\ell-1/2}}^{z_{\ell+1/2}} dz \Im(x, z), \tag{3.369}$$

are the average values of the intensity and the source function, respectively, over a given cell and

$$I_{k\pm 1/2,\ell} = \frac{1}{\Delta z_\ell}\int_{x_{\ell-1/2}}^{z_{\ell+1/2}} dz I(x_{k\pm 1/2}, z),\ I_{k,\ell\pm 1/2} = \frac{1}{\Delta x_k}\int_{x_{k-1/2}}^{x_{k+1/2}} dx I(x, z_{\ell\pm 1/2}), \tag{3.370}$$

are correspondent average values of the intensity on boundaries of the cell (k, l). Fulfilment of Eq. (3.368) guarantees that a presented scheme is a conservative one.

Intensities $I_{k-1/2,l}$ and $I_{k,l-1/2}$ are known either from boundary conditions or from the result of the calculation for the previous cell. So we need to determine only values of $I_{k-1/2,l}$, $I_{k,l-1/2}$ and $I_{k,l}$. It is not possible to evaluate three parameters from a single equation (3.368). So we need to introduce two approximate relations among these three unknown parameters. They are given as follows:

$$I_{k,\ell} = (1 - \nu_{x,k,\ell})I_{k+s(\xi)/2,\ell} + \nu_{x,k,\ell} I_{k-s(\xi)/2,\ell}, \tag{3.371}$$

$$I_{k,\ell} = (1 - \nu_{z,k,\ell})I_{k,\ell+s(\beta)/2,\ell} + \nu_{z,k,\ell} I_{k,\ell-s(\beta)/2}, \tag{3.372}$$

where $s(\xi) = sign(\xi)$, $s(\beta) = sign(\beta)$, $\nu_{x,k,\ell} \in [0, 1)$ and $\nu_{z,k,\ell} \in [0, 1)$ are weight parameters. One can use values of weight parameters as follows:

$$\nu_{x,k,\ell} = 1/(2 + h_{x,k,\ell}),\ \nu_{z,k,\ell}\frac{(h_{x,k,\ell}/h_{z,k,\ell})(1 + h_{x,k,\ell})}{2 + 2h_{x,k,\ell} + h_{x,k,\ell}^2} \text{ at } h_{x,k,\ell} \le h_{z,k,\ell}, \tag{3.373}$$

$$\nu_{x,k,\ell} = \frac{(h_{z,k,\ell}/h_{x,k,\ell})(1 + h_{z,k,\ell})}{2 + 2h_{z,k,\ell} + h_{z,k,\ell}^2} \nu_{z,k,\ell} = 1/(2 + h_{z,k,\ell} \text{ at } h_{x,k,\ell} \ge h_{z,k,\ell}. \tag{3.374}$$

Optical steps $h_{x,k,\ell}$ and $h_{z,k,\ell}$ are defined as:

$$h_{x,k,\ell} = \sigma_{ext}^{k,\ell}\Delta x_k/|\xi|,\ h_{x,k,\ell} = \sigma_{ext}^{k,\ell}\Delta z_k/|\beta|. \tag{3.375}$$

This scheme defines outside fluxes $I_{k+s(\xi)/2,\ell}$, $I_{k,\ell+s(\beta)/2}$ via entering fluxes $I_{k-s(\xi)/2,\ell}$, $I_{k,\ell-s(\beta)/2}$ in a physically correct manner in any cell of any grid. It permits to obtain discontinuous solutions and solutions with large gradients accurately. This completes a brief description of this technique. Further details are given by Nikolaeva et al. (2005).

3.9.2.3 Numerical results

Let us consider results of numerical calculations for the case presented in Fig. 3.40, where a rectangular coordinate system *xyz* is introduced. Solar light is approximated by the monodirectional source having intensity $F_0\delta(\mu - \mathrm{M})\delta(\phi - \Phi)$. Here $M = \cos\Theta$, $\mu = \cos\theta$ and the pair (Θ, Φ) gives the direction of solar light propagation in the spherical coordinate system defined by the axis z and angles (θ, ϕ). The azimuth Φ is counted with respect to the positive direction of the axis x. In this chapter only results for $\Phi = 0$ and $\Phi = \pi$ will be reported. It means that solar light enters the atmosphere from the direction of the positive values of x (then

$\Phi = \pi$) or from the direction of the negative values of x (then $\Phi = 0$). The intensity of multiply scattered light is calculated along the axis x in the zenith direction as shown in Fig. 3.40 (see the line of visualization in Fig. 3.40).

We divide the terrestrial atmosphere in two equal semi-spaces separated by a local vertical plane. One part is filled by a cloudy medium and another one by the atmospheric aerosol. The processes of molecular scattering and absorption are neglected. Scattering media are assumed to be homogeneous and infinite in planes $z = const$. We assume that there are no light scattering particles at $z > 4$ km. All downwardly propagated photons, which reach the plane $z = 0$, are assumed to be absorbed. Therefore, the contribution of the ground albedo is neglected.

Droplets in a cloud are characterized by the Cloud C.1 particle size distribution with the effective radius equal to 6 μm. The single-scattering diagram for an elementary volume of a cloudy medium is calculated at wavelength 412 nm using the Mie theory. The phase function in the aerosol medium is represented by the Henyey–Greenstein formula. The asymmetry parameter of the cloud phase function g is equal to 0.85. The value of g for the aerosol phase function is equal to 0.7. The optical thickness of cloudy and aerosol portions of the scene are 30 and 1.2, respectively. Also we have studied the variation of the reflected light as observed from a satellite for a nadir observation geometry as the function of the solar angle Θ.

Under conditions specified above, the considered problem is reduced to the 2-D problem. Both aerosol and cloud are homogeneous around z-direction on the height interval (0 km, 4 km). They are contained in rectangular boxes. We will study the upwelling light field in the zenith direction along the line of visualization shown in Fig. 3.40. Clearly, the intensity of the reflected light field must depend on the coordinate x.

The largest gradients of the reflection function $R(x)$ are expected in the area closest to the cloud boundary. Because both an aerosol and a cloud are extended to infinity along axis x, this function far from boundaries must be equal to the value, which can be obtained from the 1-D radiative transfer equation.

The results of calculations using the RADUGA code (Nikolaeva et al., 2005) are given in Figs. 3.41 and 3.42. We also show comparisons with 3-D Monte Carlo code MYSTIC in Figs. 3.41 and 3.42. The MYSTIC (Mayer, 1999) is the Monte Carlo code for the physically correct tracing of photons in cloudy atmospheres. The MYSTIC is a forward Monte Carlo code which traces photons on their individual paths through the atmosphere, similar to what is described by Cahalan et al. (1994). Radiances are calculated using a local estimate technique. In this configuration, MYSTIC has been successfully validated in the intercomparison of 3-D radiation codes (see http://climate.gsfc.nasa.gov/I3RC). MYSTIC is operated within the libRadtran package (see http://www.libradtran.org), which prepares the optical properties of the atmosphere to be used in the model. A model domain of 80 km in x was used. A large domain size is important since MYSTIC uses periodic

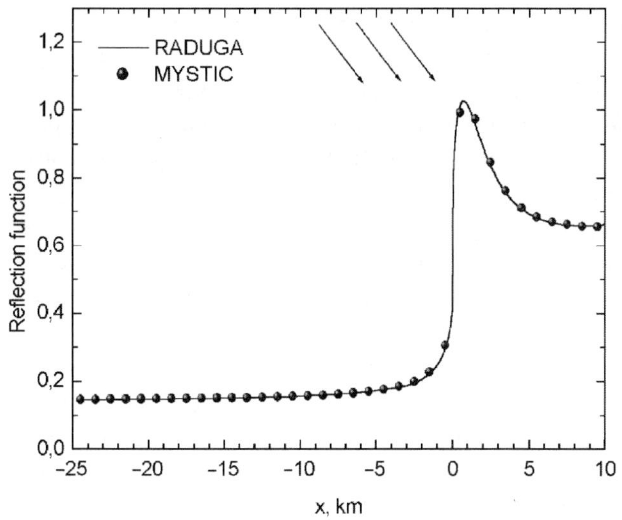

Fig. 3.41. The reflection function in the vicinity of the cloud edge in the nadir direction. Calculations are performed for the illumination from a clear sky side with the solar zenith angle 60°. The aerosol optical thickness is equal to 1.2 and the cloud optical thickness is equal to 30. Results of the RADUGA and MYSTIC codes in the direction perpendicular to the cloud band are shown (Nikolaeva et al., 2005).

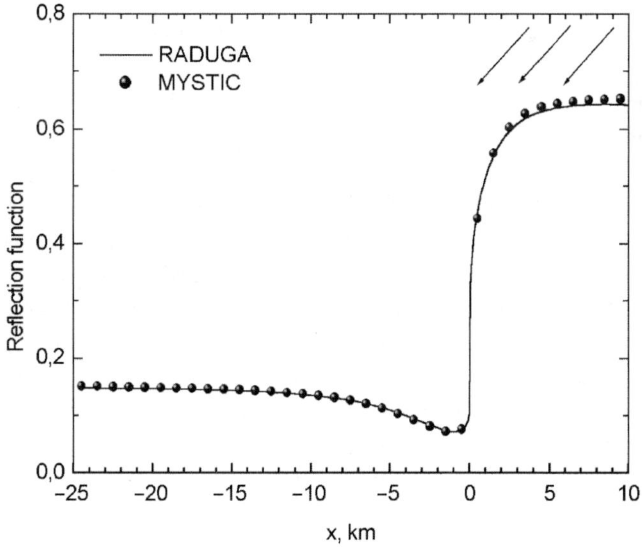

Fig. 3.42. The same as in Fig. 3.41 but for the illumination from the cloud side (Nikolaeva et al., 2005).

boundary conditions. The model resolution was set to 0.1 km; the MYSTIC results are, therefore, averages over 0.1 km bins.

It follows from results of comparisons as shown in Figs. 3.41 and 3.42 that differences are below 1%. Therefore, both the MYSTIC and RADUGA provide very accurate results as far as calculations of the light reflected by an aerosol-cloudy medium are concerned.

Some physical dependencies are clearly seen in Figs. 3.41 and 3.42. For instance, it follows from Fig. 3.42 that there is a shadow near the cloud border for the illumination from the cloud side ($\Phi = \pi$). It follows from geometrical considerations that the shadow extends from the cloud boundary up to the distance $\mathbb{Z} = Htg\vartheta_0$, where H is the cloud top altitude and ϑ_0 is the solar zenith angle. Therefore, the shadow covers a larger region for larger solar zenith angles.

Also we have a brightening effect in Fig. 3.41 due to the cloud side illumination effects ($\Phi = 0$). These two effects (*shadowing* and *brightening*) are primarily due to the direct light interaction with a scattering medium. They lead to roughening effects in 2-D–3-D transfer problems. We also observe (see Fig. 3.42) the decrease of the reflection near the border of the cloud (inside the cloud) as compared to the 1-D case. This is due to photon leaking in the area with a smaller extinction coefficient in the clear part of the scene. The increase in the aerosol phase function in Fig. 3.41 close to the cloud is due to channelling of photons from a cloud to the aerosol side. These two effects (*photons channelling* and *leaking*) lead to a smoothening of the radiative field. The four effects considered here are valid not only for a simple case studied here but also for broken cloud systems (Varnai and Marshak, 2003).

Three-dimensional effects can be parameterized taking into consideration these four fundamental effects. The parameterization of 3-D effects is of a great importance for satellite remote sensing because calculations presented here are computationally expensive and can not be included in the operational aerosol/cloud retrieval algorithms. Also, these effects cannot be neglected. In particular, if satellite pixels observe area correspondent to brightening/shadowing effects, then large biases in retrieved cloud/aerosol parameters are expected (see Figs. 3.41 and 3.42). This also points to the necessity of the development of the simultaneous aerosol-cloud (and surface) retrieval algorithm. This complex system should be considered in the retrieval process as a whole (Cahalan et al., 2001), which will allow us to obtain the most accurate estimations of the atmospheric and surface parameters from both airborne and satellite remote sensing measurements.

Chapter 4

APPLICATIONS

4.1 Optical Phenomena in Clouds

4.1.1 Corona

Sky under extended cloudiness looks grey with a uniform variation of transmitted light intensity. Isolated clouds look white on the background of a blue sky. This whiteness is primarily due to nonselective light scattering by water droplets in the visible. The solar light is composed of a mixture of waves having various wavelengths. A prism can divide the sunlight beam in a continuous colour spectrum. This is due to the change of the refractive index of the prism with the wavelength. Interestingly, clouds can also exhibit spectrally selective scattering if special conditions are met. In particular, coloured rings can appear in some observation directions. These phenomena are rather easily observed for thin clouds. Thick clouds are characterized by multiple light scattering, which reduces possible spectral variations of light scattered by single water droplets.

A typical example of an optical phenomenon associated with thin clouds is the corona. A corona appears as a sequence of coloured rings around a white light source observed through a cloud. These rings are most often observed at night in sunlight reflected from the Moon and passed through a thin cloud on its way to the detector or towards an observer's eye. The brightness of the background is low at night and the phenomenon is easily observed. Also the observation of the moonlight is not harmful for the human eye as compared to the direct observation of sunlight, even if seen through a thin cloud.

The corona can be fully described using the Mie theory for monodispersed droplets or polydispersions having very small coefficients of variance. Clouds with large variations of droplet sizes produce the colourless corona, which is due to enhanced light scattering for small scattering angles. The information on the size of droplets can in principle be obtained using measurements of the angular light distribution in the corona. In particular, smaller droplets generally produce a more extended solar corona as compared to larger droplets. A simple Fraunhofer diffraction theory can be used to understand the phenomenon. It has been shown above that angular distribution of light intensity $I(\theta)$ scattered by a single spherical droplet at small angles $\theta \ll 1$ is described by the following function:

$$I(\theta) = \frac{k^2 a^2 J_1^2(z)}{\theta^2} I_0, \tag{4.1}$$

where I_0 is the intensity of incident light, J_1 is the Bessel function, a is the radius of a droplet, and $k = 2\pi/\lambda$, λ is the wavelength, and $z = ka \sin\theta \approx ka\theta$ as $\theta \to 0$. In particular, we have at $\theta = 0$:

$$I(0) = \frac{k^4 a^4}{4}, \tag{4.2}$$

where we used the following property of the Bessel function: $J_1(x) \approx x/2$ as $x \to 0$. Also we have for the dimensionless Mie intensity $i(\theta)$ in the framework of this approximation:

$$i(\theta) = x^4 \frac{J_1^2(z)}{z^2}. \tag{4.3}$$

It follows from this equation that the intensity of scattered light in the forward direction is proportional to the squared geometrical cross-section of the particle and it is inversely proportional to the fourth power of the wavelength. So incident light with a smaller wavelength produces a brighter central core as compared to incident light having a larger wavelength. The Bessel function $J_1(z)$ (and, therefore, the function $I(\theta)$) oscillates with θ. Zeros of the Bessel function $J_1(z)$ are located at z equal to 3.8317, 7.0156, 10.1735, 13.3237, and 16.4706. Therefore, at these angles dark circles in the Fraunhofer pattern are expected. We have for the angular position θ_{min} of the first dark ring:

$$\theta_{min} = \frac{3.8317}{ka}. \tag{4.4}$$

It follows that this angle decreases with a. However, it increases with λ. These basic rules also hold for other minima and maxima. This means that the inner part of every circle in the diffraction pattern for white light illumination conditions must be bluish as compared to the reddish colour of the outer part of each diffraction ring. The first three maxima of the diffraction pattern are located at z equal to 5.1365, 8.4163, and 11.5925, respectively. It means that the angular position of the first bright ring $\theta_{\max}$ is given by

$$\theta_{\max} = \frac{5.1365}{ka} \tag{4.5}$$

or

$$\theta_{\max} = \frac{0.8175\lambda}{a}. \tag{4.6}$$

Again we see that red light maxima are shifted to outer parts of the rings.

This is actually what is observed in a corona. In particular, taking $a = 6$ μm, we obtain $\theta_{\max} = 3.1^\circ$, 3.9°, and 5.5° at $\lambda = 0.4$, 0.5, and 0.7 μm, respectively. However, it should be noted that in reality angles $\theta_{\min}$ and $\theta_{\max}$ are shifted from the values given above. This is referred to as the anomalous diffraction phenomenon. The effect is explained by the fact that the Fraunhofer diffraction gives only the first coarse approximation to the reality. As a matter of fact the Mie theory should be used for an accurate simulation of the corona. Then one has a possibility to account for contributions not only by diffracted but also by reflected and refracted light beams in a given scattering direction.

We show the phase function of a spherical water droplet calculated using Mie theory at $a = 100$ μm, $\lambda = 0.5$ μm, and $n = 1.33$ in Fig. 4.1. It follows that positions of maxima and minima coincide with those calculated using the Fraunhofer diffraction theory given above. In particular, it follows from the Fraunhofer theory that that minima must be located at 0.19°, 0.35°, 0.51°, 0.67°, and 0.83°. The positions of the first three maxima are at 0.26°, 0.42°, and 0.58° for the case studied. These numbers coincide with those obtained from Mie theory (see Fig. 4.1). So both the Fraunhofer theory and the Mie theory can be used to predict the positions of minima and maxima for large spherical particles. These positions can be used for optical droplet sizing.

Note that the polarization of scattered light is low at small scattering angles.

4.1.2 Glory

Glory or anticorona appears at scattering angles close the backward direction $\theta = \pi$. The physical mechanism behind the corona is the interference of backscattered rays. An approximate theory has been proposed by van de Hulst (1981). In particular, he showed that the scattered light intensity $I_{sca}(\theta)$ can be described as:

$$I_{sca}(\theta, a) = \left(c_1 J_0^2\left(ka(\pi - \theta)\right) + c_2 J_2^2\left(ka(\pi - \theta)\right)\right) x^2 I_0, \tag{4.7}$$

where constants c_1 and c_2 depend on the refractive index of particles and their size. Interestingly, the anticorona is much more robust to the change of the coefficient

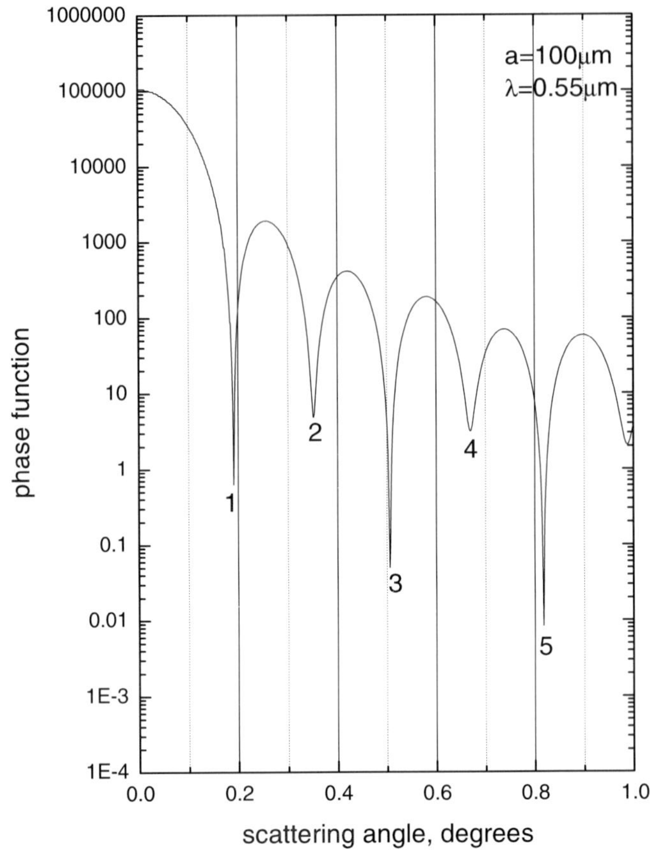

Fig. 4.1. Phase function of a water droplet having the radius 100 μm at $\lambda = 0.55$ μm. Only the small angle scattering region is shown.

of variance of the droplet size distribution as compared to the corona. In particular, it exists for relatively broad size distributions.

Let us generalize the van de Hulst's equation for the anticorona to the polydispersed case. Then it follows for the average intensity:

$$\bar{I}_{sca}(\theta) = \int_0^\infty I_{sca}(\theta, a) f(a)\, da \tag{4.8}$$

or approximately

$$\bar{I}_{sca}(\theta, a) = \left(c_1^* J_0^2\,(k\tilde{a}(\pi - \theta)) + c_2^* J_2^2\,(k\tilde{a}(\pi - \theta))\right) I_0, \tag{4.9}$$

where we used the approximation:

$$\int_{a_1}^{a_2} \phi(\zeta) f(\zeta)\, d\zeta = f(a_3) \int_{a_1}^{a_2} \phi(\zeta)\, d\zeta, \tag{4.10}$$

with $a_3 \in [a_1, a_2]$ and

$$c_j^* = k^2 \int_0^\infty a^2 c_j(a) f(a)\, da, \quad j = 1, 2. \tag{4.11}$$

Therefore, we have for the phase function:

$$p(\theta) = p(\pi) J_0^2 \left(k a_3 (\pi - \theta)\right) + q J_2^2 \left(k a_3 (\pi - \theta)\right). \tag{4.12}$$

Parameters a_3 and q can be found fitting Eq. (4.69) with the use of Mie calculations. Also, we have used the fact that $J_0(0) = 1$, $J_2(0) = 0$. The value of a_3 is generally unknown. Let us assume that $a_3 = a_{ef}$. Then it follows:

$$p(\theta) = p(\pi) J_0^2 \left(k a_{ef} (\pi - \theta)\right) + q J_2^2 \left(k a_{ef} (\pi - \theta)\right). \tag{4.13}$$

Therefore, we might expect that the phase function must depend mostly on the scaling parameter

$$z = (\pi - \theta) k a_{ef}. \tag{4.14}$$

The dependence of the phase function on the value of z at $a_{ef} = 4(2)16$ µm at $\lambda = 0.532$ µm is shown in Fig. 4.2. Points give results according to the approximate equation given above at $q = 4.0$, $p(\pi) = 0.62$. It follows that $p(\theta)$ depends mainly on the parameter z, although there is a spread depending on the effective radius. The phase function in the exact backward direction $p(\pi)$ is almost insensitive to the value of a_{ef} and close to 0.62 at $a_{ef} = 4$ µm. There is a maximum of the phase function at $z_{\max} \approx 3$ at $a_{ef} = 16$ µm. The angular position of the maximum $\theta_{\max}$ is only weakly dependent on the size of particles for a given wavelength with smaller values of $z_{\max}$ for smaller particles. It follows, therefore:

$$\theta_{\max} \approx \pi - \frac{3\lambda}{2\pi a_{ef}}. \tag{4.15}$$

This means that shorter wavelengths produce larger values of $\theta_{\max}$ making the inner region of the glory ring related to the second maximum bluish. The outer parts of the ring appear reddish. This is also confirmed by the experiment, and by Mie calculations (see Fig. 4.3). It follows from Fig. 4.2 that the value of the phase function in the maximum decreases with a_{ef} in the range of parameters studied. The accuracy of Eq. (4.13) decreases with z. The approximation can not be used at $z > 2.0$. However, the multiplication of the second term in Eq. (4.13) by $\nu = \exp(-0.07z)$ extends the applicability of the approximation till $z = 4$. Therefore, the approximation can be used, e.g., in conjunction with analytical results of laser sounding theory (Katsev et al., 1998).

We show the phase function $p(\theta)$ for several wavelengths at $a_{ef} = 6$ µm in Fig. 4.3. It follows from Figs. 4.2 and 4.3 that glory rings can be used for optical

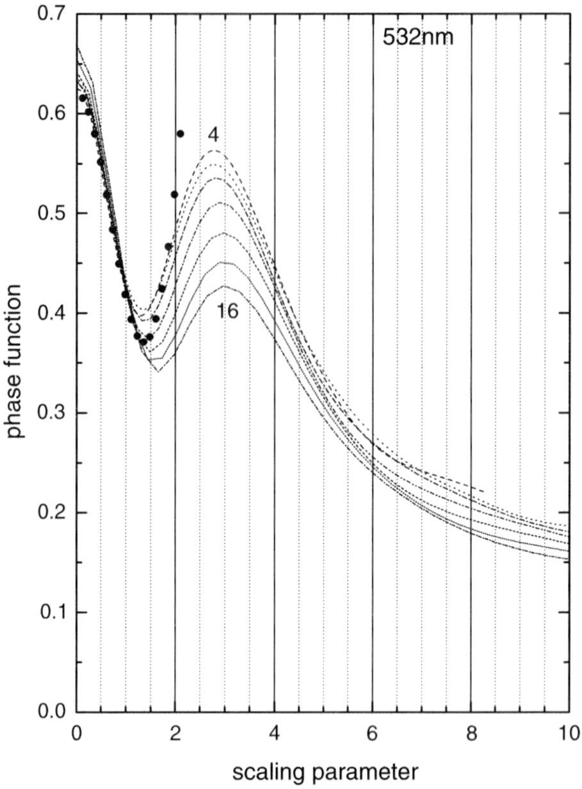

Fig. 4.2. Phase function of a water cloud at $\lambda = 532$ nm, $a_{ef} = 4, 6, 8, 10, 12, 14, 16$ μm as the function of the scaling parameter z. Points show the results using the approximate formula.

particle sizing techniques. Also we see that glory (although invisible to the human eye) exists in the infrared as well. The degree of light polarization in glory under solar light illumination conditions is shown in Fig. 4.4. It is interesting that the degree of polarization changes its sign. The neutral point, where the degree of linear polarization vanishes, depends on the wavelength. The angular position of this point θ_n is closer to the backward direction for smaller wavelengths (or larger droplets). The degree of linear polarization in the vicinity of the backward direction $\theta \approx \pi$ is positive with larger values for larger effective size parameters $x_{ef} = 2\pi a_{ef}/\lambda$ (oscillations are in predominantly in the plane perpendicular to the scattering plane). The degree of polarization is equal to zero at $\theta = \pi$. The normalized phase matrix element p_{34} shown in Fig. 4.5 describes the linear-to-circular polarization mode conversion (Kokhanovsky, 2003a). It gives the degree of circular polarization of scattered light under linearly (the azimuth 45°) polarized light illumination conditions. It follows that the circular polarization of glory for linearly

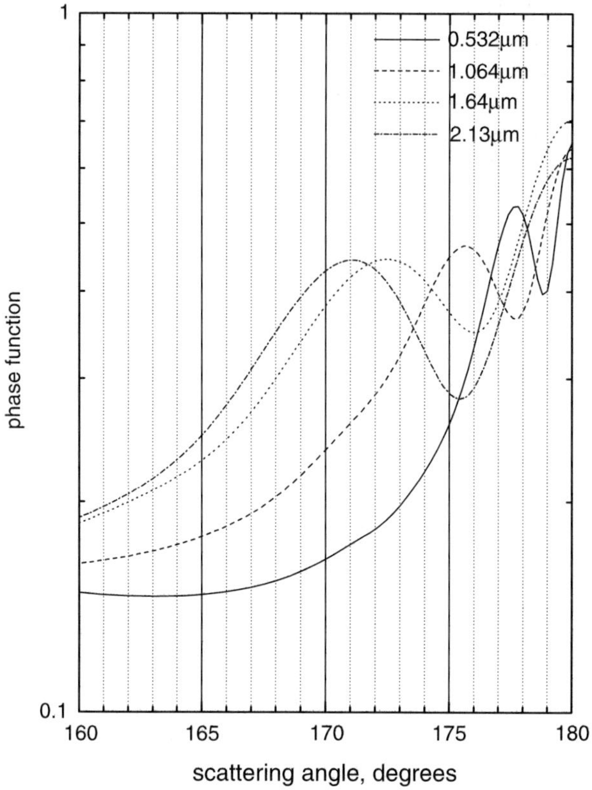

Fig. 4.3. Phase function of a water cloud at $\lambda = 0.532,\ 1.064,\ 1.64$ and 2.13 μm $a_{ef} = 6$ μm as the function of the scattering angle in the glory region.

(e.g., lidar) light illumination conditions could be quite large (especially in the infrared). The circular polarization also changes its sign in the glory region. The pulsations of circular polarization as shown in Fig. 4.5 can be used for optical particle sizing.

The normalized phase matrix element p_{44} shown in Fig. 4.6 describes the degree of circular (e.g., lidar) polarization for scattered light under circular light polarization conditions. We see that the degree of circular polarization is close to -1 for angles $\theta \to \pi$. This means that the backward scattering ($\theta \approx \pi$) does not change the absolute value of the circular polarization. However, the rotation of the electric vector is reversed. So the scattering at $\theta = \pi$ can be used in technological applications to switch the rotation direction of the electric vector. The position of the neutral point is closer to the backward scattering direction for shorter wavelengths and larger particles. This can be used for optical particle sizing using circularly polarized light beams.

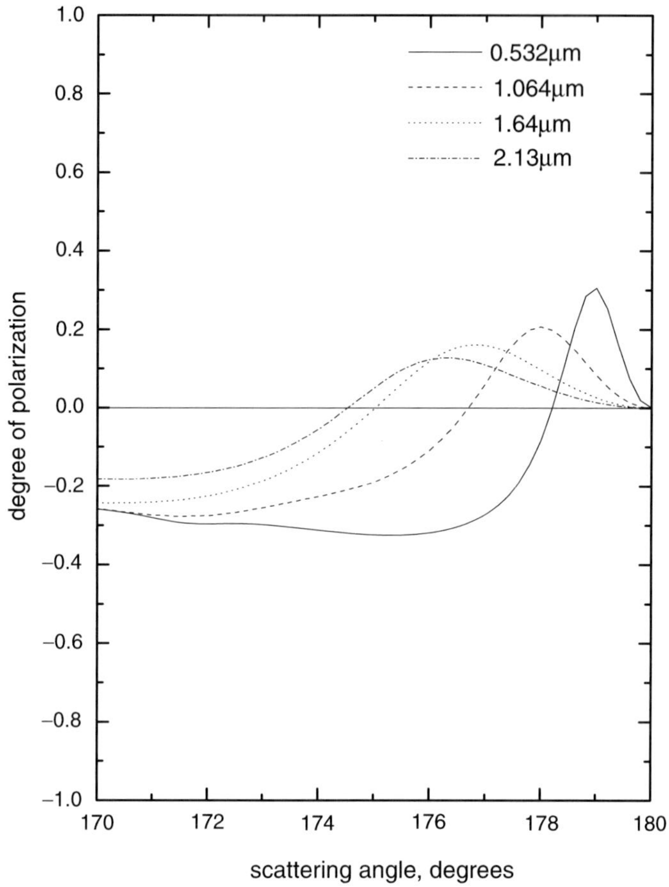

Fig. 4.4. The same as in Fig. 4.3 except for the degree of polarization.

4.1.3 The Rainbow

Perhaps the best known and most familiar optical effect associated with clouds is that of the rainbow. The rainbow occurs due to the interference of rays reflected one or several times inside the droplet. Mie theory is capable to describe main features of the rainbow with a high accuracy. However, we will begin by providing some insights using the geometrical optics approximation. In particular, simple geometrical optics calculations give the following expression for the scattering angle of a singly internally reflected ray:

$$\theta_r = \pi + 2(\varphi - 2\psi), \tag{4.16}$$

where φ is the incidence angle and ψ is the refraction angle. Let us consider the

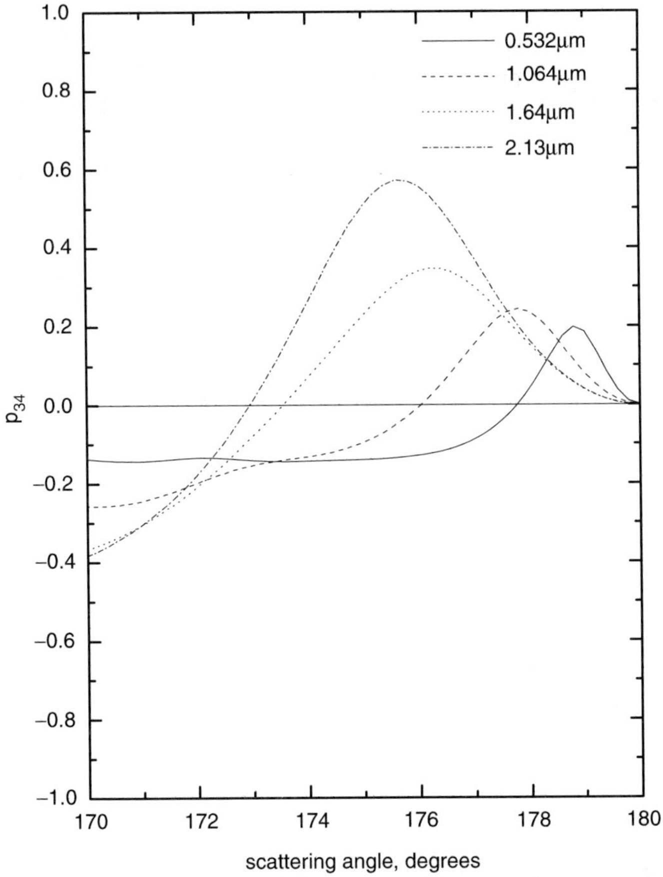

Fig. 4.5. The same as in Fig. 4.3 except for the element p_{34}.

derivative of θ_r with respect to the incidence angle. Then it follows that

$$\frac{d\theta_r}{d\varphi} = 2 - 4\frac{d\psi}{d\varphi} \tag{4.17}$$

or

$$\frac{d\theta_r}{d\varphi} = 2 - 4\frac{\cos\varphi}{n\cos\psi}, \tag{4.18}$$

where n is the refractive index and we used the Snellius law:

$$\sin\varphi = n\sin\psi. \tag{4.19}$$

It follows that θ takes a minimal value at

$$\varphi_r = \cos^{-1}\left\{\sqrt{\frac{n^2-1}{3}}\right\}. \tag{4.20}$$

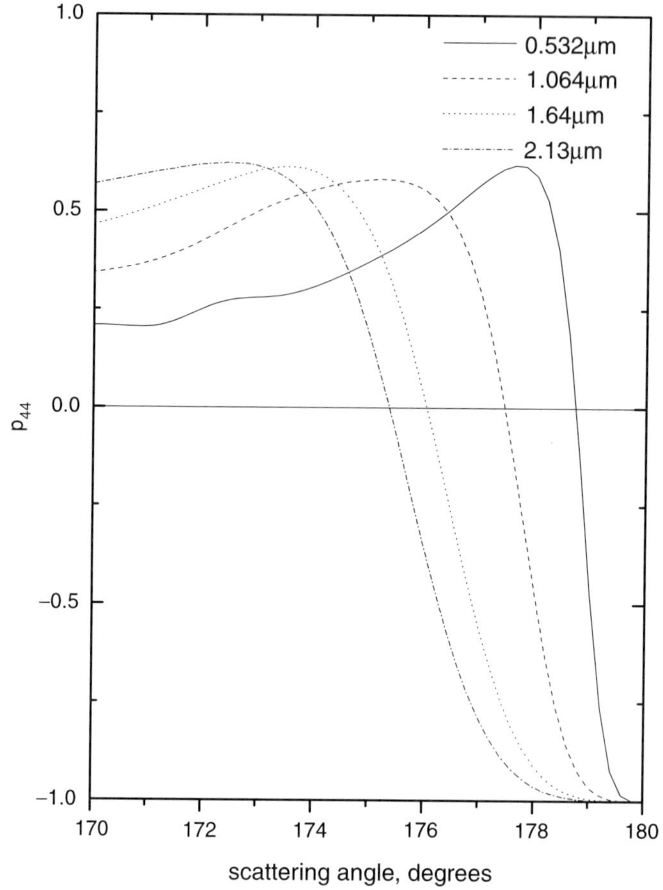

Fig. 4.6. The same as in Fig. 4.3 except for the element p_{44}.

This means that a lot of incident rays will contribute in the correspondent scattering direction. This will lead to brightening of the sky in the direction of rainbow. In particular, we have at $\lambda = 0.5$ μm and $n = 1.333$: $\varphi \approx 59°$, and, therefore, $\theta_r = 139°$. Therefore, to observe a rainbow, one should have a light source behind and look in the direction that makes a scattering angle of 139°.

Clearly, the same angle can be achieved for different azimuths. It will produce a characteristic feature known as the rainbow. As was noted above the angle φ depends on n, which means that both φ and θ depend on λ due to the wavelength dependence of n. In practice, this leads to the separation of colours in the rainbow. Because n slightly decreases with the wavelength for water in the visible, φ and θ decrease as well. Therefore, the outer parts of rainbow must be reddish in colour.

Quite often the secondary rainbow is observed. This rainbow originates due to double internally reflected light beams. Therefore, it is weaker in brightness than the primary rainbow. It follows for the scattering angle associated with two-times internally reflected rays:

$$\theta_r' = 2(\varphi - \psi). \tag{4.21}$$

Repeating the same procedure as above, we obtain that the derivative of this angle vanishes at

$$\varphi = \cos^{-1}\left\{\sqrt{\frac{n^2-1}{4}}\right\} \tag{4.22}$$

This gives that $\theta_r' = 129$ degrees at $n = 1.33$. Therefore, the secondary rainbow is observed at smaller scattering angles than the primary rainbow.

The accurate angular positions of maxima of light in a rainbow are given by the Mie theory. We show a characteristic rainbow pattern calculated with the Mie theory at $a_{ef} = 6$ μm and several wavelengths in Fig. 4.7. It follows that the rainbow maximum moves to smaller scattering angles for smaller wavelengths (or smaller droplets, see Fig. 4.8). This leads to the reddish colour of the outer band of a primary rainbow as observed in nature. For the secondary rainbow, the order of colors is opposite (see, e.g., http://www.philiplaven.com/index1.html).

The approximate theory of rainbow is given by Airy (1838) and Nussenzveig (1992). In particular, Airy has shown that the angular distribution of light in a rainbow at a given size parameter $x = ka$ is approximately proportional to the integral:

$$A(\theta) = \left\{\int_0^\infty \cos\left[\frac{\pi}{2}\left(by - y^3\right)\right]dy\right\}^2, \tag{4.23}$$

where

$$b = \sqrt[3]{\frac{12x^2}{\pi^2 c}}\,\{\theta - \theta_r\} \tag{4.24}$$

with $c = 4.89$ for the first rainbow and $c = 27.86$ for the second rainbow. Studies of this integral show that the maximum of the rainbow intensity does not coincide with θ_r but rather occurs at the somewhat larger angle $\theta_{\max}$ depending on x:

$$\theta_{\max} = \theta_r + \mathbb{C}x^{-2/3}, \tag{4.25}$$

where $\mathbb{C} = \sqrt[3]{\frac{9c\pi^2}{4}} \approx 4.77$. A similar equation, but with a different constant $\mathbb{C}$, is valid for the secondary rainbow. This dependence is confirmed by exact data shown in Fig. 4.8. It follows that the rainbow intensity at maximum increases with a_{ef}.

The thorough review of rainbow and glory theories has been conducted by Adam (2002) (see also Jackson (1999)).

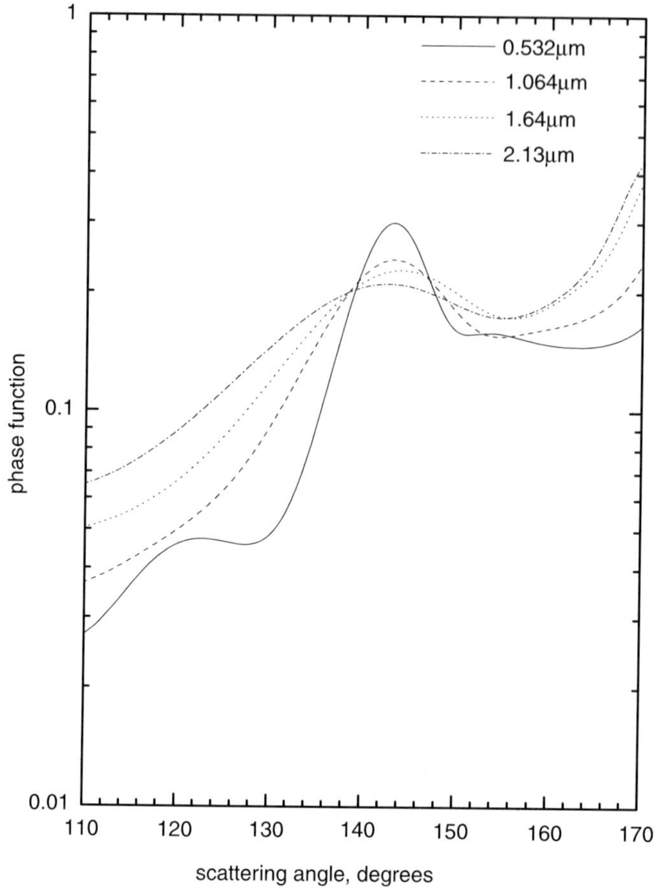

Fig. 4.7. Phase function of a water cloud at $\lambda = 0.532$, 1.064, 1.64, and 2.13μm $a_{ef} = 6$ μm as the function of the scattering angle in the rainbow region.

The degree of light polarization in a rainbow for solar light illumination conditions is quite large as shown in Fig. 4.9. It can reach 80% and even larger values depending on the size of particles and the wavelength. Oscillations occur predominantly in the plane perpendicular to the scattering plane. The reason for such a high polarization of the rainbow is due to the fact that the internal reflection of the rainbow ray in the drop occurs at the angle close to the Brewster angle equal to $\tan^{-1}(n)$.

It follows from Figs. 4.7 and 4.9 that the secondary rainbow at $\theta \approx 120°$ is better visible in the degree of polarization angular pattern as compared to the intensity angular distribution. Oscillations in Fig. 4.9 at $\theta \geq 145°$ are due to supernumery rainbows. They are sometimes observed in nature close to the outer band of the main rainbow.

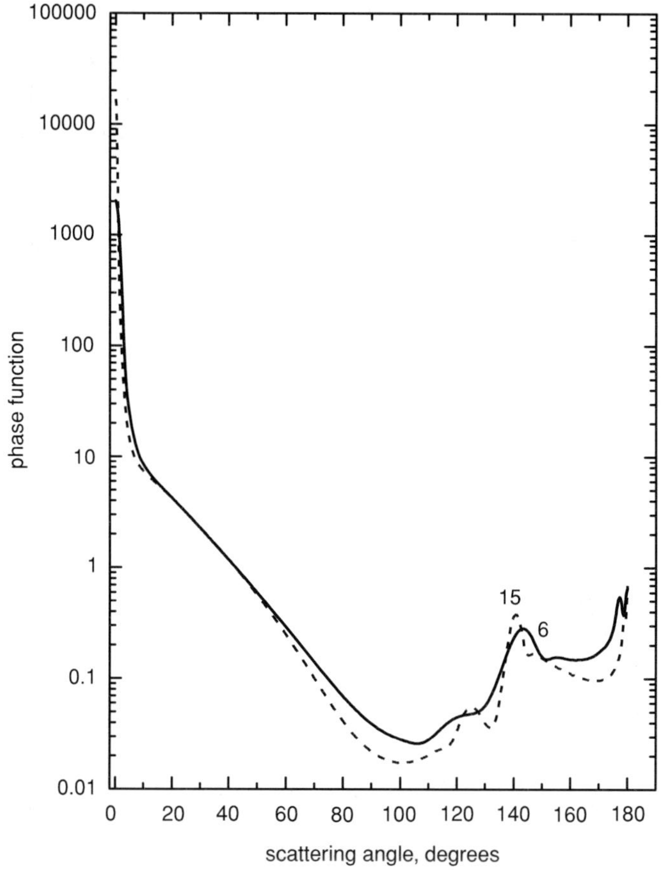

Fig. 4.8. Phase function of a water cloud at $\lambda = 0.55$ μm, $a_{ef} = 6$, 15 μm.

The wavelength dependence of the normalized phase matrix elements p_{34} and p_{44} in the vicinity of the rainbow angle is shown in Figs. 4.10 and 4.11.

4.1.4 Halo

The halo is a bright ring, which is observed around the moon or the sun. The difference from the corona is in the radius of this ring. In particular, the corona is observed for scattering angles $\theta = 0 - 5^\circ$, depending on the size of particles. The halo is observed at 22°. It should be pointed out that the Mie theory does not predict any anomaly in light scattering around 22° (see Fig. 4.8). So the halo can be explained using only the model of nonspherical particles, namely ice crystals. The most important shape of ice crystals in the terrestrial atmosphere

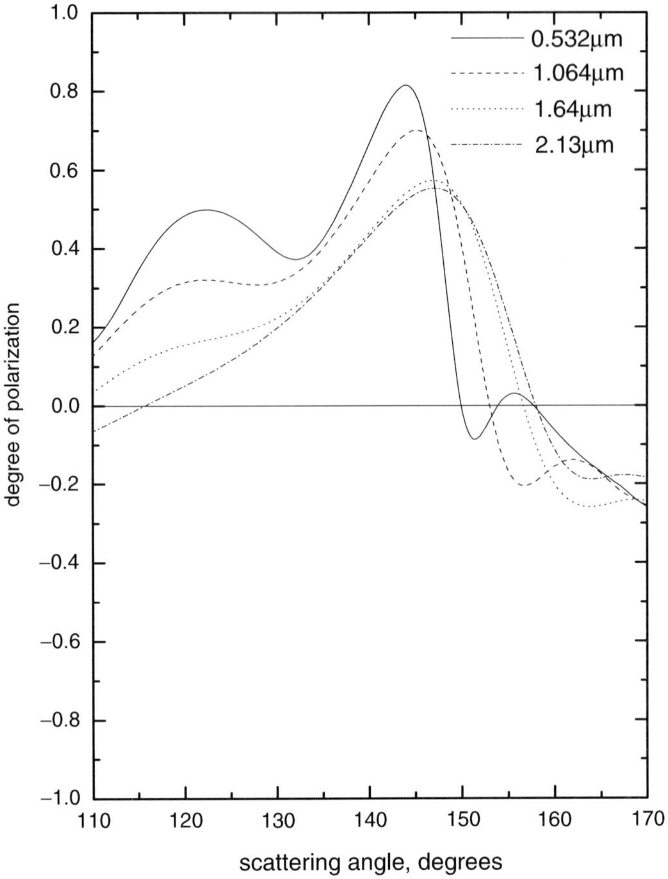

Fig. 4.9. The same as in Fig. 4.7 except for the degree of polarization.

is the hexagon. The path of a light ray in the part of crystal with the angle A is shown in Fig. 4.12. The halo corresponds to the minimal value $D_{\min}$ of the deviation angle $D = i_1 - r_1 + i_2 - r_2$. The value of D is minimal at $i_1 = i_2 \equiv i$, $r_1 = r_2 = r = \sin^{-1}\sin(i/n)$. Here n is the refractive index. Then we have $\mathrm{A} = 2r$ (see Fig. 4.12) and $D_{\min} = 2i - \mathrm{A}$. The angle of the light beam deviation for the hexagonal prism can be obtained from the Snellius law: $n\ \sin r = \sin i$, where $r = \mathrm{A}\,/\,2$ and $i = (D_{\min} + \mathrm{A})/2$. This means that

$$D_{\min} = 2\sin^{-1}\left\{\frac{n\sin \mathrm{A}}{2}\right\} - \mathrm{A}. \tag{4.26}$$

It follows from this equation that $D_{\min} = 21°54'$ for hexagons with $\mathrm{A} = 60°$ at $n = 1.31$. Also we have $D_{\min} = 45°44'$ for angles $\mathrm{A} = 90°$ at $n = 1.31$. The halos associated with angles A equal to 60° and 90° are regularly observed in the

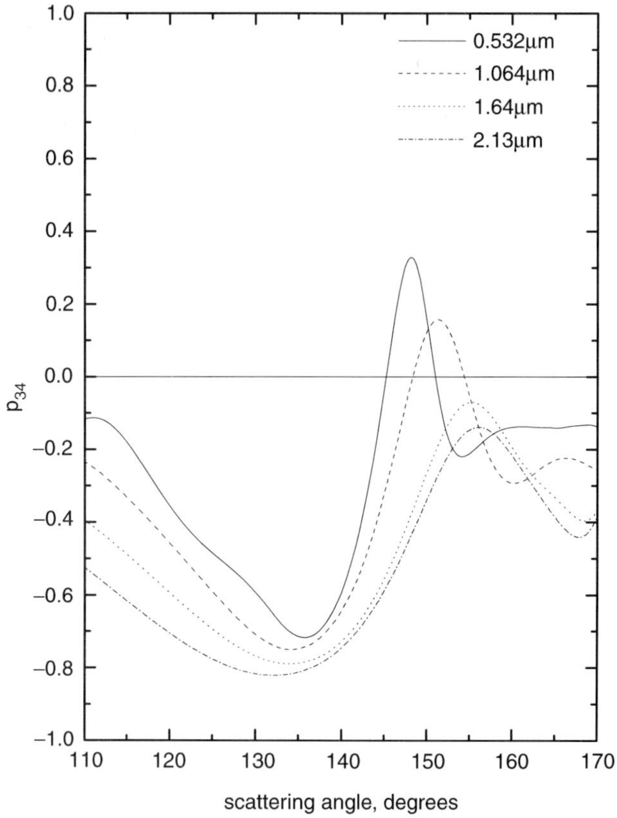

Fig. 4.10. The same as in Fig. 4.7 except for the element p_{34}.

atmosphere for randomly oriented crystals. Oriented crystals can produce the halos at other positions (Tricker, 1967). The inner part of both halos is reddish in color. This is due to the fact that $D_{\min} = 21°36'$ for red light with $n = 1.307$ and $D_{\min} = 22°22'$ for violet light with $n = 1.317$ at $A = 60°$. Also we have at $A = 90°$: $D_{\min} = 45°$ for red light with $n = 1.307$ and $D_{\min} = 47°14'$ for violet light with $n = 1.317$. Therefore, it follows that the 46° halo has an angular width twice as broad. It has a smaller intensity as compared to the halo located at $\theta = 22°$.

We conclude that a 46° halo develops when light enters one side of a columnar ice crystal and exits from either the top or bottom face of the crystal ($A = 90°$). The light is refracted twice as it passes through the ice crystal and the two refractions bend the light by 46° from its original direction. This bending produces a ring of light observed at 46° from the sun or the moon. A 22° halo develops when light enters one side of a columnar ice crystal and exits through another side ($A = 60°$). The light is refracted when it enters the ice crystal and once again when it leaves

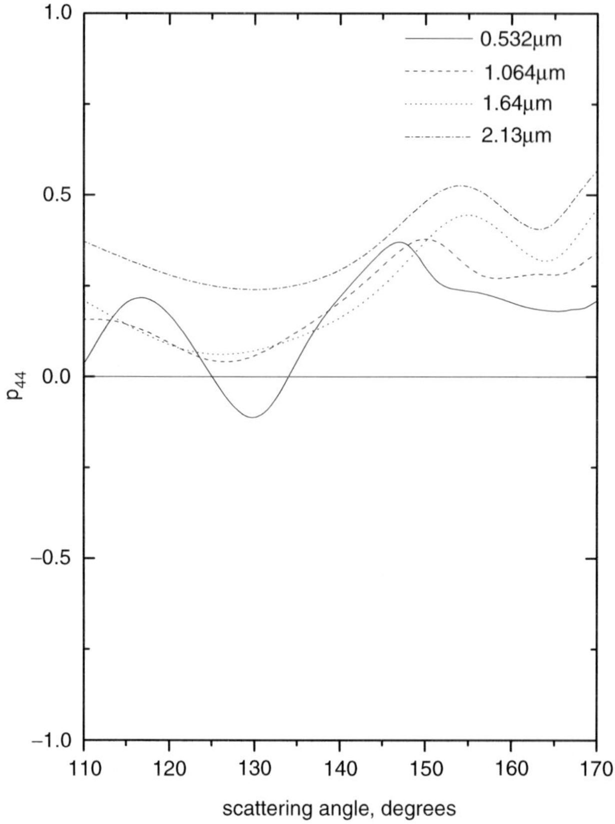

Fig. 4.11. The same as in Fig. 4.7 except for the element p_{44}.

the ice crystal. The occurrence of the 46° halo is rare (about 2% of all cases). The 22° halo is more frequent.

The halo is more easily observed for large crystals because the intensity of halos decrease with the size of crystals. A typical phase function of hexagonal ice crystals calculated using the geometrical optics approach is shown in Fig. 4.13. The halos at 22° and 46° clearly emerge from these calculations. The maximum at

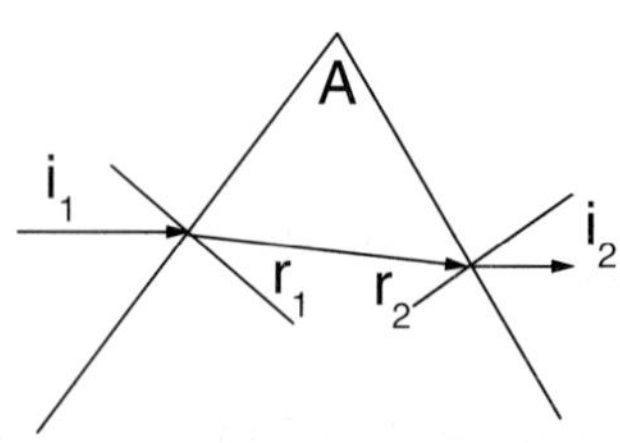

Fig. 4.12. Geometry of the problem.

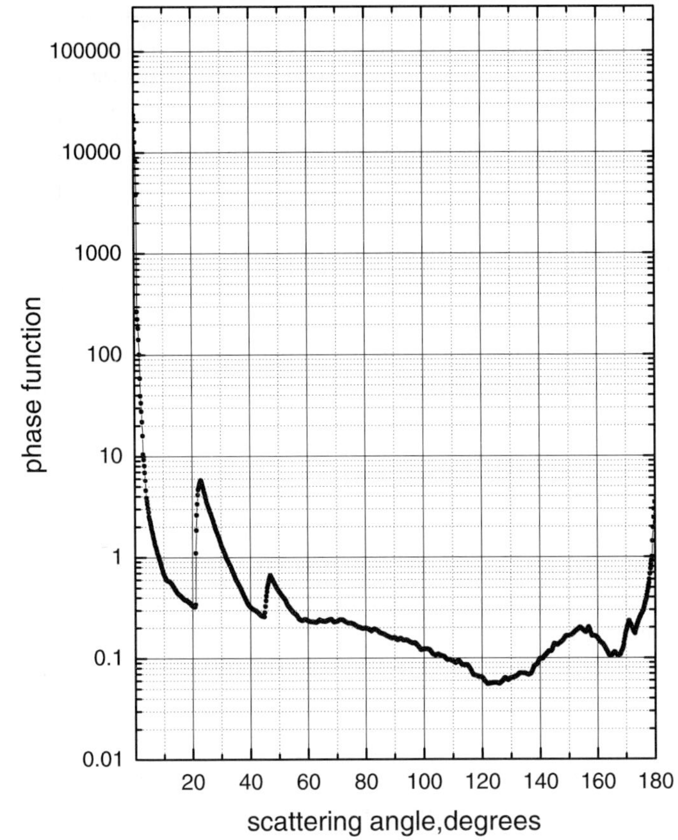

Fig. 4.13. The phase function of randomly oriented hexagonal cylinders with the effective radius 40 μm at the wavelengths 0.55 μm averaged over the size and aspect ratios distributions. The refractive index is equal to $n = 1.31$ (courtesy A. Macke).

150° is also clearly visible. The halo can be washed out by the presence of small or irregularly shaped crystals. The effects of multiple light scattering and the possible presence of supercooled water droplets also reduce the probability of halo formation. However, the halo is an atmospheric effect which occurs quite often. Some observers report a frequency of 10 times per month at certain locations.

4.2 Cloud Remote Sensing

4.2.1 Penetration Depth

The penetration depth is a parameter required in various remote sensing applications. It is defined as the length at which the intensity of incident wave is reduced by the factor $\exp(-1)$. If the absorption coefficient of a homogeneous medium under study is known, the penetration depth can easily be estimated. It could be of value to extend this notion to random media. However, this leads to a number of problems. In particular, let us take a cloud in the sky. In this case, a strongly developed multiple light scattering occurs in the medium. The downward diffuse light intensity reaches a maximum and starts to decrease preserving the angular pattern of scattered light. Brightness of this pattern decreases as $\exp(-k\tau)$, where k is the diffusion exponent and τ is the optical depth.

Space remote sensing applications require information on the thickness of an effective layer of a cloud, which interacts with incident electromagnetic radiation. This depth can be defined as the distance ℓ at which the reflection function reaches 90% of its value for the semi-infinite layer. We can state that scattering layers positioned at depths larger than ℓ only weakly influence the signal detected by an orbiting optical instrument.

The aim of this section is to present simple analytical equations, which can be used for estimations of ℓ in a cloudy atmosphere. Generally, the equations derived can be used inside and outside gaseous absorption bands. However, in the results of the numerical calculations presented, we neglect the influence of the gaseous absorption bands on the value of the penetration depth. Generally speaking, the penetration depth can be found using results of the radiative transfer calculations similar to those presented in Fig. 4.14, where we show the dependence of the reflection function on cloud optical thickness for multiple wavelengths. Data for Fig. 4.14 were obtained using the discrete ordinate method of the radiative transfer equation solution. It was assumed that droplets are characterized by the gamma particle size distribution (PSD) with the effective radius of 6 μm and the coefficient of variance of the PSD equal to 38%. Calculations were performed at $\xi = 0.5$, $\eta = 1.0$ (nadir observation). Here ξ is the cosine of the solar angle and η is the cosine of the observation angle. Obviously, the results do not depend on the relative azimuth φ for the nadir observation conditions.

It follows from Fig. 4.14 that the penetration depth decreases with wavelength. This is mostly due to the fact that water absorption generally increases with the wavelength. In particular, we find using data shown in Fig. 4.14 that the optical penetration depth $\tau_p = \ell/s$ (with $s = 1/\sigma_{ext}$ as the photon free path in a cloud) is equal to 65.5, 37.0, 20.0, 13.2, and 5.7 for wavelengths $\lambda = 900$ nm, 1240 nm, 2250 nm, and 3700 nm, respectively.

Computations as those shown in Fig. 4.14 require a radiative transfer code. Let us show that the penetration depth can also be found using a simple analytical equation. For this we will use the exponential approximation. In particular, the reflection function $R(\xi, \eta, \varphi, \tau)$ can be presented as

$$R(\xi, \eta, \varphi, \tau) = R_\infty(\xi, \eta, \varphi) - t(\tau)\exp(-x - y)u_0(\xi)u_0(\eta), \tag{4.27}$$

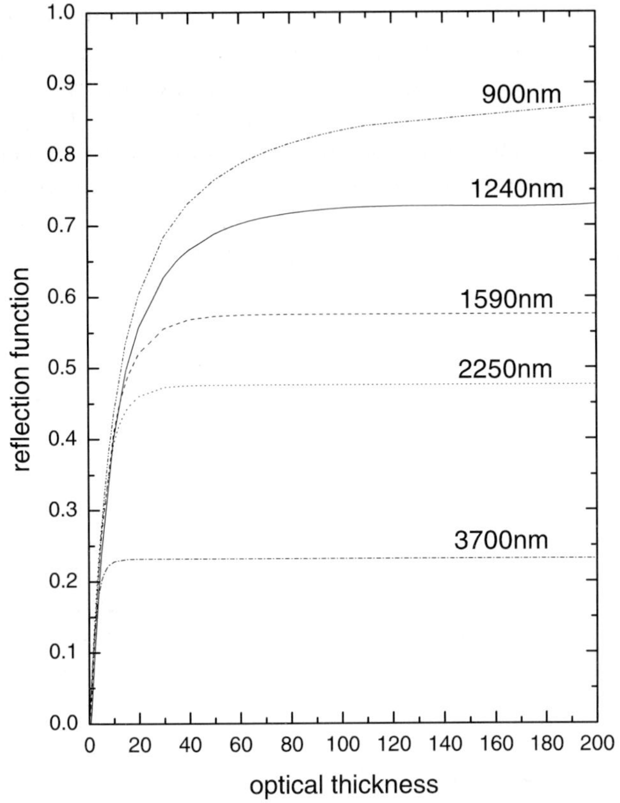

Fig. 4.14. The dependence of the reflection function on the optical thickness for several wavelengths of incident light (see details in text).

where $R_\infty(\xi, \eta, \varphi)$ is the reflection function of a semi-infinite medium having the same microphysical characteristics as a finite layer under study. The function $t(\tau)$ is the global transmittance given approximately by

$$t(\tau) = \frac{\sinh(y)}{\sinh(x + \alpha y)}. \tag{4.28}$$

Here, $x = k\tau$, $y = 4k/[3(1-g)]$, $k = \sqrt{3(1-g)(1-\omega_0)}$, $\omega_0 = 1 - \sigma_{abs}/\sigma_{ext}$, $\alpha \approx 15/14$, g is the asymmetry parameter of the phase function, σ_{abs} is the absorption coefficient and $\sigma_{ext} \equiv 1/s$ is the extinction coefficient. The accuracy of Eq. (4.27) has been thoroughly studied above.

It follows from Eq. (4.27) at $\tau = \tau_p$:

$$ut(\tau_p)\exp(-x(\tau_p) - y(1-u^*)) = b, \tag{4.29}$$

where $b = 0.1$. We used the following approximate result for the reflection function of a semi-infinite weakly absorbing medium:

$$R_{\infty}(\xi, \eta, \varphi) = R_{0\infty}(\xi, \eta, \varphi) \exp(-\mathbb{C}y), \tag{4.30}$$

where $R_{0\infty}(\xi, \eta, \varphi)$ is the reflection function of a semi-infinite medium, under the assumption that absorption of radiation in a cloud does not take place and

$$\mathbb{C} = (1 - 0.05y)\mathbb{N}, \tag{4.31}$$

$$\mathbb{N} = \frac{u_0(\xi)u_0(\eta)}{R_{0\infty}(\xi, \eta, \varphi, \tau)}. \tag{4.32}$$

Equation (4.30) is accurate to within 5% at $y \leq 1.7$. After simple algebraic derivations, it follows from Eq. (4.29):

$$\tau_p = \frac{1}{2k} \ln \{2p\mathbb{N} \sinh(y) \exp[-(1 - \mathbb{C})y] + \exp(-\alpha y)\} - \frac{2\alpha}{3(1 - g)}, \tag{4.33}$$

where $p = 1/b$. We have from Eq. (4.33) at $k = 0$ taking a limit:

$$\tau_p = \frac{4(p\mathbb{N} - \alpha)}{3(1 - g)}. \tag{4.34}$$

Equation (4.33) gives the result that we intended to gain from the very beginning. The dependence $\tau_p(\lambda)$ for a water cloud having the same microphysical characteristics and observation conditions as those used in Fig. 4.14 are shown in Fig. 4.15. The Mie theory was used to calculate ω_0, $\sigma_{ext} = 1/s$, and g in Eq. (4.33). The gaseous absorption was neglected. Symbols correspond to values of τ_p obtained from the data given in Fig. 4.14. We see that our approximate equation can be used for an accurate estimation of the sunlight penetration in clouds. Note that values obtained for $\tau < 10$ may be biased as the accuracy of Eq. (4.27) decreases in these cases. However, we do not restrict this plot to values below 2 μm to show the general trend of $\tau_p(\lambda)$. Furthermore, the point at 3.7 μm indicates that Eq. (4.33) might even be used at $\tau_p \in [5, 10]$. The following approximate result for the function $R_{0\infty}(1, \xi)$ was used in the calculations presented in Fig. 4.14:

$$R_{\infty 0}(1, \xi) = \frac{0.37 + 1.94\xi}{1 + \xi}. \tag{4.35}$$

This equation is accurate to within 5% for the nadir observation and most solar incident angles used for cloud remote sensing (oblique incident angles are excluded).

Figure 4.14 quantifies the result already mentioned above. Namely, the optical penetration depth generally decreases with the wavelength. However, at some narrow spectral intervals, the opposite is true (see, e.g., the region close to 1 μm in Fig. 4.15). The value of τ_p changes from approximately 100 at $\lambda = 0.5$ μm

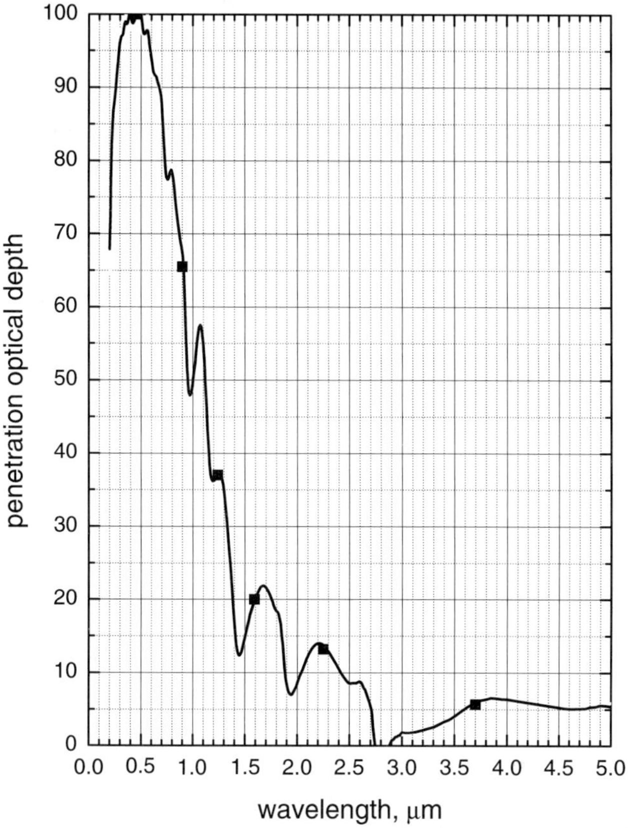

Fig. 4.15. The dependence of the penetration optical thickness on the wavelength obtained using the approximate equation for water droplets having the effective size 6 µm at the nadir observation and the solar angle equal to 60°. Symbols give the results obtained using the numerical solution of the integro-differential radiative transfer equation.

to approximately 6 at $\lambda = 3.7$ µm for the typical case shown in Fig. 4.15. Using a typical value of $\sigma_{ext} = 50$ km^{-1}, we obtain that the ℓ changes from 120 m at $\lambda = 3.7$ µm to 2000 m at $\lambda = 0.5$ µm. In fact, Fig. 4.15 is also applicable to ℓ after the rescaling the ordinate. For example, the ordinate should be multiplied by 10 at $\sigma_{ext} = 100$ km^{-1}. Then ℓ is given in meters.

It follows from Fig. 4.15 that the radiance detected at different wavelengths originates in part from different cloud depths. This sets an important question as far as cloud satellite remote sensing is concerned. Namely, the cloud liquid water path (LWP) and the effective radius of droplets are usually obtained from measurements of reflectances at multiple wavelengths. This does not lead to complications for homogeneous clouds. However, homogeneous clouds do not exist.

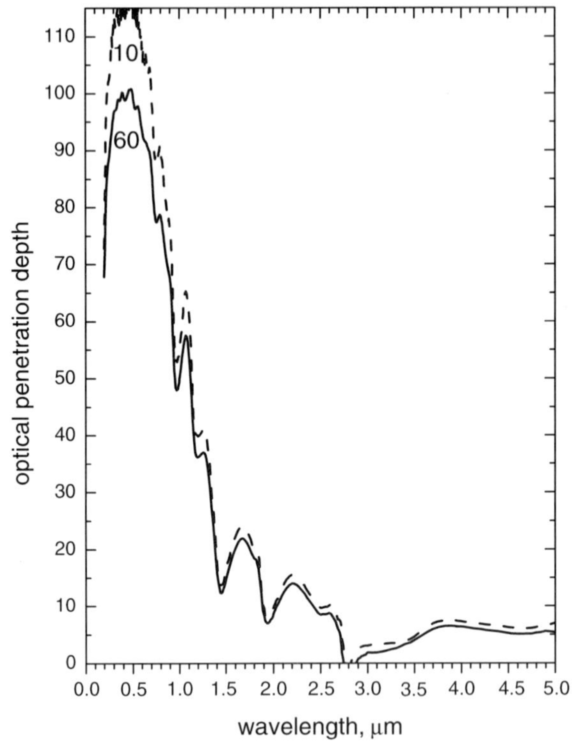

Fig. 4.16. The same as in Fig. 4.15 except the results for the solar angle 10° are also shown.

For example, the size of droplets usually increases from the bottom to the top of a cloud layer. It means that the radius of droplets obtained at 3.7μm (a_{ef}(3.7 μm)) is not necessarily representative for a whole cloud. In this case, the derivation of the LWP as the product of the optical thickness in the visible and a_{ef}(3.7 μm)(we omit a numerical multiplier) may bias the derived LWP considerably. Therefore, it is of importance to specify the wavelengths used to derive a_{ef} and the LWP while referring to their values obtained from the optical instruments onboard satellites. Generally, decreasing the wavelength will lead to smaller values of a derived a_{ef} (and also smaller values of the LWP).

It follows from Fig. 4.16 that the value of τ_p is larger for the illumination closer to the normal. This can easily be explained by the fact that photons injected into the medium along oblique angles escape more quickly and do not penetrate as deeply as photons incident on a given medium along the normal.

The spectral dependence of the optical penetration thickness is shown in Fig. 4.17 for various sizes of particles. We see that the value of τ_p decreases with the size of particles in the infrared. This can be expected from the greater light absorption by larger droplets. We also found that the optical penetration depth is slightly larger for larger droplets in the visible. This is due to larger values of g for

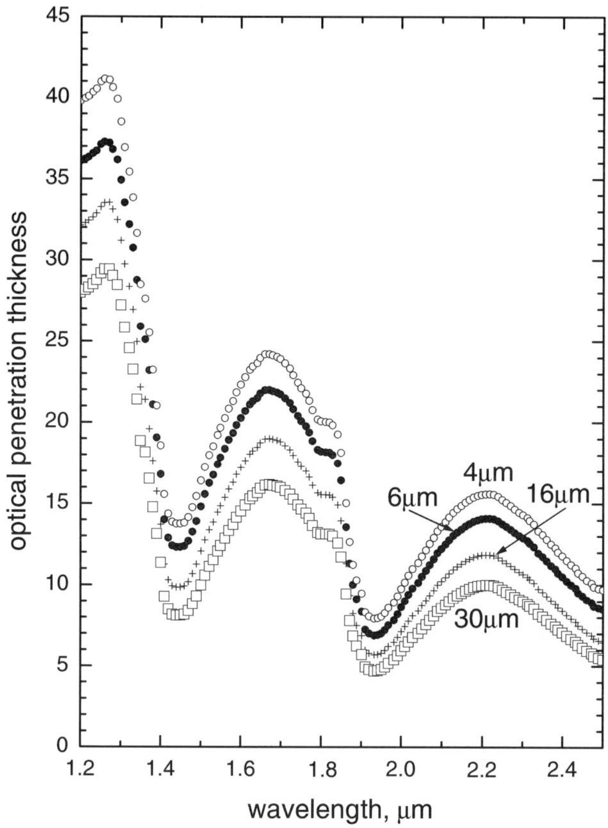

Fig. 4.17. The same as in Fig. 4.15 except the results for the several values of the effective droplet radius are shown.

clouds with larger droplets. The account for the gaseous absorption will modify the data shown in Fig. 4.16, adding an oscillating part on the general background curve depending on the gas type/concentration. However, we do not consider this contribution in any detail here.

Equation (4.33) becomes less accurate for values of the single scattering albedo (SSA) smaller than 0.95. This case may be of importance for polluted clouds and also for ice clouds having large crystals and, therefore, increased value of light absorption. In this case, the problem can be solved using the general asymptotic equation valid for cloud optical thicknesses larger than 10 and an arbitrary SSA. This equation has the following form (Nakajima and King, 1992):

$$R(\xi,\eta,\varphi,\tau) = R_\infty(\xi,\eta,\varphi)\left[1 - ml\bar{\mathbb{N}}(\xi,\eta,\varphi,\tau)\left(1 - l^2\exp(-z)\right)^{-1}\exp(-z)\right], \tag{4.36}$$

where $\bar{N}(\xi, \eta, \varphi) = u(\xi)u(\eta)/R_\infty(\xi, \eta, \varphi)$, $z = 2k\tau$, and, therefore,

$$\tau_p = (2k)^{-1} \ln \left(ap + l^2\right), \tag{4.37}$$

where

$$a = \frac{mlu(\xi)u(\eta)}{R_\infty(\xi, \eta, \varphi)}. \tag{4.38}$$

The issues related to the numerical calculation of asymptotic parameters and functions k, l, m, $u(\xi)$ and $R_\infty(\xi, \eta, \varphi)$ are discussed by Nakajima and King (1992). However, note that the essential simplicity characteristic to Eq. (4.33) is lost then.

Equation (4.27) follows from Eq. (4.36), assuming that $l = \exp(-\alpha y)$,

$$mu(\xi)u(\eta) = (1 - \exp(-2y))\, u_0(\xi)u_0(\eta) \tag{4.39}$$

as shown above. The same correspondence exists between Eqs. (4.37) and (4.33) (with account for Eq. (4.30)).

4.2.2 Cloud Optical Thickness

4.2.2.1 Retrieval procedure

Equations presented in Chapter 3 can be used for rapid estimations of the radiative and polarization characteristics of cloudy media. They can also be used to check the accuracy of new algorithms, using the fact that the numerical solution of the radiative transfer equation, and the results presented above for optically thick layers should converge as $\tau \to \infty$ and $\beta \to 0$.

The most important area of application of approximate solutions lies in the field of remote sensing (King, 1981, 1987; Kokhanovsky, 2000; Kokhanovsky, 2001; Kokhanovsky and Zege, 1996; Kokhanovsky et al., 2003; Rozenberg, 1967; Rozenberg et al., 1978). In particular, the usage of asymptotic equations (valid as $\tau_0 \to \infty$) allows us to avoid the compilation of so-called look-up tables (LUTs) in the case of optically thick clouds and, therefore, to speed up the retrieval process. Currently, LUTs are widely used in passive cloud remote sensing (Arking and Childs, 1985; Nakajima and King, 1990; Rossow et al., 1989). The minimization of differences between measured and stored (e.g., in LUTs) spectral reflectances is used as a main tool to derive cloud optical and microphysical characteristics.

The reflection function of a cloud over a reflective surface with the Lambertian albedo A can be presented as

$$R = R_b + \frac{At^2 u_0(\xi)\, u_0(\eta)}{1 - Ar}, \tag{4.40a}$$

with

$$R_b = R_{0\infty} - t u_0(\xi)\, u_0(\eta) \tag{4.40b}$$

in the visible. Here,

$$t = \frac{1}{\alpha + 0.75\tau^*} \tag{4.40c}$$

is the global transmittance with $\alpha \approx 1.07$ and $\tau^* = \tau(1-g)$.

We have for the global transmittance in the visible from Eq. (4.40a) after simple algebraic calculations:

$$t = \frac{(1 - rA)\,\Pi}{1 - rA(1+\Pi)}, \tag{4.41a}$$

where the function Π is introduced. It is given by

$$\Pi \equiv \Pi(\eta, \xi, \varphi) = \frac{R_{0\infty}(\eta, \xi, \varphi) - \hat{R}_{mes}(\eta, \xi, \varphi, \tau)}{u_0(\eta) u_0(\xi)}. \tag{4.41b}$$

The analytical results for functions $R_{0\infty}(\eta, \xi, \varphi)$ and $u_0(\eta)$ have been presented above. Thus, the global transmittance t, and correspondingly the total reflectance or the spherical albedo $r = 1 - t$, can be obtained from Eqs. (4.40) and (4.41) and knowledge of the surface albedo A and the measured value of the reflection function $\hat{R}_{mes}(\eta, \xi, \varphi, \tau)$.

For such a retrieval it is not necessary to know the optical thickness of clouds and the average size of droplets. This is an extremely important point for climate studies, where the global and temporally averaged value of the cloud spherical albedo $r = 1 - t$ is an important parameter. Usually $r < 0.8$ for natural water clouds in the visible (Danielson et al., 1969), which implies that clouds with optical thicknesses larger than 70–100 appear not very often (Trishchenko et al., 2001). The reduced reflectance in the visible can be also related to aerosol absorption in clouds (Melnikova and Mikhailov, 1994, 2000) and to the inhomogeneity and finite size of clouds (Stephens and Tsay, 1990).

Using Eq. (4.40), we obtain $t \equiv \Pi$ at $A = 0$, and $t = 0$ ($r = 1$) at $A = 1$. This shows that all photons incident on optically thick nonabsorbing clouds over surfaces with $A = 1$ survive and return back to outer space. They yield no information about actual cloud thickness. This explains why the retrieval of cloud parameters over bright surfaces (e.g., snow and ice) can be hardly performed in the visible (Platnick et al., 2001).

The information on the global transmittance t can be used to find the scaled optical thickness (King, 1987; Rozenberg et al., 1978), given by

$$\tau^* = \tau(1 - g). \tag{4.42}$$

It follows from Eq. (4.40c):

$$\tau^* = \frac{4}{3}\left[t^{-1} - \alpha\right], \tag{4.43}$$

where t is given by Eq. (4.40). The value of τ^* can be obtained although there is no information about the size of droplets and the actual optical thickness of clouds.

Equation (4.43) can be used for the retrieval of τ^* from the measurement of the reflection function at a single wavelength (King, 1987). Equation (4.42) is used for the derivation of the optical thickness τ if the value of g is known (Rossow, 1989) (approximately 0.74 for ice clouds as previously discussed). However, for warm clouds the asymmetry parameter g depends on the size of droplets even for nonabsorbing channels. Often the dependence $g(a_{ef})$ is neglected and it is assumed that $a_{ef} = 10$ μm for water clouds (Rossow and Schiffer, 1999). Then it follows from that $g = 0.86$ at $\lambda = 0.65$ μm and $a_{ef} = 10$ μm. This value of g can be used for a crude estimation of the optical thickness of liquid clouds.

Errors can be introduced if one assumes the fixed *a priori* defined value of g. It follows from at $\lambda = 0.65$ μm that $g = 0.84$–0.87 at $a_{ef} = 4$–20 μm. From Eq. (4.42), we have $\tau = \Im\tau^*$, $\Im \equiv (1-g)^{-1} \approx 6.3$–$7.6$ and $\tau \in [9.4, 11.5]$ at $\tau^* = 1.5$, depending on the value of g used. The assumption that $a_{ef} = 10$ μm yields $g = 0.86$ and $\Im = 7.2$, $\tau = 10.7$. This leads to a relative error of 7–14% in the retrieved optical thickness (i.e., a range of possible values from $\tau = 9.4$ to $\tau = 11.5$ instead of $\tau = 10.7$). This uncertainty in the optical thickness can be removed if measurements in the near infrared region of the electromagnetic spectrum are performed, enabling the size of droplets and, therefore, the asymmetry parameter g to be estimated. For this, however, we should be sure that we have a liquid and not an ice or a mixed phase cloud. Another uncertainty arises due to the possible contamination of clouds by absorbing aerosols (Asano et al., 2001, 2002). Then Eq. (4.40) is not valid and we must account for the fact that the cloud single scattering albedo (SSA) differs from one.

4.2.2.2 Hurricane Erin

Let us consider optical thickness retrieval for the Hurricane Erin, located in the western Atlantic (39.3 N, 60.4 W) on September 13th, 2001 (16:21 UTC). By definition, hurricanes contain winds in excess of 119 km per hour and large areas of heavy rainfall. Therefore, they belong to the most dangerous natural hazards. This explains the great interest in hurricane research especially in recent years (Simpson, 2002). Physical characteristics of hurricanes are usually obtained using radar remote sensing techniques (Heymsfield et al., 2001). The optical imagery serves as an important tool for a timely identification of hurricanes and for tracking their trajectories (especially where radar is not available). The optical remote sensing techniques can also be used to study physical characteristics of hurricanes like hurricane-top height, the LWP, the thermodynamic phase of particles and their size. The enhanced spatial resolution of optical imagery allows for important complimentary information as compared to microwave techniques. Clearly, some additional difficulties and problems can arise in this case. They are related mostly to the relative importance of three-dimensional (3-D) effects (e.g., in a hurricane wall), scattering by nonspherical particles, and large values of the geometrical and optical thickness of hurricanes.

Erin can be traced back to a tropical wave that emerged from Western Africa on August 30th, 2001. The hurricane took a long journey from the coast of Africa to the northern Leeward Islands and then to Greenland over the western Atlantic before it merged with high-latitude cyclonic flow on September 17th, 2001. Note that this hurricane was the first one for which a comprehensive 3-D image of the complete inner core (including the eyewall and the eye) has been created.

The retrieval procedure is performed using the SeaWiFS local area coverage imagery with the spatial resolution of about 1×1 km, taken on September 13th, 2001. The SeaWiFS instrument measures the top-of-atmosphere backscattered light intensity in eight channels (412, 443, 490, 510, 555, 670, 765, and 865 nm). Only data for the wavelength 412 nm are used in this study. This is due to a low spectral variation of the hurricane reflectance, which is due to large size of particles (as compared to the light wavelength) in a hurricane.

The map of retrieved values of transport optical thickness is given in Fig. 4.18. We see that the value of the transport optical thickness is mostly in the range of 4–10 for the hurricane studied.

We also have selected a core of a hurricane and made the retrieval of the transport optical thickness for this special case. The map obtained is shown in Fig. 4.19. The statistical distribution of the transport optical thickness for this case is given in Fig. 4.20.

One can find black areas surrounded by white color in Fig. 4.19. They correspond to negative values of the reduced optical thickness obtained owing to possible influence of 3-D effects as discussed above (e.g., shadowing effects). To determine the hurricane optical thickness, the value of the asymmetry parameter should be known. The value of g is around 0.85 for water clouds and it is close to 0.75 for ice clouds. The retrieved cloud optical thickness distribution, assuming the value of $g = 0.85$, coincides with that given in Figs. 4.18–4.20 (but scaled using the factor $B = 1/(1-g)$). Note that we have at $g = 0.85$, $B \approx 6.7$. Therefore, the spatial distribution of τ for the case given in Fig. 4.18 has a main maximum at cloud

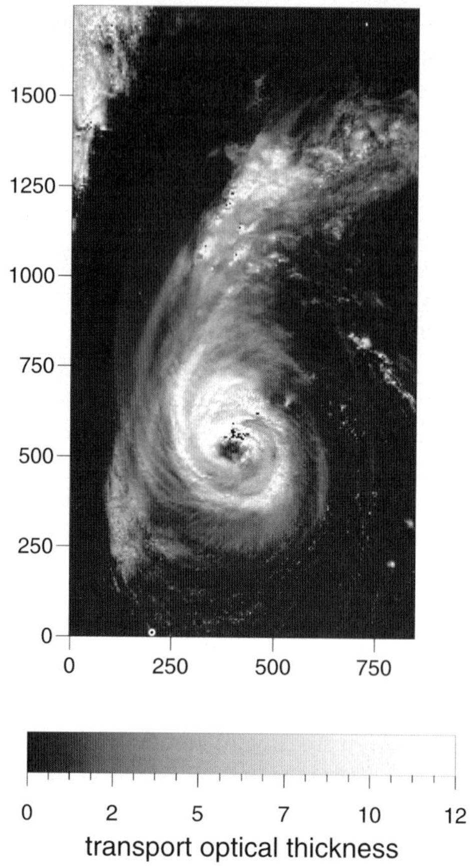

Fig. 4.18. The transport optical thickness map.

optical thickness equal to 35. Note that results at small and large values of cloud optical thickness can be biased because the accuracy of our technique decreases there. However, it works well for values of optical thickness in the range of 10–100, which are characteristic numbers for most pixels with a hurricane. We also found a highly nonuniform distribution of cloud optical thickness in the hurricane eye. The mode optical thickness was found to be equal seven in the hurricane eye.

The hurricane spherical albedo map and frequency distribution are given in Figs.4.21a and 4.21b. We see that most frequent values of a hurricane albedo is around 0.86. The distribution of the spherical albedo is similar to the well-known beta distribution, having abrupt decrease at r close to 1.0. Large values of r for a hurricane suggest that hurricanes can potentially modify planetary albedo (at least during the hurricane season). This can be also enhanced by foam-covered rough seas, which are produced by hurricanes.

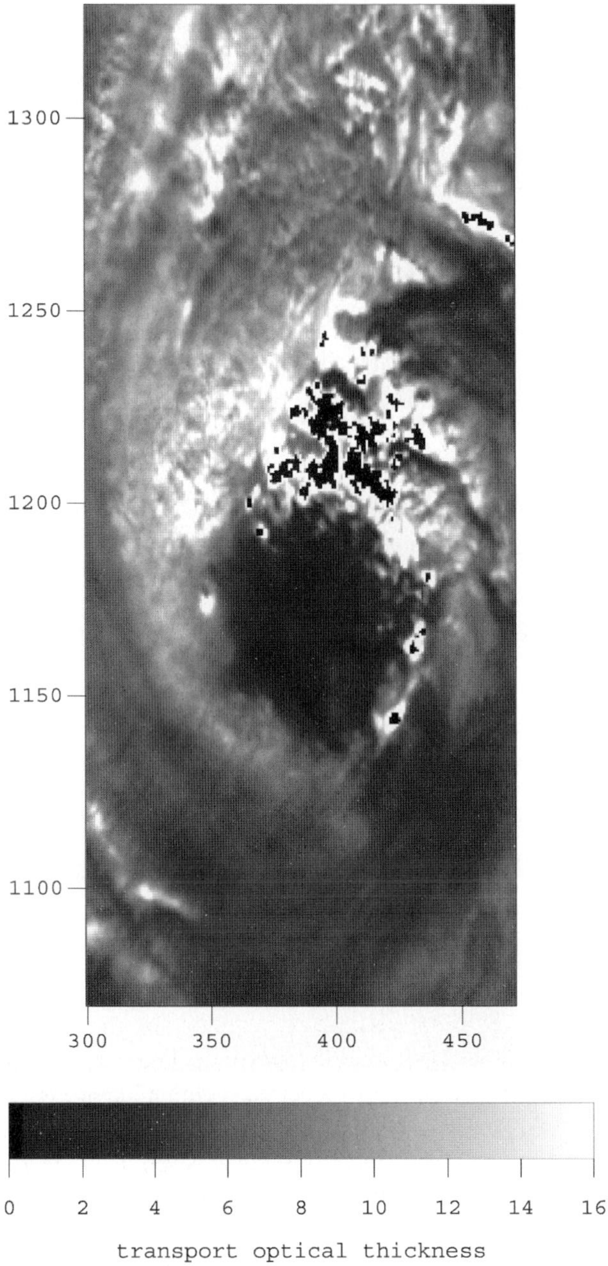

Fig. 4.19. The same as in Fig. 4.18 but for the region close to the hurricane eye.

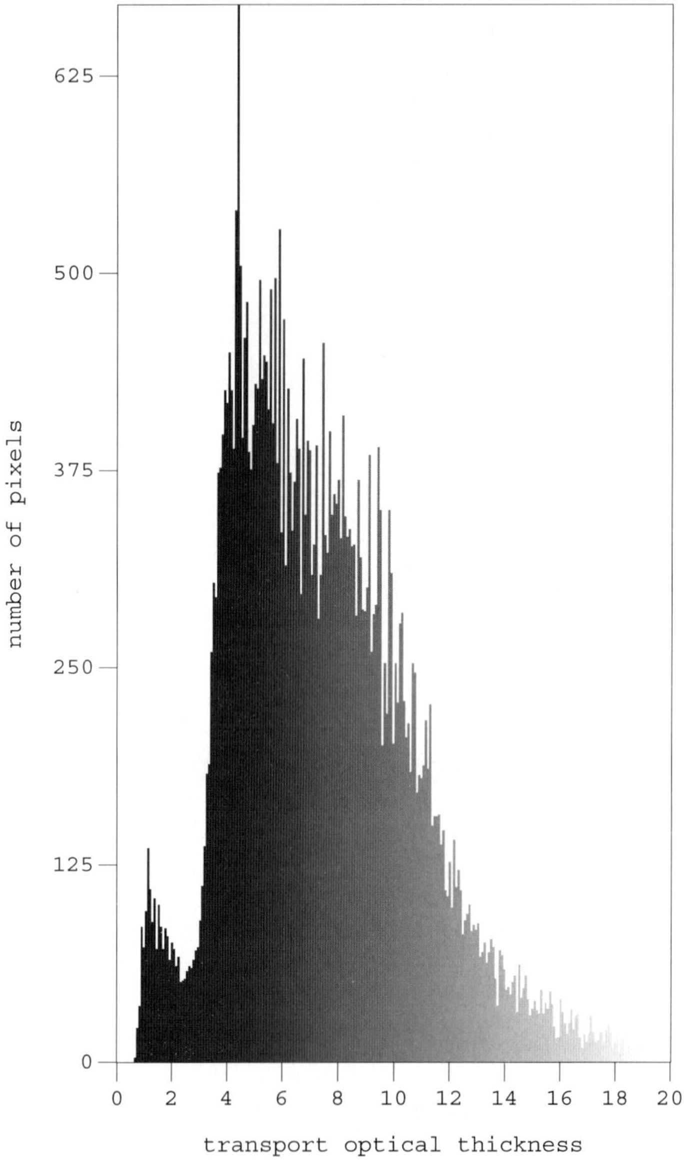

Fig. 4.20. The transport optical thickness frequency distribution.

4.2.2.3 The influence of ground albedo

The results of the cloud optical thickness retrievals are influenced by the ground reflectance. The ground reflectance is generally unknown. This may lead to considerable biases in retrievals. Sometimes minimal reflectance databases for a given

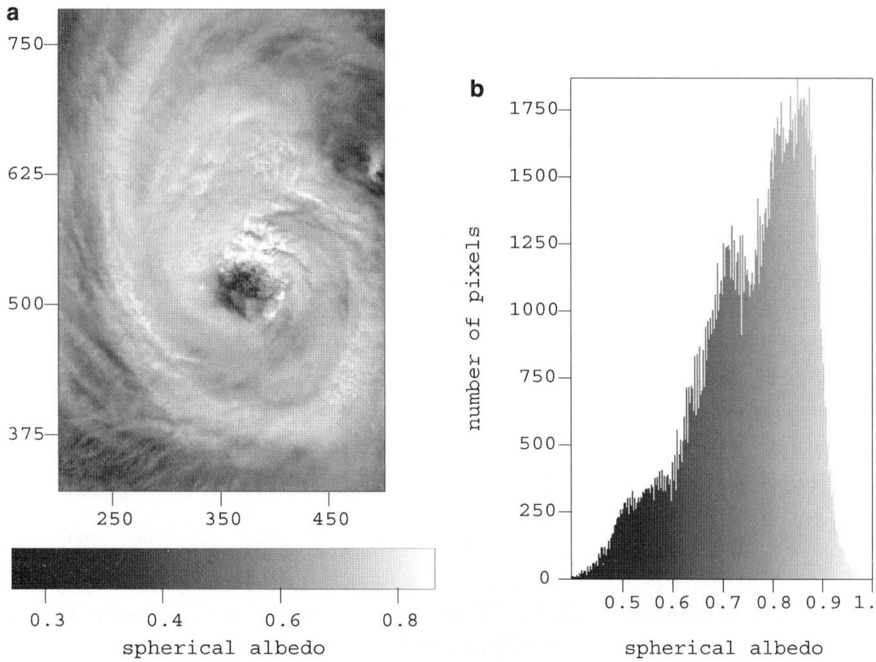

Fig. 4.21. The spherical albedo map (a) and the the spherical albedo frequency distribution (b).

location/month obtained from spaceborne observations for clear sky conditions are used to constrain the ground spectral reflectance.

It is of importance to understand in which conditions the influence of the ground reflectance can be neglected. Taking into account that the calibration error of modern spectrometers and radiometers is around 5%, we pose the following question. What is the Lambertian ground reflectance $A_{\max}$, which lead to not more than 5% differences between the cloud reflectances for a given $\tau = \tau'$ over black and not black underlying surfaces? Clearly, we can neglect the influence of the ground reflectance at $A < A_{\max}$ for a given cloud optical thickness $\tau = \tau'$.

This corresponds to the following equation:

$$1 - \frac{R_b}{R_A} = p, \tag{4.44}$$

where $p = 0.05$, R_b is the reflection function of a cloud over a black surface and R_A is the reflection function of a cloud over the ground surface having the albedo $A = A_{\max.}$ This equation can be rewritten in the following form for the Lambertian ground surface:

$$\frac{A_{\max} t^2 u_0(\mu) u_0(\mu)}{(1 - A_{\max}(1-t))\,(R_{0\infty} - t u_0(\mu) u_0\,(\mu_0))} = \frac{1-p}{p}, \tag{4.45}$$

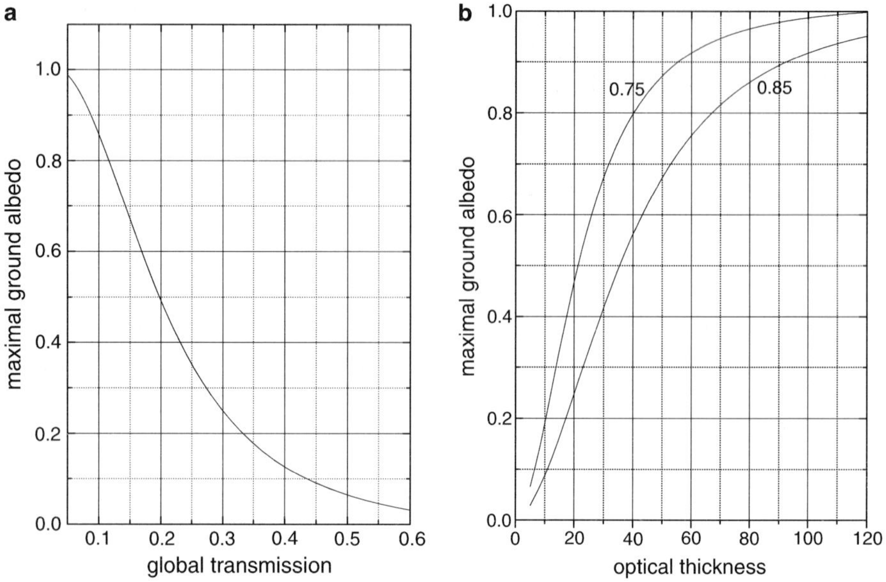

Fig. 4.22. (a) The dependence of the maximal ground albedo on the cloud transmission. (b) The dependence of the maximal ground albedo on the cloud optical thickness at $g = 0.75, 0.85$.

where we used Eq. (4.40a). So we have:

$$A_{\max} = \frac{\gamma \{Q - t\}}{\gamma (1 - t) \{Q - t\} + t^2}, \quad (4.46)$$

where $\gamma = p/(1 - p)$, $Q = R_{0\infty}(\eta, \xi, \varphi) / u_0(\eta) u_0(\xi)$. This equation allows to estimate the maximal value of the ground albedo, which can be neglected for a given p, t, and geometry. This can be easily generalized to account for the case $\omega_0 \neq 1$. Then values of $A_{\max}$ increase for a given τ.

The dependence of $A_{\max}$ on t is shown in Fig. 4.22a. It follows that for clouds having $t < 0.1$, the ground reflectance is of no importance for the geometry specified. We also show the dependence of $A_{\max}$ on cloud optical thickness in Fig. 4.22b at $g = 0.75$ (ice clouds), 0.85 (water clouds). Clearly, larger values of g lead to smaller cloud transmittances and, therefore, to smaller values of $A_{\max}$ for a given τ. This is supported by data shown in Fig. 4.22b as well. So we conclude that the account for ground reflectance is more important for water clouds as compared to ice clouds at a given τ. However, ice clouds are generally thinner as compared to water clouds and $A_{\max}$ decreases for smaller τ. So generally, surface effects cannot be neglected for ice clouds as well.

The snow ground reflectance in the visible is around 0.8. Then the ground albedo can be neglected for cloud optical thickness retrievals over snow at $\tau > \tau'$, where $\tau' = 40$ for water clouds at $\tau' = 67$ for the geometry as specified in Fig. 4.22b. The value of τ' is around 5–10 for clouds over the ocean, where the ground albedo is small (generally, below 0.05 in the visible).

4.2.3 The Size of Droplets and Crystals

As was specified above, for a correct estimation of optical thickness of clouds from space, we need to know not only the surface albedo but also the effective radius of droplets. The size of droplets can be found if the reflection function in the near-infrared is measured simultaneously (Nakajima and King, 1990). This is due to the fact that the reflection function in the infrared strongly depends on the probability of photon absorption by droplets. This probability is proportional to the effective radius of droplets, as was discussed above.

The influence of absorption and scattering of light by molecules and aerosol particles on the measured value $R(\mu, \mu_0, \varphi, \tau)$ is often neglected in the cloud retrieval algorithms. However, correction can be easily taken into account if needed (Goloub et al., 2000; Wang and King, 1997). The influence of surface reflection on the cloud reflection function, assuming that the surface is Lambertian with albedo r, is easily taken into account, leading to following results (Kokhanovsky et al., 2003):

$$\hat{R}_1\left(a_{ef}, w\right) = R_{0\infty} - \frac{t_1\left(a_{ef}, w\right)[1 - A_1]}{1 - A_1\left[1 - t_1\left(a_{ef}, w\right)\right]} u_0(\xi) u_0(\eta), \tag{4.47a}$$

$$\begin{aligned} \hat{R}_2\left(a_{ef}, w\right) = {} & R_{0\infty} \exp\left(-y\left(a_{ef}\right)\left(1 - cy\left(a_{ef}\right)\right)u\right) \\ & - \left[\exp\left(-x\left(a_{ef}, w\right) - y\left(a_{ef}\right)\right) - \frac{t_2\left(a_{ef}, w\right) A_2}{1 - A_2 r_2\left(a_{ef}, w\right)}\right] \\ & \times t_2\left(a_{ef}, w\right) u_0(\xi) u_0(\eta). \end{aligned} \tag{4.47b}$$

The subscripts ‘1’ and ‘2’ refer to wavelengths λ_1 and λ_2 in the visible and the near—infrared channels respectively. The values of A_1 and A_2 give us the surface albedos in the visible and the near-infrared. The explicit dependence of functions involved on the parameters a_{ef} and w to be retrieved is introduced in brackets. The LWP w is preferred to the optical thickness in retrieval procedures due to the independence of w on wavelength. The optical thickness is uniquely defined if a_{ef} and w are known.

Equations 4.47a and (4.47b) form a nonlinear system of two algebraic equations having two unknowns (a_{ef} and w), which can be solved by standard methods and programs. In particular, we can find the value of w from Eq. (4.47a) analytically

(Kokhanovsky et al., 2003). Substitution of this result in Eq. (4.47b) gives us a single transcendent equation for the effective radius of droplets determination. The accuracy of this semi-analytical retrieval algorithm has been studied by Kokhanovsky et al. (2003).

A similar approach can be used for the estimation of sizes of crystals in ice clouds. The simplification is possible in this case. Due to the large size of crystals one can ignore the dependence of the reflectance in the visible on the size of particles and assume that $g \approx 3/4$ (Garrett et al., 2001; Kokhanovsky, 2004a). Then it follows assuming that $A_1 = 0$:

$$R = R_{0\infty} - \frac{u_0(\xi)u_0(\eta)}{1.072 + 0.188\tau}. \tag{4.48}$$

This equation allows for a simple estimation of τ in the case of thick ice clouds over black underlying surface. The functions $u_0(\xi)$ and $R_{0\infty}$ can be found from Eqs. (3.267) and (3.256), respectively. Knowing the value of τ and also accounting for the spectral neutrality of τ for ice clouds, we can obtain the cloud SSA. The size of particles can be derived from the value of SSA assuming the shape of ice crystals. Instead of the derivation of a crystal size for a given shape model, one can retrieve the particle absorption length as discussed by Kokhanovsky and Nauss (2005).

Usually the surface albedo is not known in advance and must be assessed using a priori assumptions or measurements over the same place but at a different time, when clear conditions prevail.

4.2.4 Single Scattering Albedo

Single scattering albedo measurements in infrared are easy and can be performed using LUT approach. Also the retrieved values of a_{ef} can be used to estimate the value of SSA. The problem is much harder in the visible, where the value of SSA is close to one. For polluted clouds the SSA is associated with pollutants (e.g., soot) and not with cloud droplets. Measurements in the infrared can be used to find both a_{ef} and τ enabling the calculation of the reflectance in the visible. This calculated value of the reflection function can be compared to the measured one to quantify the load of pollutants. However, it is not easy to estimate $\beta = 1 - \omega_0 \to 0$ from reflectance measurements due to calibration problems and also model assumptions used in retrieval procedures (e.g., cloud homogeneity). Generally, cloud absorptance can be measured using two aircrafts flying below and above a cloud and measuring a transmittance and absorption. Then the difference between incident and transmitted/reflected fluxes can be used to estimate the cloud absorptance a. However, the value of a is close to zero. Therefore, large errors in the retrieved value are possible.

King (1981) proposed to use just one aircraft but flying deep inside a cloud far from the boundaries. The aircraft must have instruments to measure downward and upward light fluxes or light filed intensities $i(\pm\eta)$. Then one can derive for an infinite cloud as $\beta \to 0 : D = i(-1)/i(1) = 1 - \zeta\sqrt{\beta}$, where $\zeta = 2\sqrt{3/(1-g)}$ as it was shown above. Therefore, it follows: $\beta = (1-D)^2/\zeta^2$ as $\beta \to 0$.

One must account for the influence of cloud boundaries. Then we have for the downward propagated light intensity (van de Hulst, 1968):

$$I(\eta) \sim i(\eta) - Wi(-\eta), \tag{4.49}$$

where the second term accounts for the influence of the boundary. Clearly, it follows: $W \to 0$ with τ_B, τ_B is the optical distance from the boundary to the level of measurements. Van de Hulst (1968) found that $W = \exp(-2k(\tau_B + q))$, where $q \approx 0.71/(1-g)$. So we have for the ratio $\tilde{D} = I(-1)/I(1)$:

$$\tilde{D} = \frac{i(-1) - Wi(1)}{i(1) - Wi(-1)}, \tag{4.50}$$

where we assume that the aircraft flies exactly in the middle of cloud and $\tau_B = \tau_0/2$, where τ_0 is the cloud optical thickness. Clearly, it follows for an infinite cloud: $\tau_0 \to \infty$ and $\tilde{D} \to D$. Also we can write:

$$\tilde{D} = \frac{D - W}{1 - DW}. \tag{4.51}$$

This equation allows to find the value of β, if $\tilde{D}$ is measured. For this, however, independent measurements of the cloud asymmetry parameter g are needed (Garrett et al., 2001). The analytical approximate expression for D valid at arbitrary β has been proposed by King (1986):

$$D = \frac{(1 - \sigma s)(1 - s)}{1 + \rho s}, \tag{4.52}$$

where $\sigma = 0.979$, $\rho = 1.503$, $s = \sqrt{(1-\omega_0)/(1-\omega_0 g)}$.

4.2.5 Cloud Thermodynamic Phase

The discrimination of liquid water from ice clouds is of importance for many applications, including flight safety and Earth climate studies. The size and shape of particles in warm and ice clouds are different. This influences the energy transmitted and reflected by a cloud.

This discrimination can be performed, taking into account the difference in angular or spectral distribution of reflected light. Also one can use polarization measurements as discussed by Goloub et al. (2000). It is known that minima in the reflection function of ice clouds (e.g., near 1.5 and 2.0 μm) are moved to larger wavelengths as compared to the case of liquid droplets. This is, of course, due to the difference in spectral behaviour of the imaginary parts of the complex refractive index of liquid water and ice. Minima for liquid water also moved to larger wavelengths as compared to the absorption bands of water vapour. These different positions of minima can easily be registered with modern spectrometers (see, e.g., Dvoryashin, 2002; Knap et al., 2002).

Another possibility is to consider different angular behaviour of the reflection function for ice and water clouds at specific scattering geometries (e.g., rainbow, glory and halo scattering). In particular, the reflection function of water clouds, as distinct from ice clouds, has a maximum near the rainbow scattering angle, which also can easily be detected. This feature becomes even more pronounced if the degree of polarization [e.g., at the rainbow geometry is studied (Goloub et al., 2000; Kokhanovsky, 2003b)].

4.2.6 Cloud Top Height and Cloud Fraction

Another important characteristic of a cloud is its height. It can be retrieved using data from space-borne lidars (Winker and Trepte, 1998). Passive measurements can also be used. For instance, Yamomoto and Wark (1961) proposed the use of the oxygen A band, centred at 0.761 μm. The physical basis of this method depends on a deep minimum around 0.761 μm due to oxygen absorption. The depth of the absorption line will depend on the cloud height. Photons can hardly penetrate thick clouds and be absorbed by the oxygen in the air column below the cloud. This will increase the value of the reflection function at 0.761 μm for the case of clouds at high altitudes. The depth of the absorption line is larger for low clouds. Practical application of the method, however, is not so simple (Koelemeijer et al., 2001; Kuze and Chance, 1994; Rozanov and Kokhanovsky, 2004). First of all, the depth of line also depends on the oxygen absorption cross-section, which varies with temperature and pressure. Thus, one should use *a priori* assumptions on the temperature and pressure variation with height in the Earth's atmosphere. The generally unknown surface albedo and cloud geometrical thickness can also influence the retrieval accuracy. Other possible sources of errors are described in detail by Kuze and Chance (1994) and Koelemeijer et al. (2001).

The largest complication arises for pixels that are only partially covered by clouds. Ignoring them will lead to a big reduction of data, so to overcome this problem Koelemeijer et al. (2001) proposed an algorithm that retrieves simultaneously effective cloud top height/pressure and cloud fraction, assuming that a cloud is a Lambertian surface with a given albedo. Such an assumption leads to underestimation of cloud top heights as compared to *in situ* measurements.

Rozanov and Kokhanovsky (2004) proposed the usage of the modified exponential approximation in the cloud top height retrievals. This allows to speed up the retrieval process considerably.

Global information on cloud fraction/cover $\mathbb{Q}$ is of a considerable importance by itself. Usually the value of $\mathbb{Q}$ varies in the range 0.55–0.85, depending on the exact region under study. Globally, clouds cover around 60% of the atmosphere. This once more underlines the importance of clouds in the radiation balance and atmosphere heating rates studies.

Palle and Butler (2001) state that the global cloud cover increased during the past century. They argue against a dominating role of solar activity (via galactic cosmic rays) on cloud formation.

4.2.7 Cloud Bottom Height

The cloud bottom height is usually determined using ground-based laser measurements. However, one can also assess the cloud bottom altitude from satellite measurements, if the cloud top altitude is known (e.g., from space lidar measurements). Let us show this considering a single homogeneous cloud. The cloud reflectance spectrum is hardly sensitive to the cloud geometrical characteristics outside gaseous absorption bands. The situation is radically changed if we consider the radiative transfer in the molecular absorption line (Rozanov and Kokhanovsky, 2004; Rozanov et al., 2004; Yamomoto and Wark, 1961). Indeed, let us assume that we have a gas in a planetary atmosphere, which absorbs almost all incident radiation in a narrow band. Then the depth of this band, measured by a receiver on a satellite will depend on the cloud altitude. Gas concentrations generally decrease with the distance from the ground. Therefore, clouds at a high altitude do not allow most photons to penetrate to low atmospheric layers and be absorbed there. So the depth of a molecular line in the reflected light will decrease, if high clouds are present in the field of view of a sensor. The next question to address is the influence of cloud geometrical thickness on the reflectance spectrum $R(\lambda)$ in the gaseous absorption band. One expects that spectra $R(\lambda)$ in the gaseous absorption band for clouds having the same top heights but different cloud geometrical thicknesses will differ even if the cloud optical thicknesses τ coincide (Kokhanovsky and Rozanov, 2004). This is due to the fact that multiple light scattering will lead to large average photon path lengths in clouds as compared to a cloudless atmosphere, thereby increasing absorption. This must lead to a decrease of the reflectance for geometrically thicker clouds. Radiative transfer calculations confirm this fact.

The dependence of the spectrum $R(\lambda)$ as shown in Fig. 4.23 can be used to determine the cloud bottom height from a satellite. The main steps of the inversion technique are given below.

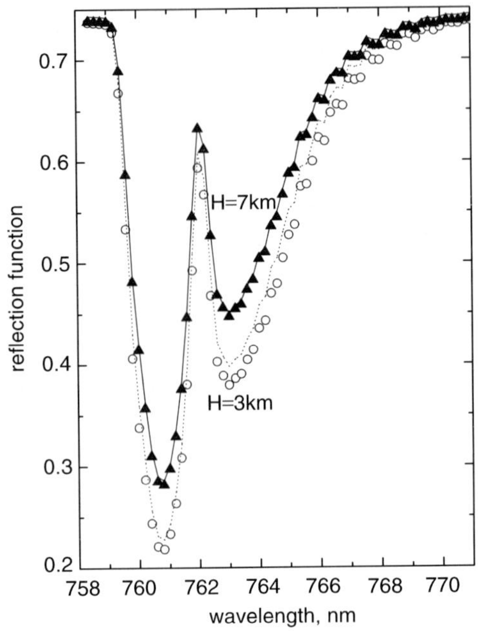

Fig. 4.23. Cloud reflection function in the oxygen A-band calculated using SCIATRAN (lines) (Rozanov et al., 2005) and asymptotic analytical theory (symbols) (Kokhanovsky and Rozanov, 2004) at the nadir observation, the solar zenith angle equal to 60°, the optical thickness equal to 50, the cloud top height equal to 9 km and cloud bottom height equal to 3 km (circles) and 7 km (triangles). All other parameters needed to calculations (e.g., atmospheric vertical profiles) coincide with those described by Kokhanovsky and Rozanov (2004). The reflectance function is averaged with the step 0.2 nm using SCIAMACHY (Bovensmann et al., 1999) response function.

Fist of all the TOA reflectance R is presented in the form of a Taylor expansion around the assumed value of the cloud bottom height equal to H_0:

$$R(H) = R(H_0) + \sum_{i=1}^{\infty} a_i (H - H_0)^i, \tag{4.53}$$

where $a_i = R^{(i)}(H_0)/i!$. Here $R^{(i)}(H_0)$ is the i-derivative of R at the point H_0. The next step is the linearization, which is a standard technique in the inversion procedures (Rozanov et al., 1998). We found that the function $R(H)$ is close to a linear one in a broad interval of the argument change (Kokhanovsky and Rozanov, 2004). Therefore, we neglect nonlinear terms in Eq. (4.53). Then it follows:

$$R = R(H_0) + R'(H_0)(H - H_0), \tag{4.54}$$

where $R' = dR/dH$. We assume that R is measured at several wavelengths $(\lambda_1, \lambda_2, \ldots \lambda_n)$ in the oxygen A-band. Then instead of the scalar quantity R we can introduce the vector $\vec{R}_{mes}$ with components $(R(\lambda_1), R(\lambda_2), \ldots R(\lambda_n))$. The same applies to other scalars in Eq. (4.53).

Therefore, Eq. (4.54) can be written in the following vector form:

$$\vec{y} = \vec{a}x \tag{4.55}$$

where $\vec{y} = \vec{R}_{mes} - \vec{R}(H_0)$, $\vec{a} = \vec{R}'(H_0)$ and $x = H - H_0$. Note that both measurement and model errors are contained in Eq. (4.55). The solution $\hat{x}$ of the inverse problem is obtained by minimizing the following cost function:

$$\Phi = \|\vec{y} - \vec{a}x\|^2, \tag{4.56}$$

where $\| \ \|$ means the norm in the Euclid space of the correspondent dimension.

The value of $\hat{x}$, where the function Φ has a minimum can be presented as

$$\hat{x} = \frac{(\vec{y}, \vec{a})}{(\vec{a}, \vec{a})} = \frac{\sum_{i=1}^{n} a_i y_i}{\sum_{i=1}^{n} a_i^2}, \tag{4.57}$$

where $(\vec{y}, \vec{a})$ denotes a scalar product in the Euclid space, and n is the number of wavelengths where the reflection function is measured.

The functions $\vec{R}(H_0)$ and $\vec{R}'(H_0)$ in Eq. (4.57) must be calculated using the radiative transfer theory with input parameters characteristic for a given atmospheric state. We use the approximate analytical theory for such a calculation. The details of this theory are given by Kokhanovsky and Rozanov (2004). Basically, the approximation has an accuracy better than 5% as compared to line-by-line calculations in the O_2 A-band (Kokhanovsky and Rozanov, 2004) for typical cloudiness with $\tau \geq 5$. The use of the exact radiative transfer theory is also possible but it leads to a huge calculation time and does not provide a better approach to the problem at hand due to all uncertainties involved (e.g., possible multi-layered cloudiness).

Therefore, knowing values of the measured spectral reflection function R_{mes} and also values of the calculated reflection function R and its derivative R' at $H = H_0$ and several wavelengths, the value of the cloud bottom height can be found from Eq. (4.57) and equality: $H = \hat{x} + H_0$. The value of H_0 can be taken equal to 0.5 km, which is a typical value for low level clouds. The main assumption in our derivation is that the dependence of R on H can be presented by a linear function on the interval x (Kokhanovsky and Rozanov, 2004).

The retrieved value of H is compared to H_0. If the difference is smaller than 100 m, the value of H is taken as a retrieved value and the inversion procedure is finished. Otherwise, the retrieved value of H is substituted in Eq. (4.56) instead of H_0 and iterations are performed until the convergence is reached.

Several additional parameters are needed in the retrieval procedure. They are the cloud optical thickness, the cloud liquid water profile, the cloud droplet radii, etc. They must be simultaneously derived or assumed using climatological values.

We have found that the cloud optical thickness τ is the most important parameter which influences the retrieval. So we find the value of τ from measurements outside the gaseous absorption band as described by Kokhanovsky and von Hoyningen-Huene (2004).

Clearly, if the forward and inverse models use the same system of equations, the inverse problem solution accurately reproduces the input parameters for the forward model in the absence of the measurement noise. We have checked this using the forward and inverse models based on the same set of analytical equations as described by Rozanov and Kokhanovsky (2004). Indeed, the cloud bottom height used as input in the retrieval scheme coincided with the cloud bottom height retrieved by the solution of the inverse problem as specified above in this case.

The next possible step is to introduce measurement errors and see the influence of these errors on the retrieval of H. However, we have chosen a different strategy. Namely, we calculate the cloud reflectance spectrum $R(\lambda)$ in the O_2 A-band using the exact radiative transfer calculations with the radiative transfer solver SCIATRAN (Rozanov et al., 2005) and use this exact spectrum in the analytical retrieval procedure described above. Because possible measurement errors are well below the accuracy of analytical equations, such an approach can be considered as a simulation of noise in the inversion procedure having as an input SCIATRAN-generated synthetic spectra.

Results of the inversion procedure described above are shown in Fig. 4.24. It follows that with the exclusion of clouds having large top altitude and small cloud bottom height (very thick clouds), which are rare cases in terrestrial atmosphere, there is a one-to-one correspondence between retrieved H_r and exact H_e cloud bottom heights. Biases $\Delta H = H_r - H_e$ are given in Fig. 4.25. It follows that biases are only weakly influenced by values of τ and generally they are below 0.5 km for clouds having the geometrical thickness below 4.5 km. Most clouds in the terrestrial atmosphere have geometrical thicknesses below 1 km. Then biases are just 0.25 km as shown in Fig. 4.25.

Therefore, we conclude that the technique presented here can be used for an accurate estimation of a cloud bottom height from a satellite.

It follows from Fig. 4.25 that the accuracy of the retrieval decreases for smaller values of H at a cloud top height equal to 9 km. This is related to the fact that analytical equations have larger errors for smaller H. Note that the probability of single clouds having small values of H and large values of the cloud top height is low. So values with large biases of the retrieved cloud bottom altitude as shown in Fig. 4.20b do not produce significant biases as far as the operational cloud retrievals are concerned.

The results given in Figs. 4.24 and 4.25 are obtained allowing for the cloud top height measurements uncertainty $\Delta h = \pm 10$ m. They are not changed appreciably even when increasing this uncertainty $10\times$ (e.g., to 100 m). The changes of biases

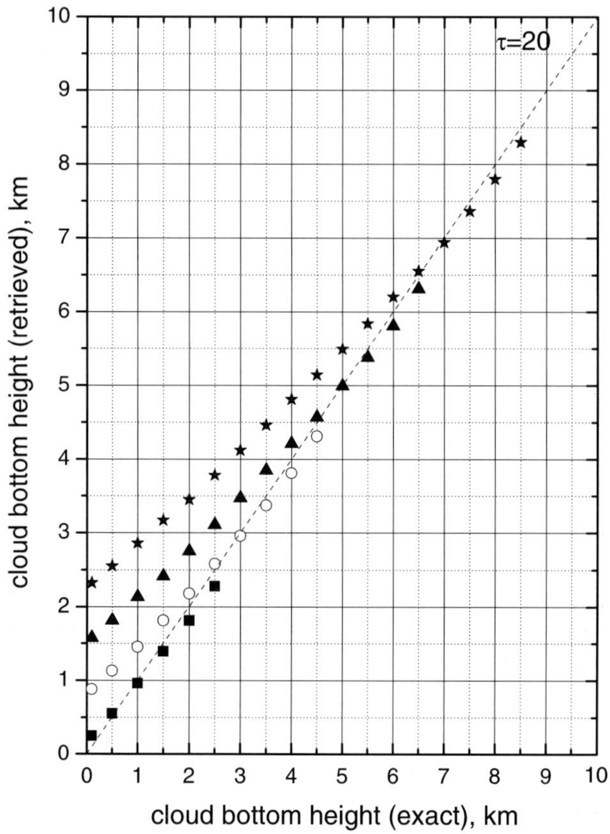

Fig. 4.24. Dependence of the retrieved cloud bottom height on the exact cloud bottom height at $\tau = 20$ for cloud top height 9 km (stars), 7 km (triangles), 5 km (circles), 3 km (squares).

as shown in Fig. 4.25 are generally below 30 m then. It means that the considered technique can be easily implemented using the lidar system. Lidars can detect the cloud boundary with the accuracy better than 20 m.

4.3 Laser Beam Propagation Through a Cloud

The studies of laser beam propagation through fogs and clouds are of importance for a number of applied problems including sounding of clouds and also for image transfer and communication problems. The aim of this section is to derive the angular distribution of a light field inside a cloud under laser beam illumination conditions along a normal to a cloud or fog layer (e.g., along axis *OZ*). Unlike most problems considered above, we need to account for the horizontal inhomogeneity

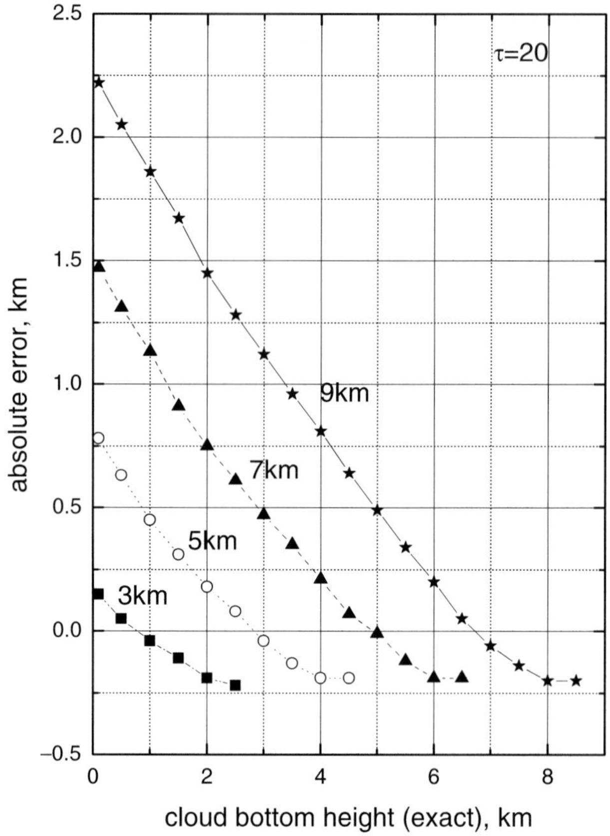

Fig. 4.25. Dependence of the absolute error of the retrieved cloud bottom height on the exact cloud bottom height at $\tau = 20$ for cloud top height 9 km (stars), 7 km (triangles), 5 km (circles), 3 km (squares). Data are obtained from Fig. 4.24.

of a multiply scattered light field. This inhomogeneity arises solely due to boundary conditions. A laser beam enters a cloud effectively at one point. So instead of uniform illumination of the cloud boundary, we have illumination of a cloud by a narrow beam. The intensity of light in a given depth depends not only on the observation direction but also on the distance of a given point from the axis OZ, which gives the illumination direction. For simplicity of derivations we will assume that a cloud is represented by a homogeneous layer. Theory is easily extended for a case of vertically inhomogeneous cloudy media.

The main equation of the problem can be written in the following form:

$$(\vec{s}\vec{\nabla})I(\vec{r},\vec{s}) + \sigma_{ext}I(\vec{r},\vec{s}) - \frac{\sigma_{sca}}{4\pi}\int_{4\pi} I(\vec{r},\vec{s}\,')p(\vec{s}\,',\vec{s})d\Omega' = 0. \qquad (4.58)$$

The task is to find the angular distribution of light field $I(\vec{r}, \vec{s})$ at a given point $\vec{r}$ in the direction specified by the vector $\vec{s}$ assuming a narrow beam illumination condition. Generally, the problem looks complicated and calls for the numerical implementation. However, an important approximate solution can be derived, if one is interested in the distribution $I(\vec{r}, \vec{s})$ close to the axis of the laser beam for vectors $\vec{s}$ directed along OZ or in the directions almost parallel to OZ. Let us show this.

We present vectors $\vec{r}$ and $\vec{s}$ as

$$\vec{r} = x\vec{e}_x + y\vec{e}_y + z\vec{e}_z, \tag{4.59}$$

$$\vec{s} = s_x\vec{e}_x + s_y\vec{e}_y + s_z\vec{e}_z, \tag{4.60}$$

where $(\vec{e}_x, \vec{e}_y, \vec{e}_z)$ are unity vectors directed along the axes OX, OY, and OZ. Axes OX and OY specify the plane perpendicular to OZ. We can write in the spherical coordinate system:

$$s_x = \sin\theta\cos\varphi, \quad s_y = \sin\theta\sin\varphi, \quad s_z = \cos\theta. \tag{4.61}$$

Here φ is the azimuthal angle and θ is the angle between the axis OZ and the observation direction. We will assume that $\theta \to 0$ in our derivations. Hence, the correspondent approximation is called the small angle approximation (SAA). Clearly, this is the generalization of the SAA considered above for a case of narrow beams.

Let us introduce the vector

$$\vec{\nabla}_\perp \equiv \vec{e}_x\frac{\partial}{\partial x} + \vec{e}_y\frac{\partial}{\partial y}. \tag{4.62}$$

Then it follows

$$(\vec{s}\vec{\nabla}_\perp)I(z, \vec{\rho}, \vec{s}) + \frac{\partial I(z, \vec{\rho}, \vec{s})}{\partial z} + \sigma_{ext}I(z, \vec{\rho}, \vec{s}) - \frac{\sigma_{sca}}{4\pi}\int_{-\infty}^{\infty} ds_x \int_{-\infty}^{\infty} ds_y I(z, \vec{\rho}, \vec{s}\,')p(\vec{s}\,' - \vec{s}) = 0, \tag{4.63}$$

where we used the fact that $ds_x ds_y = \cos\theta\sin\theta d\theta d\varphi = \cos\theta d\Omega \approx d\Omega$ and $s_z \approx 1$ as $\theta \to 0$. Also we assume that the phase function depends only on the difference vector $\vec{d} = \vec{s} - \vec{s}\,'$ and introduces the transverse vector $\vec{\rho} = x\vec{e}_x + y\vec{e}_y$. We use infinite limits of integration because the contribution of photons located at large distances from the axis OZ is low. Clearly, our assumptions are valid only if light scattering occurs predominantly in the forward direction and this is really the case for fogs and clouds due to highly extended phase functions for these cases. The approximation considered is not valid in deep layers of a cloud (e.g., at $\tau \geq 5$) because then light deviates from the axis OZ considerably.

Equation (4.63) can be solved analytically. We show this using the Fourier transforms technique. Applying the Fourier transform with respect to $\vec{\rho}$ to Eq. (4.63),

Table 4.1. Fourier transforms.

N	Fourier transform	Definition
1	$\tilde{I}(z, \vec{\nu}, \vec{s})$	$\int_{-\infty}^{\infty}\int_{-\infty}^{\infty} I(z, \vec{\rho}, \vec{s})e^{i\vec{\nu}\vec{\rho}}d\vec{\rho}$
2	$I(z, \vec{\rho}, \vec{s})$	$\frac{1}{4\pi^2}\int_{-\infty}^{\infty}\int_{-\infty}^{\infty} \tilde{I}(z, \vec{\nu}, \vec{s})e^{-i\vec{\nu}\vec{\rho}}d\vec{s}$
3	$-i\vec{s}\vec{\nu}\tilde{I}(z, \vec{\nu}, \vec{s})$	$\int_{-\infty}^{\infty}\int_{-\infty}^{\infty} (\vec{s}\vec{\nabla}_{\perp})I(z, \vec{\rho}, \vec{s})e^{i\vec{\nu}\vec{\rho}}d\vec{\rho}$
4	$\tilde{p}(\vec{q})$	$\int_{-\infty}^{\infty}\int_{-\infty}^{\infty} p(\vec{s})e^{i\vec{q}\vec{s}}d\vec{s}$
5	$\tilde{h}(\vec{\nu}) = \tilde{f}(\vec{\nu})\tilde{g}(\vec{\nu})$	$h(\vec{\beta}) = \int_{-\infty}^{\infty}\int_{-\infty}^{\infty} f(\vec{\alpha})g(\vec{\beta} - \vec{\alpha})d\vec{\alpha}$
6	$\tilde{\tilde{G}}(z, \vec{\nu}, \vec{q})$	$\int_{-\infty}^{\infty}\int_{-\infty}^{\infty} p(\vec{s})e^{i(\vec{q}-z\vec{\nu})\vec{s}}d\vec{s}$
7	$\tilde{D}(z, \vec{\nu}, \vec{s})$	$\frac{1}{4\pi^2}\int_{-\infty}^{\infty}\int_{-\infty}^{\infty} \tilde{\tilde{D}}(z, \vec{\nu}, \vec{q})e^{-i\vec{q}\vec{s}}d\vec{q}$
8	1	$\int_{-\infty}^{\infty}\int_{-\infty}^{\infty} \delta(\vec{s})e^{-i\vec{\nu}\vec{s}}d\vec{s}$

we have using definitions specified in Table 4.1:

$$\hat{\Lambda}\tilde{I}(z, \vec{\nu}, \vec{s}) - \frac{\sigma_{sca}}{4\pi}\int_{-\infty}^{\infty}\int_{-\infty}^{\infty} d\vec{s}'\tilde{I}(z, \vec{\nu}, \vec{s}')p(\vec{s} - \vec{s}') = 0, \tag{4.64}$$

where $\hat{\Lambda} \equiv (\partial/\partial z) + \sigma_{ext} - i\vec{s}\vec{\nu}$. This equation can be simplified using the substitution:

$$\tilde{I}(z, \vec{\nu}, \vec{s}) = \tilde{D}(z, \vec{\nu}, \vec{s})\exp(iz\vec{\nu}\vec{s} - \tau), \tag{4.65}$$

where $\tau = \sigma_{ext} z$. Then it follows:

$$\frac{d\tilde{D}(z, \vec{\nu}, \vec{s})}{dz} - \frac{\sigma_{sca}}{4\pi}\int_{-\infty}^{\infty}\int_{-\infty}^{\infty} \tilde{D}(z, \vec{\nu}, \vec{s}')\tilde{G}(z, \vec{\nu}, \vec{s} - \vec{s}')d\vec{s}' = 0, \tag{4.66}$$

where

$$\tilde{G}(\vec{s} - \vec{s}') = p(\vec{s} - \vec{s}')\exp(i\vec{\nu}z(\vec{s} - \vec{s}')). \tag{4.67}$$

Let us apply the Fourier transform with respect to $\vec{s}$ to Eq. (4.66). The integral in Eq. (4.66) can be transformed using the convolution property 5 in Table 4.1. Then it follows:

$$\frac{d\tilde{\tilde{D}}(z, \vec{\nu}, \vec{q})}{dz} - \frac{\sigma_{sca}}{4\pi}\tilde{\tilde{D}}(z, \vec{\nu}, \vec{q})\tilde{\tilde{G}}(z, \vec{\nu}, \vec{q}) = 0. \tag{4.68}$$

The symbol $\tilde{\tilde{D}}$ means the double Fourier transform of D (with respect to both $\vec{\rho}$ and $\vec{s}$). The expression for $\tilde{\tilde{G}}$ is given in Table 4.1. Comparing lines 3 and 5 in Table 4.1, we derive: $\tilde{\tilde{G}} \equiv \tilde{p}(\vec{q} - \vec{\nu}z)$. Therefore, it follows that Eq. (4.68):

$$\tilde{\tilde{D}}(z, \vec{\nu}, \vec{q}) = A \exp\left\{\frac{\sigma_{sca}}{4\pi} \int_0^z \tilde{p}\left(\vec{q} - \vec{\nu}z\right) dz\right\}, \tag{4.69}$$

where $A \equiv \tilde{\tilde{D}}(0, \vec{\nu}, \vec{q})$ is the integration constant. This solves the problem at hand. Indeed, the value of $\tilde{D}(z, \vec{\nu}, \vec{s})$ in Eq. (4.65) can be found using the inverse Fourier transform as specified in Table 4.1 (line 6). Then it follows for the angular distribution of light at point $\vec{\rho}$ in the direction $\vec{s}$ (see line 7 in Table 4.1):

$$I(z, \vec{\rho}, \vec{s}) = \frac{1}{4\pi^2} \int_{-\infty}^{\infty} \int_{-\infty}^{\infty} \tilde{I}(z, \vec{\nu}, \vec{s}) e^{-i\vec{\nu}\vec{\rho}} d\vec{\nu}, \tag{4.70}$$

where

$$\tilde{I}(z, \vec{\nu}, \vec{s}) = \frac{1}{4\pi^2} \int_{-\infty}^{\infty} \int_{-\infty}^{\infty} \tilde{D}(z, \vec{\nu}, \vec{q}) \exp(-i\vec{s}\vec{q} + i\vec{s}\vec{\nu}z - \tau) d\vec{q}. \tag{4.71}$$

Although the numerical integration is required to find $I(z, \vec{\rho}, \vec{s})$, a number of important results can be obtained in numerous special cases as considered by Ishimaru (1978).

Note that Eq. (4.70) can be rewritten in yet another form:

$$I(z, \vec{\rho}, \vec{s}) = \left(\frac{1}{2\pi}\right)^4 \int_{-\infty}^{\infty} \int_{-\infty}^{\infty} \int_{-\infty}^{\infty} \int_{-\infty}^{\infty} \tilde{\tilde{I}}(z, \vec{\nu}, \vec{q}) e^{-i(\vec{\nu}\rho + \vec{q}\vec{s})} d\vec{\nu} d\vec{q}, \tag{4.72}$$

where

$$\tilde{\tilde{I}}(z, \vec{\nu}, \vec{q}) = H(\vec{\nu}, \vec{q} + \vec{\nu}z) \exp\left\{-\sigma_{sca}z + \frac{\sigma_{sca}}{4\pi} \int_0^z \tilde{p}(\vec{q} - \vec{\nu}(z' - z)) dz'\right\} \tag{4.73}$$

and $H(\vec{\nu}, \vec{\kappa})$ is determined from boundary conditions:

$$H(\vec{\nu}, \vec{\kappa}) = \int_{-\infty}^{\infty} \int_{-\infty}^{\infty} \int_{-\infty}^{\infty} \int_{-\infty}^{\infty} d\vec{\rho} d\vec{s} I(0, \vec{\rho}, \vec{s}). \tag{4.74}$$

Equation (4.73) plays an important role in Fourier optics of light scattering media. In particular, assuming that $I(0, \vec{\rho}, \vec{s}) = \delta(\vec{\rho})\delta(\vec{s})$ and using property 8 in Table 4.1, we derive:

$$\tilde{\tilde{I}}(z, \vec{\nu}, \vec{q}) = \exp\left\{-\sigma_{sca}z + \frac{\sigma_{sca}}{4\pi} \int_0^z \tilde{p}(\vec{q} - \vec{\nu}(z' - z)) dz'\right\}. \tag{4.75}$$

Calculations of $\tilde{p}$ can be simplified assuming the circular symmetry of the phase function: $p(\vec{s}) \equiv p(|\vec{s}_\perp|) = p(s)$. Then it follows:

$$\tilde{p}(\kappa) = 2\pi \int_0^\infty p(s) J_0(\kappa s) s ds. \tag{4.76}$$

Therefore, we can write:

$$\tilde{\tilde{I}}(z, \vec{\nu}, \vec{q}) = \exp\left\{-\sigma_{sca} z + \frac{\sigma_{sca}}{2} \int_0^z dz' \int_0^\infty d\theta p(\theta) J_0((q - \nu z')\theta)\theta\right\}. \tag{4.77}$$

It follows from this equation at $\vec{\nu} = \vec{q} = \vec{0}$:

$$\tilde{\tilde{I}}(z, \vec{0}, \vec{0}) = \exp\left\{-\sigma_{sca} z + \frac{\sigma_{sca}}{2} \int_0^z dz' \int_0^\infty d\theta p(\theta)\theta\right\} \tag{4.78}$$

or $\tilde{\tilde{I}}(z, \vec{0}, \vec{0}) = 1$, where we accounted for the phase function normalization condition:

$$\frac{1}{2} \int_0^\infty p(\theta)\theta d\theta = 1. \tag{4.79}$$

The SAA as considered here can be generalized for the inclined incident beams with the incidence angle $\vartheta_0 \neq 0$. It can be also used to study laser backscattering in a simple approximation based on the consideration of small angle scattering in the forward direction, one single scattering backwards and the SAA in the backward direction [the quasi-single approximation (QSA)]. Details are given by Katsev et al. (1998). Note that the QSA considerably improves theoretical considerations of laser backscattering by cloudy media as compared to the traditionally used lidar equation:

$$I = Cz^{-2} \sigma_{sca}(z) \exp\left\{-2 \int_0^z \sigma_{ext}(z') dz'\right\}. \tag{4.80}$$

Here C is the lidar constant dependent on the transmitted power, pulse duration and receiver characteristics. The value of $\sigma_{sca}(z)$ gives the backscattering coefficient at the range z.

4.4 Image Transfer Through Clouds and Fogs

From the point of view of the linear systems theory a cloud is a high frequency filter. The transmission of high frequency signals (in time and space domains) is reduced considerably in comparison with low frequency signals as a result of light scattering. We introduce here the main concepts of the Fourier optics of clouds.

Any diffused source of light can be considered as a superposition of point light sources. Thus, in linear optical systems the image of such an object with the irradiance $a_0(\vec{r}\,')$ is a linear superposition of images of point sources. This can be represented as

$$a(\vec{r}) = \int_{-\infty}^{\infty}\int_{-\infty}^{\infty} S(\vec{r},\vec{r}\,')a_0(\vec{r}\,')d\vec{r}\,', \tag{4.81}$$

where the point spread function (PSF) $S(\vec{r}\,',\vec{r})$ describes the process of the transformation of the object irradiance $a_0(\vec{r}\,')$ in the initial plane to the image irradiance $a(\vec{r})$ in the image plane. The PSF is a main notion of the image transfer theory (ITT). Equation (4.81) takes a simpler form in the frequency domain:

$$a(\vec{\upsilon}) = S(\vec{\upsilon})a_0(\vec{\upsilon}), \tag{4.82}$$

where

$$a(\vec{\upsilon}) = \int_{-\infty}^{\infty}\int_{-\infty}^{\infty} a(\vec{r}\,')e^{-i\vec{\upsilon}\vec{r}\,'}d\vec{r}\,', \tag{4.83}$$

$$a_0(\vec{\upsilon}) = \int_{-\infty}^{\infty}\int_{-\infty}^{\infty} a_0(\vec{r}\,')e^{-i\vec{\upsilon}\vec{r}\,'}d\vec{r}\,', \tag{4.84}$$

$$S(\vec{\upsilon}) = \int_{-\infty}^{\infty}\int_{-\infty}^{\infty} S\left(\vec{r}\,'\right)e^{-i\vec{\upsilon}\vec{r}\,'}d\vec{r}\,', \tag{4.85}$$

and $\vec{\upsilon}$ is the space frequency. We see that the integration procedures in the Fourier space can be substituted by the multiplication. This makes all calculations much simpler.

One of the basic problems of the ITT is to determine the Fourier transform of the PSF, namely the optical transfer function (OTF) $S\left(\vec{\upsilon}\right)$ or the modulation transfer function

$$T(\vec{\upsilon}) = \frac{|S(\vec{\upsilon})|}{|S(0)|}. \tag{4.86}$$

Also we have then:

$$a(\vec{r}\,') = \frac{1}{4\pi^2}\int_{-\infty}^{\infty}\int_{-\infty}^{\infty} a(\vec{\upsilon})e^{i\vec{\upsilon}\vec{r}}d\vec{\upsilon}. \tag{4.87}$$

The OTF depends on the properties of media between an object and an image (Ishimaru, 1978; Volnistova and Drofa, 1986; Wells, 1969; Zege et al., 1991). Let us consider the case of fogs and cloudy media in the visible range. Then particles are large as compared to the wavelength and the SAA as described above can be used to solve the problem. We will study the image transfer along axis OZ perpendicular to a cloud layer. In this case the OTF is a real function. It does not depend on the azimuth for randomly oriented particles.

The SAA for the OTF follows from Eq. (4.77) at $\vec{q} = 0$. Namely, we have:

$$\tilde{\tilde{I}}(z, \nu, 0) = \exp\left\{-\sigma_{sca}z + \frac{\sigma_{sca}}{2}\int_0^z dz' \int_0^\infty d\theta p(\theta) J_0\left(\nu z'\theta\right)\theta\right\}. \quad (4.88)$$

It can be proved that $\tilde{\tilde{I}}(z, \nu, 0) \equiv S(z, \nu)$. Indeed, the double Fourier transform of the intensity $I(z, \vec{\rho}, \vec{s})$ has the following form:

$$I(z, \vec{\nu}, \vec{q}) = \int_{-\infty}^{\infty}\int_{-\infty}^{\infty}\int_{-\infty}^{\infty}\int_{-\infty}^{\infty} I(z, \vec{\rho}, \vec{s})e^{i(\vec{\nu}\vec{\rho}+\vec{q}\vec{s})}d\vec{\nu}d\vec{s}. \quad (4.89)$$

So we have at $\vec{q} = 0$:

$$I(z, \vec{\nu}, 0) = \int_{-\infty}^{\infty}\int_{-\infty}^{\infty} a(z, \vec{\rho})e^{i\vec{\nu}\vec{\rho}}d\vec{\nu}, \quad (4.90)$$

where

$$a(z, \vec{\rho}) = \int_{-\infty}^{\infty}\int_{-\infty}^{\infty} I(z, \vec{\rho}, \vec{s})d\vec{s}. \quad (4.91)$$

Comparing Eqs. (4.28) and (4.33), we conclude that $S(z, \vec{\nu}) \equiv I(z, \vec{\nu}, 0)$. So we have:

$$S(z, \nu) = \exp\left\{-\tau\left\{1 - \omega_0 B(\nu, z)\right\}\right\}, \quad (4.92)$$

where

$$B(\nu, z) = \frac{1}{2z}\int_0^z dz' \tilde{p}(\nu(z - z')). \quad (4.93)$$

Let us introduce a new variable $\varsigma = 1 - z'/z$. Then it follows:

$$B(\nu, z) = \frac{1}{2}\int_0^1 d\varsigma \tilde{p}(\nu z\varsigma). \quad (4.94)$$

We see that $B(\nu, z)$ depends on the dimensionless frequency $\kappa = \nu z$. The same is true for the OTF. Therefore, it follows:

$$S(\kappa) = \exp\{-\tau\{1 - \omega_0 B(\kappa)\}\}, \quad (4.95)$$

where

$$B(\kappa) = \frac{1}{2}\int_0^1 \tilde{p}(\kappa\varsigma)d\varsigma. \quad (4.96)$$

Equation (4.95) is of a great importance in the Fourier optics of clouds. This equation allows to derive the irradiance in the image plane if one knows the irradiance in the initial plane using Eq. (4.82). Also it follows:

$$a_0(\kappa) = S^{-1}(\kappa)a(\kappa). \tag{4.97}$$

Therefore, the initial image can be reconstructed, if the OTF is known.

It follows at $\kappa = 0$: $S(\kappa) = \exp(-\sigma_{abs}\, z)$. Therefore, only absorption processes are responsible for the OTF reduction at the zero frequency.

Let us approximate the phase function of a cloud medium as

$$p(\theta) = 4\alpha^2 \exp(-\alpha^2\theta^2). \tag{4.98}$$

Then it follows:

$$\tilde{p}(\varsigma\kappa) = \exp\left\{-\frac{\varsigma^2\kappa^2}{4\alpha^2}\right\} \tag{4.99}$$

and, therefore:

$$B(\kappa) = \frac{\alpha\sqrt{\pi}}{\kappa} erf\left[\frac{\kappa}{2\alpha}\right], \tag{4.100}$$

where the error function

$$erf(u) = \frac{2}{\sqrt{\pi}}\int_0^u \exp(-\phi^2)d\phi \tag{4.101}$$

is introduced. So we obtain the following analytical expression for the OTF:

$$S(\kappa) = \exp\left\{-\tau\left\{1 - \frac{\alpha\omega_0\sqrt{\pi}}{\kappa} erf\left[\frac{\kappa}{2\alpha}\right]\right\}\right\}. \tag{4.102}$$

In particular, it follows as $\kappa \to 0$:

$$erf\left[\frac{\kappa}{2\alpha}\right] \approx \frac{\kappa}{\alpha\sqrt{\pi}}\left[1 - \frac{\kappa^2}{12\alpha^2}\right] \tag{4.103}$$

and, therefore,

$$S(\kappa) = \exp\left\{-\tau\left[1 - \omega_0\left(1 - \frac{\kappa^2}{12\alpha^2}\right)\right]\right\}. \tag{4.104}$$

It follows for non-absorbing media: $S(\kappa) = \exp\{-\Xi\kappa^2\}$, where $\Xi = \tau/12\alpha^2$. We see that the distribution $S(\kappa)$ has the Gaussain shape at small dimensionless frequencies κ. Larger droplets in clouds are characterized by more extended phase functions. Therefore, α must be larger for larger particles. This also means that the OTF is larger for larger particles. This will lead to a better image quality for media having larger particles. Further details on the image transfer are given by Ishimaru (1978) and Zege et al. (1991).

In conclusion, let us consider the case $\vec{\nu} = 0$ in the general expression. Then it follows:

$$\tilde{\tilde{I}}(z, 0, q) = \exp\left\{-\sigma_{sca}z + \frac{\sigma_{sca}z}{2}\int_0^\infty d\theta p(\theta) J_0(q\theta)\theta\right\} \quad (4.105)$$

or

$$\tilde{\tilde{I}}(z, 0, q) = \exp\{-\sigma_{sca}z(1 - \tilde{p}(q)\}, \quad (4.106)$$

where

$$\tilde{p}(q) = \frac{1}{2}\int_0^\infty p(\theta) J_0(q\theta)\theta d\theta. \quad (4.107)$$

Using Eq. (4.32), we obtain:

$$I(z, 0, \vec{s}) = \int_{-\infty}^{\infty}\int_{-\infty}^{\infty} I(z, \vec{s}) e^{i\vec{q}\vec{s}} d\vec{s}, \quad (4.108)$$

where

$$I(z, \vec{s}) = \int_{-\infty}^{\infty}\int_{-\infty}^{\infty} I(z, \vec{\rho}, \vec{s}) d\vec{\rho}. \quad (4.109)$$

$\tilde{\tilde{I}}(z, 0, q)$ gives the angular spectrum of transmitted light. According to Zege et al. (1991), this function coincides with the mutual coherence function (MCF) $\Gamma(z, q)$:

$$\Gamma(z, q) = \exp\{-\sigma_{sca}z(1 - \tilde{p}(q))\}. \quad (4.110)$$

So one can study also the coherence loss due to scattering processes in cloudy media using approximate results presented here. By definition, the inverse Fourier transform of $\Gamma(z, q)$ gives the angular distribution of light transmitted through a scattering layer having the geometrical thickness z. Further discussions of this topic are given by Zege and Kokhanovsky (1994).

4.5 Clouds and Climate

The weather could be highly variable at any given place. However, meteorological characteristics (e.g., surface temperature T_s) are quite stable if one considers their average values for long periods of time (e.g., month, year, 100 years). The climate is the averaged weather. It is argued that the human activity leads to the climate change on a global scale. For instance, there is a strong correlation between the increase of the surface temperature T_s and the increase of the concentration c of CO_2 in the atmosphere. It is believed that c increases due to human activity and this leads to the increase of T_s. There are also other human factors, which influence the climate (in particular, the extensive land use and industrial production). All this constitutes a major scientific problem of modern time, namely, the climate change and its influence on the biosphere. Another issue is how we may change human behaviour to minimize its possibly disastrous influence on climate (e.g., melting of ice at northern latitudes). Clouds influence climate in many ways. First of all they reflect a large portion of incident solar energy back to outer space cooling the system. So the change of cloud albedo, which is influenced by cloud optical thickness and in a smaller extent by cloud thermodynamic state and size of particles is very important. Clouds also warm the system protecting the escape of energy from our planet at night. This is especially true for high clouds. Therefore, we see that not only the cloud thicknesses but also their altitudes are of importance for climate studies. Polluted clouds (e.g., containing soot) also warm the Earth-atmosphere system reflecting less solar energy back to outer space. However, pollution can also lead to larger droplet concentrations and, therefore, to smaller particles. This leads to the increase of reflectance by cloudy media.

The energy balance equation can be written in the following form:

$$(1-r)\frac{E_0}{4} = \sigma T^4, \tag{4.111}$$

where r is the global average albedo, σ is the Stefan–Botzman constant, E_0 is the solar constant and T_{ef} is averaged radiation temperature. The coefficient 1/4 shows that the illuminated surface (πa^2) is $4\times$ smaller than the emission surface ($4\pi a^2$). Here a is the radius of Earth. Basically, this equation means that energy transmitted through the top-of-atmosphere escapes to outer space through thermal radiative effects. This equation is only approximately valid because it ignores many important processes. However, Eq. (4.53) clearly shows that the temperature T_{ef} is influenced mainly by the albedo of the planet r:

$$T_{ef} = \sqrt[4]{\frac{1-r}{4\sigma}E_0}. \tag{4.112}$$

In turn, the albedo of the planet is mainly determined by its cloud cover and cloud thickness. This simple example shows the importance of clouds in the climate problem. Note that the surface temperature T_{ef} averaged on large time intervals (e.g., 1 year) is highly correlated with the product $x = \sigma T_{ef}^4$. In particular, it was found that (in °C): $T_s = 0.5\sigma T_{ef}^4 - 104$. Using values $E_0 = 1360\frac{W}{m^2}$, $r = 0.3$, we obtain that 1% increase in r will result in the cooling of the surface by approximately 0.5°C.

Therefore, it is of great importance to monitor the albedo of cloudless atmosphere and also clouds on a global scale. In particular, the albedo of cloud fields depends on

- cloud cover;
- cloud horizontal inhomogeneity;
- cloud optical thickness;
- cloud thermodynamic state;
- the size and shape of particles;
- the pollution of clouds by finely dispersed matter (e.g., soot, dust aerosol) and
- cloud altitudes.

Therefore, there is a need to monitor all these parameters on a global scale using observations from satellites. Such observations make a substantial contribution to the most important scientific problem of modern time—studies of climate change.

Chapter A

APPENDIX A. REFRACTIVE INDICES

Table A.1. Water (Hale and Querry, 1973).

λ (μm)	n	χ	λ (μm)	n	χ
0.2	1.396	1.1×10^{-7}	2.0	1.306	1.1×10^{-3}
0.225	1.373	4.9×10^{-8}	2.2	1.296	2.89×10^{-4}
0.250	1.362	3.35×10^{-8}	2.4	1.279	9.56×10^{-4}
0.275	1.354	2.35×10^{-8}	2.6	1.242	3.17×10^{-3}
0.300	1.349	1.6×10^{-8}	2.65	1.219	6.7×10^{-3}
0.325	1.346	1.08×10^{-8}	2.70	1.188	0.019
0.350	1.343	6.5×10^{-9}	2.75	1.157	0.059
0.375	1.341	3.5×10^{-9}	2.80	1.142	0.115
0.400	1.339	1.86×10^{-9}	2.85	1.149	0.185
0.425	1.338	1.3×10^{-9}	2.90	1.201	0.268
0.450	1.337	1.02×10^{-9}	2.95	1.292	0.298
0.475	1.336	9.35×10^{-9}	3.00	1.371	0.272
0.500	1.335	1.00×10^{-9}	3.05	1.426	0.240
0.525	1.334	1.32×10^{-9}	3.10	1.467	0.192
0.550	1.333	1.96×10^{-9}	3.15	1.483	0.135
0.575	1.333	3.60×10^{-9}	3.20	1.478	0.0924
0.600	1.332	1.09×10^{-9}	3.25	1.467	0.0610
0.625	1.332	1.39×10^{-8}	3.30	1.450	0.0368
0.650	1.331	1.64×10^{-8}	3.35	1.432	0.0261
0.675	1.331	2.23×10^{-8}	3.40	1.420	0.0195
0.700	1.331	3.35×10^{-8}	3.45	1.410	0.0132
0.725	1.330	9.15×10^{-8}	3.50	1.400	0.0094
0.750	1.330	1.56×10^{-7}	3.6	1.385	0.00515
0.775	1.330	1.48×10^{-7}	3.7	1.374	0.00360
0.800	1.329	1.25×10^{-7}	3.8	1.364	0.00340

(continued)

Table A.1. *(Continued)*

λ (μm)	n	χ	λ (μm)	n	χ
0.825	1.329	1.82×10^{-7}	3.9	1.357	0.00380
0.850	1.329	2.93×10^{-7}	4.0	1.351	0.00460
0.875	1.328	3.91×10^{-7}	4.1	1.346	0.00562
0.900	1.328	4.86×10^{-7}	4.2	1.342	0.00688
0.925	1.328	1.06×10^{-6}	4.3	1.338	0.00845
0.950	1.327	2.93×10^{-6}	4.4	1.334	0.0103
0.975	1.327	3.48×10^{-6}	4.5	1.332	0.0134
1.0	1.327	2.89×10^{-6}	4.6	1.330	0.0147
1.2	1.324	9.89×10^{-6}	4.7	1.330	0.0157
1.4	1.321	1.38×10^{-4}	4.8	1.330	0.0150
1.6	1.317	8.55×10^{-5}	4.9	1.328	0.0137
1.8	1.312	1.15×10^{-4}	5.0	1.325	0.0124

Table A.2. Ice (Warren, 1984 (with updates given by S. G. Warren in 1995)).

λ, μm	n	χ	λ, μm	n	χ
2.00363E − 01	1.39309E + 00	3.74825E − 08	3.35598E − 01	1.32739E + 00	7.27663E − 09
2.04227E − 01	1.38776E + 00	3.15556E − 08	3.42071E − 01	1.32624E + 00	6.81271E − 09
2.08166E − 01	1.38244E + 00	2.69447E − 08	3.48668E − 01	1.32512E + 00	6.25559E − 09
2.12181E − 01	1.37829E + 00	2.35935E − 08	3.55393E − 01	1.32430E + 00	5.55574E − 09
2.16273E − 01	1.37512E + 00	2.12678E − 08	3.62247E − 01	1.32353E + 00	4.86137E − 09
2.20445E − 01	1.37196E + 00	1.96115E − 08	3.69234E − 01	1.32269E + 00	4.19643E − 09
2.24696E − 01	1.36873E + 00	1.79143E − 08	3.76355E − 01	1.32190E + 00	3.66810E − 09
2.29030E − 01	1.36551E + 00	1.64961E − 08	3.83614E − 01	1.32110E + 00	3.31362E − 09
2.33447E − 01	1.36231E + 00	1.53601E − 08	3.91013E − 01	1.32030E + 00	3.08786E − 09
2.37950E − 01	1.35913E + 00	1.45812E − 08	3.98554E − 01	1.31954E + 00	2.86107E − 09
2.42539E − 01	1.35591E + 00	1.39906E − 08	4.06241E − 01	1.31884E + 00	2.62143E − 09
2.47217E − 01	1.35268E + 00	1.35202E − 08	4.14076E − 01	1.31819E + 00	2.40419E − 09
2.51985E − 01	1.35007E + 00	1.32798E − 08	4.22062E − 01	1.31760E + 00	2.22129E − 09
2.56845E − 01	1.34832E + 00	1.32311E − 08	4.30202E − 01	1.31701E + 00	2.07637E − 09
2.61798E − 01	1.34656E + 00	1.31632E − 08	4.38500E − 01	1.31643E + 00	1.93435E − 09
2.66848E − 01	1.34479E + 00	1.30619E − 08	4.46957E − 01	1.31588E + 00	1.64345E − 09
2.71994E − 01	1.34301E + 00	1.29180E − 08	4.55577E − 01	1.31535E + 00	1.53439E − 09
2.77240E − 01	1.34123E + 00	1.27077E − 08	4.64364E − 01	1.31484E + 00	1.53875E − 09
2.82587E − 01	1.33946E + 00	1.23852E − 08	4.73320E − 01	1.31434E + 00	1.57953E − 09
2.88037E − 01	1.33772E + 00	1.19507E − 08	4.82449E − 01	1.31387E + 00	1.67349E − 09
2.93593E − 01	1.33594E + 00	1.15029E − 08	4.91754E − 01	1.31341E + 00	1.80232E − 09
2.99255E − 01	1.33413E + 00	1.10568E − 08	5.01238E − 01	1.31296E + 00	1.93731E − 09
3.05027E − 01	1.33294E + 00	1.00918E − 08	5.10905E − 01	1.31253E + 00	2.15069E − 09
3.10910E − 01	1.33183E + 00	9.17977E − 09	5.20759E − 01	1.31212E + 00	2.28030E − 09
3.16906E − 01	1.33075E + 00	8.55242E − 09	5.30803E − 01	1.31172E + 00	2.56955E − 09
3.23018E − 01	1.32965E + 00	8.06212E − 09	5.41040E − 01	1.31132E + 00	2.94838E − 09
3.29248E − 01	1.32853E + 00	7.69293E − 09	5.51475E − 01	1.31094E + 00	3.13612E − 09

Table A.2. *(Continued)*

λ, μm	n	χ	λ, μm	n	χ
5.62111E − 01	1.31057E + 00	3.33761E − 09	1.27811E + 00	1.29650E + 00	1.33375E − 05
5.72953E − 01	1.31021E + 00	3.66707E − 09	1.30276E + 00	1.29599E + 00	1.31723E − 05
5.84003E − 01	1.30987E + 00	4.35903E − 09	1.32788E + 00	1.29546E + 00	1.32000E − 05
5.95266E − 01	1.30952E + 00	5.31240E − 09	1.35349E + 00	1.29491E + 00	1.40043E − 05
6.06747E − 01	1.30918E + 00	6.49113E − 09	1.37960E + 00	1.29431E + 00	1.57585E − 05
6.18449E − 01	1.30886E + 00	8.29497E − 09	1.40620E + 00	1.29365E + 00	2.28869E − 05
6.30377E − 01	1.30854E + 00	1.04633E − 08	1.43333E + 00	1.29290E + 00	1.01737E − 04
6.42535E − 01	1.30823E + 00	1.27042E − 08	1.46097E + 00	1.29230E + 00	2.88632E − 04
6.54928E − 01	1.30792E + 00	1.53949E − 08	1.48915E + 00	1.29178E + 00	5.28659E − 04
6.67559E − 01	1.30762E + 00	1.83140E − 08	1.51787E + 00	1.29124E + 00	5.02587E − 04
6.80434E − 01	1.30732E + 00	2.10267E − 08	1.54714E + 00	1.29062E + 00	4.27116E − 04
6.93557E − 01	1.30702E + 00	2.56793E − 08	1.57698E + 00	1.28991E + 00	3.25113E − 04
7.06934E − 01	1.30673E + 00	3.26536E − 08	1.60740E + 00	1.28901E + 00	2.59410E − 04
7.20568E − 01	1.30645E + 00	4.04497E − 08	1.63840E + 00	1.28813E + 00	2.35842E − 04
7.34466E − 01	1.30616E + 00	4.56762E − 08	1.67000E + 00	1.28718E + 00	2.00162E − 04
7.48631E − 01	1.30588E + 00	5.73064E − 08	1.70221E + 00	1.28615E + 00	1.70685E − 04
7.63070E − 01	1.30560E + 00	7.51223E − 08	1.73504E + 00	1.28504E + 00	1.49527E − 04
7.77787E − 01	1.30532E + 00	9.81886E − 08	1.76850E + 00	1.28375E + 00	1.38725E − 04
7.92788E − 01	1.30504E + 00	1.22278E − 07	1.80261E + 00	1.28248E + 00	1.30382E − 04
8.08078E − 01	1.30476E + 00	1.38832E − 07	1.83737E + 00	1.28104E + 00	1.23000E − 04
8.23663E − 01	1.30448E + 00	1.43732E − 07	1.87281E + 00	1.27939E + 00	1.56453E − 04
8.39549E − 01	1.30420E + 00	1.50726E − 07	1.90893E + 00	1.27748E + 00	4.88735E − 04
8.55741E − 01	1.30392E + 00	2.00786E − 07	1.94575E + 00	1.27599E + 00	1.16465E − 03
8.72246E − 01	1.30363E + 00	2.79390E − 07	1.98328E + 00	1.27462E + 00	1.62574E − 03
8.89069E − 01	1.30335E + 00	3.86335E − 07	2.02153E + 00	1.27320E + 00	1.56339E − 03
9.06216E − 01	1.30305E + 00	4.34792E − 07	2.06052E + 00	1.27146E + 00	1.28836E − 03
9.23694E − 01	1.30277E + 00	4.87391E − 07	2.10026E + 00	1.26933E + 00	8.26460E − 04
9.41509E − 01	1.30246E + 00	5.60162E − 07	2.14076E + 00	1.26659E + 00	4.59263E − 04
9.59668E − 01	1.30217E + 00	7.49369E − 07	2.18205E + 00	1.26341E + 00	2.96285E − 04
9.78177E − 01	1.30185E + 00	1.08198E − 06	2.22414E + 00	1.26002E + 00	2.25004E − 04
9.97042E − 01	1.30154E + 00	1.52851E − 06	2.26703E + 00	1.25605E + 00	2.21324E − 04
1.01627E + 00	1.30122E + 00	2.15364E − 06	2.31076E + 00	1.25172E + 00	3.25536E − 04
1.03587E + 00	1.30089E + 00	2.33000E − 06	2.35532E + 00	1.24687E + 00	4.86439E − 04
1.05585E + 00	1.30055E + 00	2.04429E − 06	2.40075E + 00	1.24122E + 00	5.93097E − 04
1.07622E + 00	1.30020E + 00	1.76609E − 06	2.44705E + 00	1.23469E + 00	6.81003E − 04
1.09697E + 00	1.29984E + 00	1.70900E − 06	2.49425E + 00	1.22677E + 00	7.89017E − 04
1.11813E + 00	1.29948E + 00	1.80866E − 06	2.54236E + 00	1.21713E + 00	8.49488E − 04
1.13969E + 00	1.29910E + 00	2.24330E − 06	2.59139E + 00	1.20466E + 00	9.40740E − 04
1.16168E + 00	1.29870E + 00	3.16177E − 06	2.64137E + 00	1.18723E + 00	1.94166E − 03
1.18408E + 00	1.29829E + 00	5.15251E − 06	2.69231E + 00	1.16478E + 00	4.58211E − 03
1.20692E + 00	1.29787E + 00	8.00677E − 06	2.74424E + 00	1.13285E + 00	1.11083E − 02
1.23019E + 00	1.29743E + 00	1.13169E − 05	2.79717E + 00	1.08780E + 00	2.74474E − 02
1.25392E + 00	1.29698E + 00	1.30171E − 05	2.85111E + 00	1.02346E + 00	7.08075E − 02

(continued)

Table A.2. *(Continued)*

λ, μm	n	χ	λ, μm	n	χ
2.90610E+00	9.62625E−01	1.90406E−01	3.87041E+00	1.37485E+00	7.34151E−03
2.96215E+00	1.00242E+00	3.36656E−01	3.94505E+00	1.36697E+00	8.68508E−03
3.01928E+00	1.07280E+00	4.87479E−01	4.02114E+00	1.36029E+00	9.90104E−03
3.07751E+00	1.32783E+00	6.24110E−01	4.09870E+00	1.35416E+00	1.10951E−02
3.13687E+00	1.55554E+00	4.85772E−01	4.17775E+00	1.34871E+00	1.29144E−02
3.19737E+00	1.64935E+00	2.91863E−01	4.25832E+00	1.34353E+00	1.53598E−02
3.25904E+00	1.61095E+00	1.54136E−01	4.34045E+00	1.33921E+00	1.95646E−02
3.32189E+00	1.55778E+00	8.36852E−02	4.42416E+00	1.33833E+00	2.49185E−02
3.38596E+00	1.51404E+00	4.40622E−02	4.50949E+00	1.34078E+00	2.90321E−02
3.45126E+00	1.47661E+00	2.38255E−02	4.59646E+00	1.34609E+00	2.68426E−02
3.51783E+00	1.44807E+00	1.42235E−02	4.68512E+00	1.34473E+00	1.95120E−02
3.58568E+00	1.42618E+00	9.44774E−03	4.77548E+00	1.34122E+00	1.57984E−02
3.65483E+00	1.40937E+00	7.66333E−03	4.86758E+00	1.33641E+00	1.36552E−02
3.72532E+00	1.39547E+00	6.91620E−03	4.96146E+00	1.33119E+00	1.23514E−02
3.79717E+00	1.38423E+00	6.70954E−03	5.05715E+00	1.32573E+00	1.22845E−02

Chapter B

APPENDIX B. PHASE FUNCTIONS

Table B.1. Phase functions of clouds at wavelength 0.5 μm (g is the asymmetry parameter).

θ	Water cloud (Kokhanovsky, 2004a) $g = 0.8552$	Ice cloud (Liou, 1992) $g = 0.75$
0.0	3.386×10^3	1.083×10^5
0.1	3.361×10^3	6.037×10^4
0.2	3.286×10^3	3.231×10^4
0.3	3.166×10^3	1.809×10^4
0.4	3.005×10^3	9.985×10^3
0.5	2.813×10^3	5.477×10^3
0.6	2.597×10^3	3.210×10^3
0.7	2.366×10^3	2.106×10^3
0.8	2.130×10^3	1.502×10^3
0.9	1.894×10^3	1.095×10^3
1.0	1.667×10^3	7.875×10^2
1.1	1.451×10^3	5.550×10^2
1.2	1.253×10^3	3.855×10^2
1.3	1.074×10^3	2.758×10^2
1.4	9.133×10^2	2.027×10^2
1.5	7.728×10^2	1.563×10^2
1.6	6.511×10^2	1.267×10^2
1.7	5.470×10^2	1.069×10^2
1.8	4.587×10^2	9.266×10
1.9	3.846×10^2	8.151×10
2.0	3.229×10^2	7.210×10
3.0	7.524×10	2.223×10
4.0	3.437×10	9.666
5.0	2.146×10	5.198

(*continued*)

Table B.1. (*Continued*)

θ	Water cloud (Kokhanovsky, 2004a) $g = 0.8552$	Ice cloud (Liou, 1992) $g = 0.75$
6.0	1.538×10	3.208
7.0	1.212×10	2.182
8.0	9.934	1.598
9.0	8.871	1.236
10	8.316	1.013
11	7.378	8.296×10^{-1}
12	6.786	7.494×10^{-1}
13	6.371	6.753×10^{-1}
14	5.912	6.220×10^{-1}
15	5.714	5.681×10^{-1}
16	5.319	5.248×10^{-1}
17	5.018	4.883×10^{-1}
18	4.755	4.598×10^{-1}
19	4.493	4.409×10^{-1}
20	4.183	4.227×10^{-1}
21	3.982	1.935
22	3.738	5.333
23	3.471	6.137
24	3.322	6.043
25	3.092	4.660
26	2.915	3.665
27	2.744	2.955
28	2.586	2.404
29	2.422	1.982
30	2.275	1.638
31	2.138	1.342
32	2.0128	1.130
33	1.864	9.85×10^{-1}
34	1.768	8.54×10^{-1}
35	1.643	7.24×10^{-1}
36	1.548	6.26×10^{-1}
37	1.443	5.55×10^{-1}
38	1.347	5.55×10^{-1}
39	1.261	4.45×10^{-1}
40	1.179	4.12×10^{-1}
41	1.104	4.03×10^{-1}
42	1.026	3.74×10^{-1}
43	9.604×10^{-1}	3.57×10^{-1}
44	8.976×10^{-1}	4.09×10^{-1}
45	8.312×10^{-1}	5.44×10^{-1}
46	7.795×10^{-1}	7.44×10^{-1}
47	7.260×10^{-1}	8.88×10^{-1}
48	6.746×10^{-1}	8.78×10^{-1}
49	6.313×10^{-1}	8.10×10^{-1}

Table B.1. (*Continued*)

θ	Water cloud (Kokhanovsky, 2004a) $g = 0.8552$	Ice cloud (Liou, 1992) $g = 0.75$
50	5.839×10^{-1}	7.15×10^{-1}
51	5.467×10^{-1}	6.56×10^{-1}
52	5.055×10^{-1}	5.91×10^{-1}
53	4.731×10^{-1}	5.09×10^{-1}
54	4.370×10^{-1}	4.52×10^{-1}
55	4.098×10^{-1}	4.16×10^{-1}
56	3.773×10^{-1}	3.89×10^{-1}
57	3.504×10^{-1}	3.72×10^{-1}
58	3.287×10^{-1}	3.58×10^{-1}
59	3.022×10^{-1}	3.47×10^{-1}
60	2.808×10^{-1}	3.46×10^{-1}
61	2.606×10^{-1}	3.44×10^{-1}
62	2.423×10^{-1}	3.43×10^{-1}
63	2.239×10^{-1}	3.44×10^{-1}
64	2.076×10^{-1}	3.43×10^{-1}
65	1.931×10^{-1}	3.40×10^{-1}
66	1.785×10^{-1}	3.38×10^{-1}
67	1.648×10^{-1}	3.37×10^{-1}
68	1.532×10^{-1}	3.36×10^{-1}
69	1.419×10^{-1}	3.36×10^{-1}
70	1.322×10^{-1}	3.36×10^{-1}
71	1.214×10^{-1}	3.35×10^{-1}
72	1.134×10^{-1}	3.34×10^{-1}
73	1.049×10^{-1}	3.33×10^{-1}
74	9.713×10^{-2}	3.32×10^{-1}
75	9.033×10^{-1}	3.31×10^{-1}
76	8.429×10^{-2}	3.29×10^{-1}
77	7.781×10^{-2}	3.27×10^{-1}
78	7.263×10^{-2}	3.25×10^{-1}
79	6.780×10^{-2}	3.24×10^{-1}
80	6.780×10^{-2}	3.22×10^{-1}
81	6.285×10^{-2}	3.21×10^{-1}
82	5.887×10^{-2}	3.18×10^{-1}
83	5.510×10^{-2}	3.14×10^{-1}
84	5.155×10^{-2}	3.11×10^{-1}
85	4.834×10^{-2}	3.11×10^{-1}
86	4.564×10^{-2}	3.08×10^{-1}
87	4.056×10^{-2}	3.03×10^{-1}
88	3.822×10^{-2}	2.98×10^{-1}
89	3.640×10^{-2}	2.95×10^{-1}
90	3.442×10^{-2}	2.91×10^{-1}
91	3.293×10^{-2}	2.89×10^{-1}
92	3.153×10^{-2}	2.87×10^{-1}

(*continued*)

Table B.1. (*Continued*)

θ	Water cloud (Kokhanovsky, 2004a) $g = 0.8552$	Ice cloud (Liou, 1992) $g = 0.75$
93	3.021×10^{-2}	2.85×10^{-1}
94	2.926×10^{-2}	2.83×10^{-1}
95	2.851×10^{-2}	2.82×10^{-1}
96	2.768×10^{-2}	2.81×10^{-1}
97	2.699×10^{-2}	2.78×10^{-1}
98	2.659×10^{-2}	2.75×10^{-1}
99	2.612×10^{-2}	2.72×10^{-1}
100	2.560×10^{-2}	2.68×10^{-1}
101	2.521×10^{-2}	2.62×10^{-1}
102	2.506×10^{-2}	2.56×10^{-1}
103	2.459×10^{-2}	2.53×10^{-1}
104	2.460×10^{-2}	2.48×10^{-1}
105	2.453×10^{-2}	2.42×10^{-1}
106	2.467×10^{-2}	2.35×10^{-1}
107	2.463×10^{-2}	2.27×10^{-1}
108	2.482×10^{-2}	2.21×10^{-1}
109	2.524×10^{-2}	2.16×10^{-1}
110	2.571×10^{-2}	2.11×10^{-1}
111	2.659×10^{-2}	2.05×10^{-1}
112	2.788×10^{-2}	1.99×10^{-1}
113	2.942×10^{-2}	1.96×10^{-1}
114	3.175×10^{-2}	1.89×10^{-1}
115	3.386×10^{-2}	1.79×10^{-1}
116	3.654×10^{-2}	1.64×10^{-1}
117	3.916×10^{-2}	1.42×10^{-1}
118	4.138×10^{-2}	1.27×10^{-1}
119	4.363×10^{-2}	1.18×10^{-1}
120	4.548×10^{-2}	1.11×10^{-1}
121	4.701×10^{-2}	1.04×10^{-1}
122	4.785×10^{-2}	9.93×10^{-2}
123	4.835×10^{-2}	9.86×10^{-2}
124	4.873×10^{-2}	9.79×10^{-2}
125	4.792×10^{-2}	9.70×10^{-2}
126	4.757×10^{-2}	9.67×10^{-2}
127	4.717×10^{-2}	9.68×10^{-2}
128	4.714×10^{-2}	9.76×10^{-2}
129	4.757×10^{-2}	9.97×10^{-2}
130	4.933×10^{-2}	1.01×10^{-1}
131	5.242×10^{-2}	1.03×10^{-1}
132	5.760×10^{-2}	1.04×10^{-1}
133	6.597×10^{-2}	1.02×10^{-1}
134	7.796×10^{-2}	1.02×10^{-1}
135	9.477×10^{-2}	9.96×10^{-2}
136	1.150×10^{-1}	1.02×10^{-1}

Table B.1. (*Continued*)

θ	Water cloud (Kokhanovsky, 2004a) $g = 0.8552$	Ice cloud (Liou, 1992) $g = 0.75$
137	1.435×10^{-1}	1.12×10^{-1}
138	1.756×10^{-1}	1.21×10^{-1}
139	2.109×10^{-1}	1.27×10^{-1}
140	2.467×10^{-1}	1.35×10^{-1}
141	2.773×10^{-1}	1.44×10^{-1}
142	2.959×10^{-1}	1.54×10^{-1}
143	2.988×10^{-1}	1.66×10^{-1}
144	2.886×10^{-1}	1.78×10^{-1}
145	2.622×10^{-1}	1.91×10^{-1}
146	2.289×10^{-1}	2.04×10^{-1}
147	1.980×10^{-1}	2.13×10^{-1}
148	1.734×10^{-1}	2.25×10^{-1}
149	1.613×10^{-1}	2.43×10^{-1}
150	1.546×10^{-1}	2.63×10^{-1}
151	1.554×10^{-1}	2.87×10^{-1}
152	1.564×10^{-1}	3.04×10^{-1}
153	1.565×10^{-1}	3.08×10^{-1}
154	1.561×10^{-1}	3.09×10^{-1}
155	1.534×10^{-1}	3.07×10^{-1}
156	1.507×10^{-1}	2.99×10^{-1}
157	1.473×10^{-1}	2.80×10^{-1}
158	1.472×10^{-1}	2.63×10^{-1}
159	1.425×10^{-1}	2.52×10^{-1}
160	1.449×10^{-1}	2.36×10^{-1}
161	1.435×10^{-1}	2.13×10^{-1}
162	1.398×10^{-1}	1.95×10^{-1}
163	1.421×10^{-1}	1.77×10^{-1}
164	1.394×10^{-1}	1.66×10^{-1}
165	1.399×10^{-1}	1.57×10^{-1}
166	1.444×10^{-1}	1.58×10^{-1}
167	1.440×10^{-1}	1.69×10^{-1}
168	1.488×10^{-1}	1.90×10^{-1}
169	1.491×10^{-1}	2.35×10^{-1}
170	1.511×10^{-1}	2.71×10^{-1}
171	1.700×10^{-1}	2.84×10^{-1}
172	1.696×10^{-1}	2.88×10^{-1}
173	1.795×10^{-1}	2.63×10^{-1}
174	2.040×10^{-1}	2.67×10^{-1}
175	2.400×10^{-1}	3.04×10^{-1}
176	2.984×10^{-1}	3.81×10^{-1}
177	4.177×10^{-1}	5.76×10^{-1}
178	4.992×10^{-1}	7.98×10^{-1}
179	3.968×10^{-1}	1.01
180	6.779×10^{-1}	1.18

REFERENCES

Adam, J. A., 2002, The mathematical physics of rainbows and glories, *Phys. Rep., 356*(4–5), 229–365.

Airy, G. B., 1838, On the intensity of light in the neighborhood of a caustic, *Trans. Camb. Phil. Soc.*, *6*, 397–403.

Ambartsumian, V. A., 1943, On diffuse reflection of light from turbid media, *Doklady AN SSSR*, *8*, 257–265.

Arking, A. and J. D. Childs, 1985, Retrieval of clouds cover parameters from multispectral satellite images, *J. Appl. Meteorol.*, *24*, 323–333.

Asano, S., A. Uchiyama, A. Yamazaki, and M. Tanizono, 2001, Effects of aerosols on retrieval of the microphysical properties of water clouds from the airborne solar spectral reflectance measurements, *Proceedings of SPIE, 4150*, Optical Remote Sensing of the Atmosphere and Clouds II (Y. Sasano, J. Wang, and T. Hayasaka, eds.), pp. 208–216.

Asano, S., A. Uchiyama, A. Yamazaki, J.-F. Gayet, and M. Tanizono, 2002, Two case studies of winter continental-type water and mixed-phase stratocumuli over the sea. 2. Absorption of solar radiation, *J. Geophys. Res.*, *107*, 10.1029/2001JD001108.

Auer A. H. and D. L. Veal, 1970, The dimension of ice crystals in natural clouds, *J. Atmos. Sci.*, *27*, 919–926.

Ayvazyan, G. M., 1991, *The Propagation of Millimeter and Sub-millimeter Waves in Clouds*, Gidrometeoizdat, Leningrad.

Babenko, V. A., V. N. Kuzmin, L. G. Astafyeva, 2003, *Electromagnetic Scattering in Dispersed Media: Inhomgeneous and Anisotropic Particles*, Springer-Praxis, Berlin.

Barker, H. W., 1996, A parameterization for computing grid-averaged solar fluxes for inhomogeneous marine boundary layer clouds, I, methodology and homogeneous biases, *J. Atmos. Sci., 53*, 2304–2316.

Barker, H. W.,1996, A parameterization for computing grid-averaged solar fluxes for inhomogeneous marine boundary layer clouds, II, validation using satellite data, *J. Atmos. Sci.*, *53*, 2304–2316.

Barker, H. W. and Q. Fu, 2000, Assessment and optimization of the Gamma-weighted two—stream approximation, *J. Atmos. Sci.*, *57*, 1181–1188.

Bovensmann, H., J. P. Burrows, M. Buchwitz, J. Frerick, S. Noël, V. V. Rozanov, K. V. Chance and A. P. H. Goede, 1999, SCIAMACHY: mission objectives and measurement modes, *J. Atmos. Sci.*, *56*, 127–150.

Cahalan, R. F., W. Ridgway, W. J. Wiscombe, Harshvardhan, and S. Gollmer, 1994, Independent pixel and Monte Carlo estimates of stratocumulus albedo, *J. Atmos. Sci.*, *51*, 3776–3790.

Cahalan, R. F., L. Oreopoulos, G. Wen, A. Marshak, S. C. Tsay, and T. P. DeFelice, 2001, Cloud characterization and clear-sky correction from Landsat-7, *Remote Sens. Environ.*, *78*, 83–98.

Chandrasekhar, S., 1950, *Radiative Transfer*, Oxford Press, Oxford.

Danielson, R. E., D. R., Moore, and H. C. van de Hulst, 1969, The transfer of visible radiation through clouds, *J. Atmos. Sci.*, *26*, 1078–1087.

Davis, A. and A. Marshak, 2002, Space-time characteristics of light transmitted through dense clouds: a Green's function analysis. *J. Atmos. Sci.*, *59*, 2713–2727.

Davis, A., A. Marshak, R. Cahalan, and W. Wiscombe, 1997, The Landsat scale break in stratocumulus as a three-dimensional radiative transfer effect: implication for cloud remote sensing, *J. Atmos. Sci.*, *54*, 241–260.

Debye, P., 1909, Der Lichtdruck auf Kugeln von beliebigem Material, *Ann. Physik*, *30*, 57–136.

Deirmendjian, A., 1969, *Electromagnetic Scattering on Spherical Polydispersions*, Elsevier, Amsterdam.

de Rooij, W. A., 1985, *Reflection and Transmission of Polarized Light by Planetary Atmospheres*, PhD thesis, Free University of Amsterdam, Amsterdam.

Deschamps, P.-Y., F.-M. Breon, M. Leroy, A. Podaire, A. Bricaud, J.-C. Buriez, and G. Séze, 1994, The POLDER Mission: instrument characteristics and scientific objectives, *IEEE Trans.*, *GE 32*, 598–614.

Domke, H., 1978a, Linear Fredholm integral equations for radiative transfer problems in finite plane-parallel media. I. Imbedding in an infinite medium, *Astronaut. Nachr.*, *299*, 87–93.

Domke, H., 1978b, Linear Fredholm integral equations for radiative transfer problems in finite plane-parallel media. II. Imbedding in a semi-infinite medium, *Astronaut. Nachr.*, *299*, 95–102.

Dvoryashin, S. V., 2002, Remote determination of the ratio between the coefficients of water and ice absorption in clouds in the 2.15–2.35 μm spectral range, *Izvestiya, Atmos. Ocean Phys.*, *38*, 523–528.

Evans, K. F., 1998, The spherical harmonics discrete ordinate method for three-dimensional atmospheric radiative transfer, *J. Atmos. Sci.*, *55*, 429–446.

Feigelson, E. M., ed., 1981, *Radiation in a Cloudy Atmosphere*, Gidrometeoizdat, Leningrad, p. 280.

Fomin, B. A. and I. P. Mazin, 1998, Model for an investigation of radiative transfer in cloudy atmosphere, *Atmos. Res.*, *47–48*, 127–153.

Fresnel, A., 1866, *Oeuvres complètes d'Augustin Fresnel*, v. 1–3.

Garrett, T. J., P. V. Hobbs, and H. Gerber, 2001, Shortwave, single — scattering properties of arctic ice clouds, *J. Geophys. Res.*, *106*(D14), 15155–15172.

Germogenova, T. A. and N. V. Konovalov, 1974, Asymptotic characteristic solutions of transport equation for the inhomogeneous slab problem, *J. Appl. Math. Comp. Phys.*, *14*, 928–946.

Ghosh, S., P. R. Jonas, and R. Wood, 2000, Some impact of pollutants on the development and optical properties of stratocumulus clouds, *Q. J. R. Meteorol. Soc. (Royal Meteorological Society)*, *569*, 2851–2872.

Goloub, P., M. Herman, H. Chepfer, J. Riedi, G. Brogniez, P. Couvert, G. Séze, 2000, Cloud thermodynamical phase classification from the POLDER spaceborne instrument, *J. Geophys. Res. D*, *105*, 14747–14759.

Hale, G. M. and M. R. Querry, 1973, Optical constacts of water in the 200-nm to 200-μm wavelength region, *Appl. Opt.*, *12*, 555–563.

Han, Q. Y., W. B. Rossow, and A. A. Lacis, 1994, Near global survey of effective droplet radii in liquid water clouds using ISCCP data, *J. Climate*, *7*, 465–497.

Hansen, J. E., and L. D. Travis (1974), Light scattering in planetary atmospheres, *Space Sci. Rev.*, *16*, 527–610.

Hansen, J. E. and J. Hovenier, 1974, Interpretation of the polarization of Venus, *J. Atmos. Sci.*, *31*, 1137–1160.

Hapke, B., 1993, *Theory of the Reflectance and Emmitance Spectroscopy*, University Press, Cambridge.

Heymsfield, G. M., J. B. Halverson, J. Simpson, L. Tian, T. P. Bui, 2001, ER-2 Doppler Radar Investigations of the Eyewall of Hurricane Bonnie during the Convection and Moisture Experiment-3, *Journal of Applied Meteorology, 40*, 8, 1310–1330.

Ishimaru, A., 1978, *Wave Propagation and Scattering in Random Media*, Academic Press, New York.

Jackson, J. D., 1999, From Alexander of Aphrodisias to Young and Airy, *Phys. Rep.*, *320*, 27–36.

Katsev, I. L., E. P. Zege, A. S. Prikhach, and I. N. Polonsky, 1998, Efficient technique to determine backscattered light power for various atmospheric and oceanic sounding and imaging systems, *J. Opt. Soc. Am.*, *A14*, 1338–1346.

Khrgian, A. H. and I. P. Mazin, 1952, On the droplet size distributions in clouds, *Proc. Cent. Aerol. Obs.*, *7*, 56–61.

King, M. D., 1981, A method for determining the single scattering albedo of clouds through observation of the internal scattered radiation field, *J. Atmos. Sci.*, *38*, 2031–2044.

King, M. D., 1987, Determination of the scaled optical thickness of clouds from reflected soar radiation measurements, *J. Atmos. Sci.*, *44*, 1734–1751.

King, M. D. and Harshvardhan, 1986, Comparative accuracy of selected multiple scattering approximations, *J. Atmos. Sci.*, *43*, 784–801.

King, M. D., et al., 1992, Remote sensing of cloud, aerosol, and water vapour properties from the moderate resolution imaging spectrometer (MODIS), *IEEE Trans.,GE 30*, 2–27.

Knap, W. H., P. Stammes, and R. B. A. Koelemeijer, 2002, Cloud thermodynamic phase determination from near-infrared spectra of reflected sunlight, *J. Atmos. Sci., 59,* 83–96.

Koelemeijer, R. B. A., P. Stammes, J. W. Hovenier, and J. F. De Haan, 2001, A fast method for retrieval of cloud parameters using oxygen a band measurements from GOME, *J. Geophys. Res.*, *106*, 3475–3490.

Kokhanovsky, A. A. and E. P. Zege, 1996, The determination of the effective radius of drops and liquid water path of water clouds from satellite measurements, *Earth Res. Space*, *2*, 33–44.

Kokhanovsky, A. A., 2000, Determination of the effective radius of drops in water clouds from polarization measurements, *Phys. Chem. Earth B*, *5–6*, 471–474.

Kokhanovsky, A. A., 2001, Simple relationships between radiative and microphysical characteristics of cloudy media, *Proceedings of the International Radiation Symposium*, St Petersburg, Russia, 24–29 July 2000 (W. L. Smith and Y. M. Timofeyev, eds.), pp. 315–318.

Kokhanovsky, A. A., 2002, A simple approximate formula for the reflection function of a homogeneous semi-infinite turbid medium, *JOSA*, *A19*, 957–960.

Kokhanovsky, A. A. and V. V. Rozanov, 2003, The reflection function of optically thick weakly absorbing turbid layers: a simple approximation, *J. Quant. Spectrosc. Radiat. Transf.*, *77*, 165–175.

Kokhanovsky, A. A., et al., 2003, A semi-analytical cloud retrieval algorithm using backscattering radiation in 04.–2.4 μm spectral range,*J. Geophys. Res. D*, *108*, 10.1029/2001JD001543.

Kokhanovsky, A. A., 2003a, *Polarization Optics of Random Media*, Springer-Praxis, Berlin.

Kokhanovsky, A. A., 2003b, The influence of the horizontal inhomogeneity on radiative characteristics of clouds: an asymptotic case study, *IEEE Trans.Geosci. Remote Sens.*, *41*, 817–825.

Kokhanovsky, A. A. and V. V. Rozanov, 2004, Simple approximate solutions of the radiative transfer equation for a cloudy atmosphere, *Remote Sensing of Clouds and the Atmosphere IX, Proceedings of SPIE,* v. 5571 (K. P. Schafer et al., eds.), SPIE, Bellingham, WA, pp. 86–93.

Kokhanovsky, A. A. and W. von Hoyningen-Huene, 2004, Optical properties of a hurricane, *Atmos. Res., 69*, 165–183.

Kokhanovsky, A. A., et al., 2004, The determination of cloud altitudes using SCIAMACHY onboard ENVISAT, *IEEE Trans. Geosci. Remote Sens. Lett.*, *1*, 211–214.

Kokhanovsky, A. A., 2004a, *Light Scattering Media Optics*, Springer-Verlag, Berlin.

Kokhanovsky, A. A., 2004b, Reflection of light from nonabsorbing semi-infinite cloudy media: a simple approximation, *J. Quant. Spectrosc. Radiat. Transf.*, *85*, 25–33.

Kokhanovsky, A. A., 2005, Reflection of light from particulate media with irregularly shaped particles, *Journal of Quantitative Spectroscopy and Radiative Transfer*, Volume 96, 1–10.

Kokhanovsky A. A. and T. Nauss, 2005, Satellite-based retrieval of ice cloud properties using a semianalytical algorithm, J. Geophys. Res., 110, D19206, doi:10.1029/2004JD005744.

Kondratyev, K. Y., 1998, *Multidimensional Global Change*, Wiley-Praxis, London.

Kondratyev, K. Y. and V. I. Binenko, 1984, *Impact of Cloudiness on Radiation and Climate*, Gidrometeoizdat, Leningrad.

Konovalov, N. V., 1975, On range of applicability of asymptotical formulae for calculations of monochromatic radiation in a nonuniform optically thick plane parallel layer, *Izv. AN SSSR,FAO*, *11*, 1263–1271.

Korolev, A., G. A. Isaac, and J. Hallett, 2000, Ice particle habits in stratiform clouds, *Q. J. R. Meteorol. Soc.*, *126*, 2873–2902.

Kuo, L., D. Labrie, and P. Chylek, 1993, Refractive indices of water and ice in the 0.65–2.5 μm spectral range, *Appl. Opt.*, *32*, 3531–3540.

Kuze, A. and K. V. Chance, 1994, Analysis of cloud top height and cloud coverage from satellites using the O_2 A and B bands, *J. Geophys. Res.*, *99*, 14481–14491.

Landolt-Bornstein, 1988, Numerical data and functional relationships in science and technology, *Group V: Geophysics and Space Research. V.5: Meteorology. Subvolume b. Physical and Chemical Properties of the Air* (G. Fischer, ed.), Springer-Verlag, Berlin, p. 570.

Liou, K. N., 1992, *Radiation and Cloud Processes in Atmosphere*, Oxford University Press, Oxford.

Liou, K. N., 2002, *An Introduction to Atmospheric Radiation*, Academic Press, New York.

Loeb, N. G. and R. Davies, 1996, Observational evidence of plane parallel model biases: apparent dependence of cloud optical depth on solar zenith angle, *J. Geophys. Res.*, *101*, 1621–1634.

Lutomirski, R. F., A. P. Ciervo, and J. H. Gainford, 1995, Moments of multiple scattering, *Appl. Opt.*, *34*, 7125–7136.

Macke, A. and F. Tzschihholz, 1992, Scattering of light by fractal particles: a qualitative estimate exemplary for two-dimensional triadic Koch island, *Physica A*, *191*, 159–170.

Macke, A., J. Mueller, and E. Raschke, 1996, Scattering properties of atmospheric ice crystals, *J. Atmos. Sci.*, *53*, 2813–2825.

Magano, C. and C. V. Lee, 1966, Meteorological classification of natural snow crystals, *J. Fac. Sci. Hokkaido Univ.*, *7*, 321–362.

Marchuk, G. I., G. A. Mikhailov, and M. A. Nazarliev, 1980, *The Monte-Carlo Methods in Atmospheric Optics*, Spriger-Verlag, Berlin.

Markel, V. A., 2002, The effects of averaging on the enhancement factor for absorption of light by carbon particles in microdroplets of water, *J. Quant. Spectrosc. Radiat. Transf.*, *72*, 765–774.

Marsh, N. D. and Svensmark, H., 2000, Cosmic rays, clouds and climate, *Phys. Rev. Lett.*, *85*, 5004–5007.

Marshak, A. and A. Davis, 2005, *3D Radiative Transfer in Cloudy, Atmospheres*, Berlin: Springer.

Marshak, A., A. Davis, W. Wiscombe, and R. Cahalan, 1998, Radiative effects of sub-mean free path liquid water variability observed in stratiform clouds, *J. Geophys. Res.*, *103*(D16), 19557–19568, 10.1029/98JD01728.

Mason, B. J., 1975, *Clouds, Rain, and Rainmaking*, Cambridge University Press, Cambridge.

Masunaga H., T. Y. Nakajima, T. Nakajima, M. Kachi, R. Oki, and S. Kuroda, 2002, Physical properties of maritime low clouds as retrieved by combined use of Tropical Rainfall Measurement

Mission Microwave Imager and Visible/Infrared Scanner: Algorithm, *J. Geophys. Res.*, *107*(D10), doi:10.1029/2001JD000743.

Mayer B., 1999, I3RC phase 1 results from the MYSTIC Monte Carlo model, *I3RC Workshop Proceedings*, Tucson, Arizona, November 17–19.

McGraw, R., S. Nemesure, and S. E. Schwartz, 1998, Properties and evolution of aerosols with size distributions having identical moments, *J. Aerosol Sci.*, *29*, 761–772.

Melnikova, I. N. and V. V. Mikhailov, 1994, Spectral scattering and absorption coefficients in strati derived from aircraft measurements, *J. Atmos. Sci.*, *51*, 925–931.

Melnikova, I. N. and V. V. Mikhailov, 2000, Vertical profile of spectral optical parameters of stratus clouds from airborne radiative measurements, *J. Geophys. Res.*, *D105*, 23255–23272.

Melnikova, I. N. and I. N. Minin, 1977, On the monochromatic radiative transfer in cloud layers, Izvestiya, *Atmos. Oceanic Phys.*, *13*(3), 254–263.

Melnikova, I. N. and A. V. Vasilyev, 2005, *Short-Wave Solar Radiation In The Earth's Atmosphere: Calculation, Observation, Interpretation*, Springer, Berlin.

Mie, G., 1908, Beitrage zur optik truber Medien speziell kolloidaler Metallosungen, *Ann. Phys.*, *25*, 377–445.

Miles, N. L., J. Verlinde, and E. E. Clothiaux, 2000, Cloud droplet size distributions in low-level stratiform clouds, *J. Atmos. Sci.*, *57*, 295–311.

Min, Q. and M. Duan, 2004, A successive order of scattering model for solving vector radiative transfer in the atmosphere, *J. Quant. Spectrosc. Radiat. Transf.*, *87*, 243–259.

Min, E., et al., 2004, Retrievals of thin cloud optical depth from a multifilter rotating shadowband radiometer, *J. Geophys. Res.*, *D109*, doi:10.1029/2003JD003964.

Minin, I. N., 1988, *Radiative Transfer Theory in Planetary Atmospheres*, Nauka, Moscow.

Mishchenko, M. I., J. M. Dlugach, E. G. Yanovitskij, and N. T. Zakharova, 1999, Bidirectional reflectance of flat, optically thick particulate layers: an efficient radiative transfer solution and applications to snow and soil surfaces. *J. Quant. Spectrosc. Radiat. Transf.* *63*, 409–432.

Mishchenko, M. I., et al., 2002, *Absorption, Scattering, and Emission of Light by Small Particles,* Cambridge University Press, Cambridge.

Mitchel, D. L. and W. P. Arnott, 1994, A model predicting the evolution of ice particle spectra and radiative properties of cirrus clouds. II. Dependence of absorption and extinction on ice morphology,*J. Atmos. Sci.*, *51*, 817–832.

Muinonen, K., et al., 1996, Light scattering by Gaussian random particles: ray optics approximation, *J. Quant. Spectrosc. Radiat. Transf.*, *55*, 577–601.

Nakajima, T. and M. Tanaka, 1988, Algorithms for radiative intensity calculations in moderately thick atmospheres using a truncation approximation, *J. Quant. Spectrosc. Radiat. Transf.*, *40*, 51–69.

Nakajima, T. and M. D. King, 1990, Determination of the optical thickness and effective particle radius of clouds from reflected solar radiation measurements. Part 1. Theory, *J. Atmos. Sci.*, *47*, 1878–1893.

Nakajima, T., M. D. King, and J. Spinhirne, 1991, Determination of the optical thickness and effective particle radius of clouds from reflected solar radiation measurements. Part 2. Marine stratocumulus observations, *J. Atmos. Sci.*, *47*, 1878–1893.

Nakajima, T. and M. D. King, 1992, Asymptotic theory for optically thick layers: application to the discrete ordinates method, *Appl. Opt.*, *31*, 7669–7683.

Nikolaeva, O. V., L. P. Bass, T. A. Germogenova, A. A. Kokhanovsky, V. S. Kuznetsov, and B. Mayer, 2005, The influence of neighbouring clouds on the clear sky reflectance studied with the 3–D transport code RADUGA, *J. Quant. Spectrosc. Radiat. Transf.*, *94*, 405–424.

Nussenzveig, H. M., 1992, *Diffracrtion Effects in Semiclassical Scattering*, Cambridge University Press, London.

Okada, et al., 2001, Shape of atmospheric mineral particles collected in three Chinese arid-regions, *Geophys. Res. Lett.*, *28*, 3123–3126.

Palle, E. and C. J. Butler, 2001, Sunshine records from Ireland, cloud factors and possible links to solar activity and climate, *Int. J. Climatol.*, *21*, 709–729.

Peltoniemi, J. I., et al., 1989, Scattering of light by stochastically rough particles, *Appl. Opt., 28*, 4088–4095.

Pincus, R. and S. A. Klein, 2000, Unresolved spatial variability and microphysical process rates in large-scale models, *J. Geophys. Res., 105*(D22), 27059–27065.

Pinsky, N. B. and A. P. Khain, 2002, Effects of in-cloud nucleation and turbulence on droplet spectrum formation in cumulus clouds, *Q. J. R. Meteorol. Soc.*, *128*, 501–534.

Platnick, S., 2001, Approximations for horizontal photon transport in cloud remote sensing problems, *J. Quantum Spectr. and Radiative Transfer, 68*, 75–99.

Platnick, S., M. D. King, S. A. Ackerman, W. P. Menzel, B. A. Baum, J. C. Rièdi, and R. A. Frey, 2003, The MODIS cloud products: algorithms and examples from Terra. *IEEE Trans. Geosci. Remote Sens.*, *41*, 459–473.

Prishivalko, A. P., 1983: *Optical and Thermal Fields Inside Light Scattering Particles*, Minsk: Nauka I Tekhnika.

Prishivalko, A. P., V. A. Babenko, and V. N. Kuzmin, 1984, *Scattering and Absorption of Light by Inhomogeneous and Anisotropic Spherical Particles*, Nauka i Tekhnika, Minsk.

Pruppacher, H. R. and J. D. Klett, 1978, *Microphysics of Clouds and Precipitation*, D. Reidel Publishing Company, Boston.

Rossow, W. B., 1989, Measuring cloud properties from space: a review, *J. Climate*, *2*, 419–458.

Rossow, W. B., L. C. Garder, and A. A. Lacis, 1989, Global, seasonal cloud variations from satellite radiance measurements. Part I: sensitivity of analysis. *J. Climate*, *2*, 419–458.

Rossow, W. B. and R. A. Schiffer, 1999, Advances in understanding clouds from ISCCP, *Bull. Am. Meteorol. Soc.*, *80*, 2261–2288.

Rothman, L. S., et al., 2003, The HITRAN molecular spectroscopic database: edition of 2000 including updates through 2001, *J. Quant. Spectrosc. Radiat. Transf.*, *82*, 5–44.

Rozanov, V. V. and A. A. Kokhanovsky, 2004, Semianalytical cloud retrieval algorithm as applied to the cloud top altitude and the cloud geometrical thickness determination from top-of-atmosphere reflectance measurements in the oxygen A band, *J. Geophys. Res.*, *109*, doi:10.1029/2003JD004104.

Rozanov, V. V., A. A. Kokhanovsky, and J. P. Burrows, 2004, The determination of cloud top altitudes using GOME reflectance spectra: multi-layered cloud systems, *IEEE Trans. Geosci. Remote Sens.*, *42*, 1009–1017.

Rozanov, V. V., T. Kurosu, and J. P. Burrows, 1998, Retrieval of atmospheric constituents in the UV-Visible: a new quasi-analytical approach for the calculation of weighting functions, *J. Quant. Spectrosc. Radiat. Transf.*, *60*, 277–299.

Rozanov, A. V., V. V. Rozanov, M. Buchwitz, A. A. Kokhanovsky, and J. P. Burrows, 2005, SCIATRAN 2.0 – a new radiative transfer model for geophysical applications in the 175–2400nm spectral range, *Adv. Space Res.*, in press.

Rozenberg, G. V., 1967, Physical foundations of light scattering media spectroscopy, *Soviet Physics Uspekhi, 91*, 569–608 .

Rozenberg, G. V., 1973, Coherence, observability, and the photometric aspect of beam optics. *Appl. Opt.*, *12*, 2855–2863.

Rozenberg, G. V., M. S. Malkevitch, V. S. Malkova, and V. I. Syachinov, 1978, The determination of optical characteristics of clouds from measurements of the reflected solar radiation using data from the Sputnik "KOSMOS-320", *Izvestiya Acad. Sci. USSR, Fizika Atmos. Okeana, 10*, 14–24.

Scheirer, R. and A. Macke, 2001, On the accuracy of the independent column approximation in calculating the downward fluxes in the UVA, UVB, and PAR spectral ranges, *J. Geophys. Res.*, *106*(D13), 14301–14312, 10.1029/2001JD900130.

Siewert, C. E., 2000, A discrete-ordinates solution for radiative – transfer models that include polarization effects, *J. Quant. Spectrosc. Radiat. Transf.*, *64*, 227–254.

Simpson, R., ed., 2002, *Hurricane! Coping with Disaster*, American Geophysical Union, Washington.

Sobolev, V. V., 1975, *Light Scattering in Planetary Atmospheres*, Moscow: Nauka.

Sobolev, V. V., 1984, Integral relations and asymptotic expressions in the theory of radiative transfer, *Astrofizika*, *20*, 123–132.

Stephens, G. L. and S.-C. Tsay, 1990, On the cloud absorption anomaly, *Q. J. R. Meteorol. Soc.*, *116*, 671–704.

Stokes, G. G., 1852, On the composition and resolution of streams of polarized light from different sources, *Trans. Camb. Phil. Soc.*, *9*, 399–416.

Stokes, G. G., 1862, On the intensity of the light reflected from or transmitted through a pile of plates,*Proc. R. Soc. Lond. 11*, 545–556.

Stratton, J. A., 1941, *Electromagnetic Theory*, McGraw-Hill Book Company, New York.

Svensmark, H., 1998, Influence of cosmic rays on Earth's climate, *Phys. Rev. Lett.*, *81*, 5027–5030.

Svensmark, H. and Friis-Christensen, E., 1997, Variations of cosmic ray flux and global cloud coverage. A missing link in solar-climate relationships. *J. Atmos. Solar-Terrestrial Phys. 59*, 1225–1232.

Thomas, G. E. and K. Stamnes, 1999, *Radiative Transfer in the Atmosphere and Ocean*, Cambridge University Press, Cambridge.

Tricker, R. A. R., 1970, *Introduction to Meteorological Optics*, London: Elsevier.

Trishchenko, A. P., Z. Li, F.-L. Chang, and H. Barker, 2001, Cloud optical depth and TOA fluxes: Comparison between satellite and surface retrievals from multiple platforms, *Geophys. Res. Lett.*, *28*, 979–982.

Twomey, S., 1977, *Atmospheric Aerosols*, Elsevier, London.

Umow, N., 1905, Chromatische depolarisation durch Lichtzerstreung, *Phys. Z.*, *6*, 674–676.

van de Hulst, H. C., 1968, Radiative transfer in thick atmospheres with an arbitrary scattering function, *Bull. Astron. Neth.*, *20*, 77–90.

van de Hulst, H. C., 1980, *Multiple Light Scattering: Tables, Formulas and Applications*, New York: Academic Press.

van de Hulst, H. C., 1981, *Light Scattering by Small Particles*, Dover, New York.

Varnai T., and A. Marshak, 2003, A method for analyzing how various parts of clouds influence each other's brightness, *J. Geophys. Res.*, *108*(D22), 4706, doi:10.1029/2003JD003561.

Volnistova, L. P. and A. S. Drofa, 1986, Quality of image transfer through light scattering media, *Opt. Spektrosk.*, *61*, 116–121.

Volten, H., 2001, *Light Scattering by Small Planetary Particles. An Experimental Study,* PhD thesis, Amsterdam Free University, Amsterdam.

Volten, H., O. Muñoz, E. Rol, J. F. de Haan, W. Vassen, J. W. Hovenier, K. Muinonen, and T. Nousiainen, 2001, Scattering matrices of mineral aerosol particles at 441.6 and 632.8 nm, *J. Geophys. Res.*, *106*(D15), 17375–17401.

Wang, M. and M. D. King, 1997, Correction of Rayleigh scattering effects in cloud optical thickness retrievals, *J. Geophys. Res.*, *102*(D22), 25915–25926.

Warner, J., 1973, The microstructure of cumulus clouds: Part IV. The effect on the droplet spectrum of mixing between cloud and environment, *J. Atmos. Sci.*, *30*, 256–261.

Warren, S. G., 1984, Optical constants of ice from the ultraviolet to the microwave, *Appl. Opt.*, *23*, 1206–1225.

Wauben, W. M. F., 1992, *Multiple Scattering of Polarized Radiation in Planetary Atmospheres*, PhD thesis, Free University of Amsterdam, Amsterdam.

Wells, W. H., 1969, Loss of resolution in water as a result of multiple small-angle scattering, *J. Opt. Soc. Am.*, *59*, 686–691.

Winker, D. M. and C. R. Trepte, 1998, Laminar cirrus observed near the tropical tropopause by LITE, *Geophys. Res. Lett.*, *25*, 3351–3354.

Yamomoto, G. and D. Q. Wark, 1961, Discussion of the letter by R. A. Hanel, 'Determination of cloud altitude from a satellite', *J. Geophys. Res.*, *66*, 3596.

Yang, P., K. N. Liou, K. Wyser, and D. Mitchell, 2000, Parametrization of the scattering and absorption parameters of individual ice crystals, *J. Geophys. Res.*, *105*, 4699–4718.
Yang, P., B.-C. Gao, B. A. Baum, W. J. Wiscombe, Y. X. Hu, S. L. Nasiri, P. F. Soulen, A. J. Heymsfield, G. M. McFarquhar, and L. M. Miloshevich, 2001, Sensitivity of cirrus bidirectional reflectance to vertical inhomogenity of ice crystal habits and size distributions for two Moderate-Resolution Imaging Spectroradiometer (MODIS) bands, *J. Geophys. Res.*, *106*(D15), 17267–17291.
Yanovitskij, E. G., 1997, *Light Scattering in Inhomogeneous Atmospheres*, Springer-Verlag, New York.
Zege, E. P., A. P. Ivanov, and I. L. Katsev, 1991, *Image Transfer Through a Scattering Medium*, Springer-Verlag, New York.
Zege, E. P., and A. A. Kokhanovsky, 1994, Analytical solution for optical transfer function of a light scattering medium with large particles, *Appl. Opt.*, *33*, 6547–6554.

ATMOSPHERIC AND OCEANOGRAPHIC SCIENCES LIBRARY

1. F.T.M. Nieuwstadt and H. van Dop (eds.): *Atmospheric Turbulence and Air Pollution Modelling.* 1982; rev. ed. 1984 ISBN 90-277-1365-6; Pb (1984) 90-277-1807-5
2. L.T. Matveev: *Cloud Dynamics.* Translated from Russian. 1984 ISBN 90-277-1737-0
3. H. Flohn and R. Fantechi (eds.): *The Climate of Europe: Past, Present and Future.* Natural and Man-Induced Climate Changes: A European Perspective. 1984 ISBN 90-277-1745-1
4. V.E. Zuev, A.A. Zemlyanov, Yu.D. Kopytin, and A.V. Kuzikovskii: *High-Power Laser Radiation in Atmospheric Aerosols.* Nonlinear Optics of Aerodispersed Media. Translated from Russian. 1985 ISBN 90-277-1736-2
5. G. Brasseur and S. Solomon: *Aeronomy of the Middle Atmosphere.* Chemistry and Physics of the Stratosphere and Mesosphere. 1984; rev. ed. 1986 ISBN (1986) 90-277-2343-5; Pb 90-277-2344-3
6. E.M. Feigelson (ed.): *Radiation in a Cloudy Atmosphere.* Translated from Russian. 1984 ISBN 90-277-1803-2
7. A.S. Monin: *An Introduction to the Theory of Climate.* Translated from Russian. 1986 ISBN 90-277-1935-7
8. S. Hastenrath: *Climate Dynamics of the Tropics*, Updated Edition from *Climate and Circulation of the Tropics.* 1985; rev. ed. 1991 ISBN 0-7923-1213-9; Pb 0-7923-1346-1
9. M.I. Budyko: *The Evolution of the Biosphere.* Translated from Russian. 1986 ISBN 90-277-2140-8
10. R.S. Bortkovskii: *Air-Sea Exchange of Heat and Moisture During Storms.* Translated from Russian, rev. ed. 1987 ISBN 90-277-2346-X
11. V.E. Zuev and V.S. Komarov: *Statistical Models of the Temperature and Gaseous Components of the Atmosphere.* Translated from Russian. 1987 ISBN 90-277-2466-0
12. H. Volland: *Atmospheric Tidal and Planetary Waves.* 1988 ISBN 90-277-2630-2
13. R.B. Stull: *An Introduction to Boundary Layer Meteorology.* 1988 ISBN 90-277-2768-6; Pb 90-277-2769-4
14. M.E. Berlyand: *Prediction and Regulation of Air Pollution.* Translated from Russian, rev. ed. 1991 ISBN 0-7923-1000-4
15. F. Baer, N.L. Canfield and J.M. Mitchell (eds.): *Climate in Human Perspective.* A tribute to Helmut E. Landsberg (1906-1985). 1991 ISBN 0-7923-1072-1
16. Ding Yihui: *Monsoons over China.* 1994 ISBN 0-7923-1757-2
17. A. Henderson-Sellers and A.-M. Hansen: *Climate Change Atlas.* Greenhouse Simulations from the Model Evaluation Consortium for Climate Assessment. 1995 ISBN 0-7923-3465-5
18. H.R. Pruppacher and J.D. Klett: *Microphysics of Clouds and Precipitation*, 2nd rev. ed. 1997 ISBN 0-7923-4211-9; Pb 0-7923-4409-X
19. R.L. Kagan: *Averaging of Meteorological Fields.* 1997 ISBN 0-7923-4801-X
20. G.L. Geernaert (ed.): *Air-Sea Exchange: Physics, Chemistry and Dynamics.* 1999 ISBN 0-7923-5937-2
21. G.L. Hammer, N. Nicholls and C. Mitchell (eds.): *Applications of Seasonal Climate Forecasting in Agricultural and Natural Ecosystems.* 2000 ISBN 0-7923-6270-5

ATMOSPHERIC AND OCEANOGRAPHIC SCIENCES LIBRARY

22. H.A. Dijkstra: *Nonlinear Physical Oceanography*. A Dynamical Systems Approach to the Large Scale Ocean Circulation and El Niño. 2000 ISBN 0-7923-6522-4
23. Y. Shao: *Physics and Modelling of Wind Erosion*. 2000 ISBN 0-7923-6657-3
24. Yu.Z. Miropol'sky: *Dynamics of Internal Gravity Waves in the Ocean*. Edited by O.D. Shishkina. 2001 ISBN 0-7923-6935-1
25. R. Przybylak: *Variability of Air Temperature and Atmospheric Precipitation during a Period of Instrumental Observations in the Arctic*. 2002 ISBN 1-4020-0952-6
26. R. Przybylak: *The Climate of the Arctic*. 2003 ISBN 1-4020-1134-2
27. S. Raghavan: *Radar Meteorology*. 2003 ISBN 1-4020-1604-2
28. H.A. Dijkstra: *Nonlinear Physical Oceanography*. A Dynamical Systems Approach to the Large Scale Ocean Circulation and El Niño. 2nd Revised and Enlarged Edition. 2005 ISBN 1-4020-2272-7
29. X. Lee, W. Massman and B. Law (eds.): *Handbook of Micrometeorology*. A Guide for Surface Flux Measurement and Analysis. 2004 ISBN 1-4020-2264-6
30. A. Gelencsér: *Carbonaceous Aerosol*. 2005 ISBN 1-4020-2886-5
31. A. Soloviev and L. Roger : *The Near-Surface Layer of the Ocean*. Structure, Dynamics and Applications. 2006 ISBN 1-4020-4052-0
32. G.P. Brasseur and S. Solomon: *Aeronomy of the Middle Atmosphere*. Chemistry and Physics of the Stratosphere and Mesosphere. 2005 ISBN 1-4020-3284-6